Lecture Notes in Computer Science 8721

Commenced Publication in 1973
Founding and Former Series Editors:
Gerhard Goos, Juris Hartmanis, and Jan van Leeuwen

T0213755

András Horváth   Katinka Wolter (Eds.)

Computer Performance Engineering

11th European Workshop, EPEW 2014
Florence, Italy, September 11-12, 2014
Proceedings

 Springer

Volume Editors

András Horváth
Università di Torino
Dipartimento di Informatica
Corso Svizzera 185, 10149 Turin, Italy
E-mail: horvath@di.unito.it

Katinka Wolter
Freie Universität Berlin
Institute of Computer Science
Takustr. 9, 14195 Berlin, Germany
E-mail: katinka.wolter@fu-berlin.de

ISSN 0302-9743   e-ISSN 1611-3349
ISBN 978-3-319-10884-1   e-ISBN 978-3-319-10885-8
DOI 10.1007/978-3-319-10885-8
Springer Cham Heidelberg New York Dordrecht London

Library of Congress Control Number: 2014947117

LNCS Sublibrary: SL 2 – Programming and Software Engineering

Typesetting: Camera-ready by author, data conversion by Scientific Publishing Services, Chennai, India

Printed on acid-free paper

Springer is part of Springer Science+Business Media (www.springer.com)

Preface

It is our pleasure to present the proceedings of EPEW 2014, the 11th European Workshop on Performance Engineering held during September 11–12, 2014, in Florence, Italy. This annual workshop series aims to gather academic and industrial researchers working on all aspects of performance engineering. The accepted papers reflect the diversity of modern performance engineering. A number of papers tackle theoretical modeling issues in queuing theory and the analysis of Markov chains. Others address practical problems of resource allocation, communications networks, or even car parking systems and bus services.

The program of EPEW 2014 comprised 18 papers selected from 30 submissions. Each paper was peer reviewed by three reviewers from the Program Committee (PC). After the collection of reviews the PC members carefully discussed the quality of the papers for one week before deciding about acceptance. We therefore owe special thanks to the members of the PC for their work in the reviewing process and the subsequent discussion panels.

EPEW 2014 was honored to have two keynote speakers. Luca Bortolussi explored combination of statistical machine tools and stochastic model checking. Armando Tacchella considered quantitative model checking for analysis and repair of stochastic control policies.

We wish to express our gratitude to the staff of the University of Florence for hosting the workshop, the EasyChair team for having allowed us to use their conference system, and Springer for the continued editorial support of this workshop series. Finally, we thank the authors of all the papers for their contribution.

July 2014   András Horváth
                                                                   Katinka Wolter

Organization

Program Committee

Gianfranco Balbo	University of Turin, Italy
Simonetta Balsamo	Università Ca' Foscari di Venezia, Italy
Marta Beltran	Rey Juan Carlos University, Spain
Marco Bernardo	University of Urbino, Italy
Jeremy Bradley	Imperial College London, UK
Domenico Cotroneo	University of Naples Federico II, Italy
Tadeusz Czachorski	IITiS PAN, Polish Academy of Sciences, Poland
Dieter Fiems	Ghent University, Belgium
Jean-Michel Fourneau	Université de Versailles St. Quentin, France
Stephen Gilmore	School of Informatics Edinburgh, UK
Andras Horvath	University of Turin, Italy
Carlos Juiz	University of the Balearic Islands, Spain
Helen Karatza	Aristotle University of Thessaloniki, Greece
Leila Kloul	Université de Versailles St. Quentin, France
William Knottenbelt	Imperial College London, UK
Samuel Kounev	Karlsruhe Institute of Technology (KIT), Germany
Fumio Machida	NEC Corporation, Japan
Andrea Marin	University of Venice, Italy
Rivalino Matias Jr.	Universidade Federal de Uberlandia, Brazil
Luca Muscariello	Orange Labs, France Telecom, France
Roberto Pietrantuono	University of Naples Federico II, Italy
Philipp Reinecke	Freie Universität Berlin, Germany
Anne Remke	University of Twente, The Netherlands
Jacques Resing	Eindhoven University of Technology, The Netherlands
Sabina Rossi	Università Ca' Foscari di Venezia, Italy
Nahum Shimkin	Technion - Israel Institute of Technology, Israel
Markus Siegle	University of the Armed Forces Munich, Germany
Mark Squillante	IBM Research, USA
Yutaka Takahashi	Kyoto University, Japan
Miklos Telek	Budapest University of Technology and Economics, Hungary
Mirco Tribastone	University of Southampton, UK
Petr Tuma	Charles University, Prague
Benny Van Houdt	University of Antwerp, Belgium
Maria Grazia Vigliotti	Imperial College London, UK
Jean-Marc Vincent	Laboratoire LIG, France

Katinka Wolter   Freie Universität Berlin, Germany
Armin Zimmermann   Technische Universität Ilmenau, Germany

Additional Reviewers

Frattini, Flavio   Lera, Isaac
Gouberman, Alexander   Spinner, Simon
Gómez, Beatriz   Stojic, Ivan
Horvath, Illes   V. Kistowski, Jóakim
Khouja, Mehdi   Walter, Jürgen
Kolesnichenko, Anna

Abstracts of Invited Talks

Quantitative Model Checking
for Analysis and Repair of Stochastic Control
Policies

Armando Tacchella

Dipartimento. di Informatica, Bioingegneria, Robotica e Ingegneria dei Sistemi
(DIBRIS)
Università degli Studi di Genova, Via Opera Pia, 13 – 16145 Genova – Italy
Armando.Tacchella@unige.it

Abstract. Given a system operating in an environment modeled as a
Markov decision process, it is possible to synthesize (optimal) stochastic
control policies in a variety of ways. In particular, if the model of the environment is not available, a policy can be obtained using Reinforcement
Learning (RL) techniques, wherein trial-and-error interaction between
the system and the environment is leveraged. RL methods have shown
robust and efficient learning on a variety of robot-control problems —
see, e.g., [1]. However, while policies learned by reinforcement are satisfactory according to utility-based measures, they may fail to meet other
requirements, e.g., safety. In this direction RL methods *per se* cannot provide adequate guarantees. In the words of [2]: "The asymptotic nature of
guarantees about RL performances makes it difficult to bound the probability of damaging the controlled robot and/or the environment". How
to guarantee that, given a control policy synthesized by RL, such policy
will have a very low probability of yielding undesirable behaviors? Our
answer leverages Probabilistic Model Checking techniques — see, e.g., [3]
– by describing robot-environment interactions using Markov chains, and
the related safety properties using probabilistic logic. Both the encoding
of the interaction models and their verification can be fully automated,
and only properties have to be manually specified. Our research goes
beyond automating verification, to consider the problem of automating
repair, i.e., if the policy is found unsatisfactory, how to fix it without
manual intervention. In this talk, we detail how to automate the analysis
of policies using probabilistic model checking techniques. Our methodology includes algorithms to repair control policies until they satisfy a set
of safety requirements. We describe theoretical and empirical evidence
about the effectiveness of our methodology. Alternative approaches and
similar results available in the literature are also discussed.

References

1. Bagnell, J.A., Schaal, S.: Special issue on Machine Learning in Robotics (Editorial).
The International Journal of Robotics Research 27(2), 155–156 (2008)

2. Gillula, J.H., Tomlin, C.J.: Guaranteed Safe Online Learning via Reachability: tracking a ground target using a quadrotor. In: ICRA, pp. 2723–2730 (2012)
3. Kwiatkowska, M., Norman, G., Parker, D.: Stochastic model checking. Formal Methods for Performance Evaluation, 220–270 (2007)

Machine Learning Meets Stochastic Model Checking*

Luca Bortolussi[1,2,3]

[1] Dept. of Computer Science, Saarland University, Germany
[2] DMG, University of Trieste, Italy
[3] CNR/ISTI, Pisa, Italy

Introduction. Performance modelling is more and more concerned with the modelling of complex systems, in which the interactions of many heterogeneous entities produce complex emergent system-level behaviours. Examples span from sensor networks to cloud computing and to smart cities. When the interest is in predictive modelling, as is the case in performance, these models are necessarily quantitative, e.g. expressed as Continuous Time Markov Chains (CTMC) or as other kinds of stochastic processes. Their quantitative nature is reflected in their dependence on several parameters, which are often known with a considerable margin of uncertainty. Practically, this means that we most likely can provide a bounded interval supposed to contain the true value of a parameter, but not the true value itself. We will refer to this class of CTMC as Uncertain CTMC.

Problem and Methodology. The problem we address is, given an uncertain CTMC model, how to reason about it with automatic tools. Our approach is statistical, combining state-of-the-art Statistical Machine Learning tools, specifically designed to tackle uncertainty, with classic tools from formal methods, mainly Model Checking.

The first problem we face is how to estimate the satisfaction probability of a linear temporal logic property under such uncertainty [3]. We will show how we can reconstruct the functional dependency of the satisfaction probability on unknown parameters using Machine Learning techniques based on Gaussian Processes, which offer a flexible framework for regression and classification. We dubbed this approach Smoothed Model Checking, as it relies on smoothness properties of CTMCs.

An alternative way to deal with uncertainty is to try to eliminate it exploiting available observations of the real system modelled. As it is often easier to observe and capture qualitative properties, rather than performing precise measurements, we tackled the problem of parameter estimation from observations of truth values of temporal logic formulae [2]. Also in this case, Gaussian Processes and Bayesian Optimisation play a central role in the solution of this problem. In

---

* Work in collaboration with Guido Sanguinetti, from the University of Edinburgh. L.B. acknowledges partial supported from EU-FET project QUANTICOL (nr. 600708).

a similar way we can tackle the twin problem of system design [2, 1], which consists in finding optimal parameter values to enforce a desired behaviour, again given by a linear temporal property.

References

1. Bartocci, E., Bortolussi, L., Nenzi, L., Sanguinetti, G.: On the robustness of temporal properties for stochastic models. Electronic Proceedings in Theoretical Computer Science 125, 3–19 (2013)
2. Bortolussi, L., Sanguinetti, G.: Learning and designing stochastic processes from logical constraints. In: Joshi, K., Siegle, M., Stoelinga, M., D'Argenio, P.R. (eds.) QEST 2013. LNCS, vol. 8054, pp. 89–105. Springer, Heidelberg (2013)
3. Bortolussi, L., Sanguinetti, G.: Smoothed model checking for uncertain continuous time Markov chains. CoRR ArXiv, 1402.1450 (2014)

Table of Contents

Cloud Performance Modelling

Queueing and Fluid Models

Performance of Computation and Programming

Fitting

Urban Traffic Modelling

Decision Making

Markovian Models, Above and Beyond

Optimal Hiring of Cloud Servers

Andrew Stephen McGough[1] and Isi Mitrani[2]

[1] School of Engineering and Computing Sciences, Durham University, DH1 3LE, U.K.
stephen.mcgough@durham.ac.uk
[2] School of Computing Science, Newcastle University, NE1 7RU, U.K.
isi.mitrani@ncl.ac.uk

Abstract. A host uses servers hired from a Cloud in order to offer certain services to paying customers. It must decide dynamically when and how many servers to hire, and when to release them, so as to minimize both the job holding costs and the server costs. Under certain assumptions, the problem can be formulated in terms of a semi-Markov decision process and the optimal hiring policy can be computed. Two situations are considered: (a) jobs are submitted in random batches and servers can be hired for arbitrary periods of time; (b) jobs arrive singly and servers must be hired for fixed periods of time. In both cases, the optimal policies are compared with some simple and easily implementable heuristics.

1 Introduction

This paper focuses on certain special, and important, dynamic scheduling problems that arise in the market for computer services. It presents a general optimization methodology and applies it in situations where detailed exact and approximate solutions can be developed.

A host offers certain services which involve running user jobs. It does not own servers, but hires them on a temporary basis from a Cloud provider. The host must decide dynamically when, and how many, servers to hire. The objective is to manage optimally the long-term trade-offs between the operating costs (which depend on the number of servers hired), and the Quality-of-Service, or 'holding' costs (which are proportional to the number of jobs present).

Two distinct models are considered. In the first, jobs are submitted in batches of random size and at random intervals. Servers may be hired and released at arbitrary moments of time, hence the hiring decisions can be taken at the instants when new batches arrive. In the second model, servers must be hired for reasonably long fixed periods of time, e.g. by the hour. Hiring decisions are therefore assumed to take place at discrete moments in time, while jobs arrive and depart singly and in continuous time. The second model is perhaps closer to current practice, but the first one may come into its own since some Cloud providers are beginning to offer servers for very short-term hire, e.g. by the minute.

We show how, under certain assumptions, these dynamic optimization problems can be solved by formulating them in terms of semi-Markov decision processes and applying a policy improvement algorithm. The optimal hiring policy

A. Horváth and K. Wolter (Eds.): EPEW 2014, LNCS 8721, pp. 1–15, 2014.

can then be computed in a finite number of iterations. Although that computation is efficient, it may sometimes be too expensive to be carried out on-line. We therefore propose simple and easily implementable heuristic policies for both models. In numerical experiments, the performance of the heuristics is compared to that of the optimal policy.

An example of a company using Cloud servers is Cycle Computing[1], which acts as a broker offering virtual High-Throughput HTCondor [15] clusters in the Cloud. Different service facilities are also provided by interfaces such as e-Science Central [8], whereby access to Cloud computing resources is offered to users in a transparent manner. However, at present these systems do not make any attempt to optimize their operating policies.

The main distinguishing feature of the present study is that we carry out a rigorous dynamic optimization of the systems considered. That is, we consider operational decisions that depend not only on the system parameters, but also on the changing system state. Moreover, the optimization takes into account the transition probabilities between states, and hence covers a long-term system trajectory. This does not appear to have been done before.

Being able to determine the optimal operating policy is valuable, even when good heuristics exist. One may suspect that a simple heuristic policy will perform well, but the only way to quantify such a statement is to compute the optimal policy and carry out a proper comparison.

1.1   Related Work

The existing approaches to the server hiring problem are, on the whole, concerned with static policies. That is, the hiring decisions are based on knowing or estimating the characteristics of user demand. Those decisions change only when the demand parameters change. On the other hand, a dynamic policy reacts to random changes in the system state, even if the demand characteristics remain the same. In general, dynamic policies are more efficient than static ones, as we shall see when presenting our numerical results.

Mazzucco *et al.* [10] have used workload estimation in order to determine the optimal number of servers to hire. By assuming that impatient users will abandon job requests (common for HTTP) an Erlang-C problem is converted into an Erlang-A problem and a solution is obtained by a binary search algorithm. This work is extended in [12] to evaluate the number of Virtual Machines (VMs) required by a Software-as-a-Service (SaaS) provider using an Infrastructure-as-a-Service (IaaS) backend. Bodík *et al.* [2] use statistical machine learning to estimate the workload in the next epoch. Like other approaches, this requires additional servers to be provisioned in case the estimate is low.

Another static version of the server hiring problem was considered by Lampe *et al.* [9], who examined the optimal placement of a fixed set of jobs, with given run times and resource requirements, onto different Cloud servers. An exact formulation based on Binary Integer Programming and an approximate algorithm

---

[1] http://www.cyclecomputing.com

using bin-packing techniques were proposed. A similar problem involving work-flows was addressed by Byun *et al.* [3,4]. In this instance, the servers are not different, but the jobs must satisfy a set of precedence constraints. Again, the aim is to minimize the cost of executing a given workflow on the Cloud. An approximate scheduling algorithm is proposed.

Chaisiri *et al.* [6] attempt to exploit the lower costs of future reservations in order to minimize the overall cost of hiring Cloud resources. They use stochastic and deterministic programming techniques, coupled with sample-average approximations or Benders decomposition. This study has some dynamic aspects. However, the actual demand process is not modelled and therefore the costs of waiting cannot be taken into account.

The server hiring problem is distantly related to other server allocation topics, for which a large body of literature exists. These topics include the trade-offs between performance and power consumption in a service center. In Mazzucco *et al.* [11] and Mitrani [13], certain dynamic server allocation policies were analysed, but no attempt was made to find the optimal policy. The maximization of throughput and the minimization of waiting or response time were considered in Urgaonkar *et al.* [17], Chandra *et al.* [5] and Bennani and Menascé [1].

The general Semi-Markov decision process and the algorithm for computing the optimal policy are described in section 2. The applications of the theory to the models with batch arrivals and with fixed hiring periods are presented in sections 3 and 4, respectively. Section 5 introduces the heuristic and shows the results of some numerical experiments. Some directions for further research are outlined in the conclusion – section 6.

2 Semi-Markov Decision Processes

Consider a finite-state system which is observed at random points in time, t_i ($i = 0, 1, \ldots$). These instants are called 'decision epochs' and the intervals between them are 'decision intervals'. If at time t_i the system is in state j ($j = 1, 2, \ldots, J$), an action, or decision, a_j, is taken. That action may influence the length of the ensuing decision interval, $t_{i+1} - t_i$, and also the system state at the next epoch. However, neither the decision interval nor the next state depend on anything that happened prior to t_i. Such a process is called a 'semi-Markov decision process'. The actions taken in the various states constitute a 'stationary policy', if for all states j, whenever the state j is observed, the same action, a_j, is taken, regardless of current time and past history.

The system incurs costs which depend on the states it passes through and on the decisions taken in those states. Let $Z_A(t)$ be the total cost incurred up to time t under a stationary policy A. Then the long-run average cost of policy A per unit time is defined as the limit:

$$g(A) = \lim_{t \to \infty} \frac{1}{t} E[Z_A(t)] . \tag{1}$$

That quantity, which does not depend on the initial state, is the optimization criterion. The object is to find a policy A that minimizes $g(A)$.

The evolution of the process under the control of a stationary policy A is governed by the succession of states at decision epochs, the decisions made at those epochs and the costs incurred during the decision intervals. Let $p_{j,k}(A)$ be the transition probability that the system will be in state k at the next decision epoch, given that the current state is j and the policy is A; $j, k = 1, 2, \ldots, J$. Also, denote by $c_j(A)$ the average cost incurred during a decision interval, given the current state j and policy A. Finally, let $\tau_j(A)$ be the average length of the decision interval, given the current state and policy.

The long-run average cost of policy A, $g(A)$, can be computed by introducing certain quantities called 'relative values', v_j, $j = 1, 2, \ldots, J$, (Tijms [16]). These relative values, together with $g(A)$, satisfy a set of simultaneous linear equations:

$$v_j = c_j(A) - \tau_j(A)g(A) + \sum_{k=1}^{J} p_{j,k}(A)v_k \; ; \; j = 1, 2, \ldots, J , \qquad (2)$$

with $c_j(A)$, $p_{j,k}(A)$ and $\tau_j(A)$ as defined above.

In this set, there are J equations with $J + 1$ unknowns. However, if the same constant, c, is added to all relative values v_j, the value of $g(A)$ would not change (since for each j, the sum of $p_{j,k}(A)$ with respect to k is 1). Therefore, the solution of (2) can be made unique by choosing an arbitrary state, m, and setting $v_m = 0$. The optimal policy can be determined by the following 'policy improvement' algorithm.

1. Choose some stationary policy A.
2. Compute $g(A)$ and v_j by solving (2).
3. For each j, find action a^* that minimizes the right-hand side of equation (2):

$$\min_a \left[c_j(A) - \tau_j(a)g(A) + \sum_{k=1}^{J} p_{j,k}(a)v_k \right] ,$$

   where $g(A)$ and v_k keep the values already computed.
4. If new actions a^* are the same as the old ones for all states, i.e. new policy A^* is the same as A, stop. Otherwise repeat from step 2, replace A with A^*.

This algorithm terminates after a finite number of iterations, producing an optimal policy and the corresponding long-run average cost, g.

An efficient and stable method for solving the set of equations (2) is to use Gauss-Seidel iterations, starting with $v_j = 0$ for all j. Convergence is assured because the coefficients in the right-hand sides of (2), being probabilities, do not exceed 1. If that method is adopted, then the complexity of computing the optimal policy is on the order of $O(J^2 SI)$, where J is the size of the state space, S is the number of iterations in the Gauss-Seidel solution and I is the number of iterations in the policy-improvement algorithm.

3   Batch Arrivals

The first system we examine is one where user demands arrive at the host's site in a Poisson stream with rate λ. Consecutive demands consist of batches of jobs whose sizes are i.i.d. random variables with an arbitrary distribution. Let b_j be the probability that a batch contains j jobs ($j = 1, 2, \ldots, \ldots$). The average batch size is denoted by b:

$$b = \sum_{j=1}^{\infty} jb_j \ . \tag{3}$$

A job's runtime, on any available server, is distributed exponentially with mean $1/\mu$. Thus, the total offered load at the site is $\rho = \lambda b/\mu$. When all available servers are busy, jobs wait in a common FIFO queue. Servers may be hired and released at any moment.

In this model, the decision epochs are the instants just after the arrival of a new batch. The system state at a decision epoch is the total number, j, of jobs present. That number may include jobs from previous batches that are still waiting or are in service. The decision taken at a decision epoch is the number of servers, n, that are hired from a Cloud provider and will be available to serve jobs. That number may include previously hired servers, plus any newly hired ones, or minus any servers whose hire is terminated at this decision epoch.

Each job present incurs a holding cost of c_1 per unit time spent in the system. These costs reflect the importance attached to fast service. In addition, each hired server incurs a cost of c_2 per unit time. This is predicated on the assumption that the host is dealing with a Cloud that allows hire and release at arbitrary moments, with charges proportional to the duration of hire. A different hire regime will be modeled in the next section.

Thus, the total cost incurred per unit of time during which there are j jobs present and n servers hired is $c_1 j + c_2 n$.

Note that in this model the decision interval does not depend on the current state or on the decision taken. The average length of that interval is the average interarrival time between batches: $\tau = 1/\lambda$.

Since the algorithms available for determining the optimal policy require that the state space is finite, we assume that there is an upper bound, J, for the number of jobs that may be present. If an incoming batch would cause that bound to be exceeded, some or all of its jobs are rejected. That condition is not too restrictive: under any policy that does not allow the queue to saturate, one can choose J sufficiently large so that the probability of rejecting jobs is negligible. However, the numerical complexity of the solution increases with J.

To write equations (2) for a given policy A in the present model, we need expressions for $c_j(n)$ and $p_{j,k}(n)$, where n is the number of servers hired in state j under policy A. We start with the costs. Let $T_j(n)$ be the total average time that the j jobs currently present spend in the system during the decision period, given that n servers are available to serve them. There are two cases to consider:

1. If $j \leq n$, all jobs present are being served. The contribution of each job to $T_j(n)$ is the average minimum of its remaining service time and the remaining decision period. Hence, in this case,

$$T_j(n) = \frac{j}{\lambda + \mu} \ ; \ j = 1, 2, \ldots, n \ . \tag{4}$$

2. If $j > n$, then n jobs are being served and $j - n$ are waiting. The next event to occur is either a service completion, with probability $n\mu/(\lambda + n\mu)$, or an arrival of a new batch, with probability $\lambda/(\lambda+n\mu)$. The average interval until that event is $1/(\lambda + n\mu)$, and there are j jobs present during it. If the next event is a service completion, then the decision period continues with $j - 1$ jobs present; otherwise it terminates and there is no further contribution to $T_j(n)$. This provides a recurrence relation,

$$T_j(n) = \frac{j}{\lambda + n\mu} + \frac{n\mu}{\lambda + n\mu} T_{j-1}(n) \; ;$$

$$j = n+1, n+2, \ldots, J \,. \tag{5}$$

Equation (4), together with the recurrences (5), allow the holding times $T_j(n)$ to be computed easily for all j and n. The average cost, $c_j(n)$, incurred during a decision period is the sum of the holding cost and the server cost:

$$c_j(n) = c_1 T_j(n) + c_2 n\frac{1}{\lambda} \,. \tag{6}$$

Before addressing the transition probabilities $p_{j,k}(n)$, consider the probability, $q_{j,k}(n)$, that there will be k jobs present *just before* the next decision epoch, given that there are j jobs now and n servers are available. That is the probability that $j - k$ jobs are completed during the decision interval. There are three distinct cases:

1. If $j < n$, more servers become idle with each departing job. In order that k jobs are left at the end of the decision period, the latter must terminate when there are k busy servers. Hence,

$$q_{j,k}(n) = \left[\prod_{i=k+1}^{j} \frac{i\mu}{\lambda + i\mu} \right] \frac{\lambda}{\lambda + k\mu} \; ; \quad k = 0, 1, \ldots, j \,, \tag{7}$$

where an empty product is equal to 1 by definition.

2. If $j \geq n$ and $k \geq n$, then $q_{j,k}(n)$ is the probability that exactly $j - k$ jobs are completed by n busy servers before the decision period terminates:

$$q_{j,k}(n) = \left[\frac{n\mu}{\lambda + n\mu} \right]^{j-k} \frac{\lambda}{\lambda + n\mu} \; ; \quad k = n, n+1, \ldots, j \,. \tag{8}$$

3. If $j \geq n$ and $k < n$, then of the $j - k$ completions that must take place before the end of the observation period, $j - n + 1$ occur while n servers are busy and $n - 1 - k$ with gradually diminishing number of busy servers:

$$q_{j,k}(n) = \left[\frac{n\mu}{\lambda + n\mu} \right]^{j-n+1} \left[\prod_{i=k+1}^{n-1} \frac{i\mu}{\lambda + i\mu} \right] \frac{\lambda}{\lambda + k\mu} \; ;$$

$$k = 0, 1, \ldots, n-1 \,. \tag{9}$$

Now we can obtain the transition probabilities from state j to state k, $p_{j,k}(n)$, by remarking that the number of jobs present after the arrival of the next batch is the convolution of the number left over at the end of the decision interval and the number contained in the new batch. Hence,

$$p_{j,k}(n) = \sum_{i=0}^{m} q_{j,i}(n) b_{k-i} \; ; \quad k = 1, 2, \ldots, J - 1 , \tag{10}$$

where $m = \min(j, k - 1)$. The exception to that pattern is destination state J, which may be reached after rejecting some new arrivals:

$$p_{j,J}(n) = \sum_{i=0}^{j} q_{j,i}(n) \sum_{s=J-i}^{\infty} b_s . \tag{11}$$

All quantities necessary for setting up equations (2), and hence for applying the policy improvement algorithm, are now available.

N.B. The reason for assuming that the batch interarrival intervals are distributed exponentially was the tractability of the expressions for $c_j(n)$ and $p_{j,k}(n)$. It would be possible to relax that assumption, e.g. by replacing the exponential with a phase-type distribution. However, the resulting expressions would be considerably more complicated.

4 Fixed Hiring Periods

We now address a system where a server must be hired for a sizeable minimum period of time, τ. Amazon, for example, hires servers by the hour. Although in principle one could initiate a hire at any time, it is reasonable, and more tractable, to use the instants $0, \tau, 2\tau, \ldots$, as decision epochs (i.e., the length of the decision interval is τ). Assume that jobs arrive singly during a decision interval, in a Poisson stream with rate λ. Their service times are again distributed exponentially, with mean $1/\mu$.

Thus, if there are j jobs in the system at a decision epoch, and n servers are hired, then during an interval of length τ the queue behaves as a transient $M/M/n/J$ queue (J is the bound on the number of jobs present), with initial state j. To define our decision process, we need the transition probabilities, $p_{j,k}(n)$, that there will be k jobs at time τ, given that there were j jobs at time 0 and n servers were hired.

Denote by $P(t) = [p_{j,k}(t)]$, $j, k = 0, 1, \ldots, J$, the transient transition probability matrix for the $M/M/n/J$ queue over the interval $(0,t)$. Clearly, $P(0) = I$, where I is the $(J+1) \times (J+1)$ identity matrix. We are interested in computing the j'th row of $P(\tau)$.

Let G be the generator matrix for the $M/M/n/J$ queue:

$$
G = \begin{bmatrix}
-\lambda & \lambda \\
\mu_1 & -(\lambda + \mu_1) & \lambda \\
& & \ddots \\
& & & -(\lambda + \mu_{J-1}) & \lambda \\
& & & \mu_J & -\mu_J
\end{bmatrix}, \tag{12}
$$

where $\mu_i = \min(i, n)\mu$. The matrix $P(t)$ is given by the matrix-exponential:

$$
P(t) = e^{Gt} . \tag{13}
$$

If the solution algorithms are implemented in Matlab, this matrix exponentiation can be performed by the built-in function $expm(G * t)$, which is stable and fast. If that is not available, one could employ the 'uniformization' technique, which involves replacing the continuous-time Markov process with an equivalent discrete-time Markov chain using the parameter $\gamma = \lambda + n\mu$ (e.g., see [14]). The generator matrix G is replaced by the matrix:

$$
Q = I + \frac{G}{\gamma} ,
$$

where I is the identity matrix. Then $P(t)$ is given by the series:

$$
P(t) = \sum_{i=0}^{\infty} Q^i \frac{(\gamma t)^i}{i!} e^{-\gamma t} . \tag{14}
$$

This expression provides an efficient way of computing $P(t)$ because (a) Q is a stochastic matrix, so the elements of Q^i remain uniformly bounded for all i (since the rows always sum up to 1), and (b) the Poisson probabilities that appear in (14) converge rapidly to 0. Hence, the infinite series can be truncated on the right, and possibly on the left, resulting in a finite sum:

$$
P(t) = \sum_{i=\ell}^{r} Q^i \frac{(\gamma t)^i}{i!} e^{-\gamma t} , \tag{15}
$$

where ℓ and r are chosen so that the two omitted tails are negligible (see [7]).

It remains to determine the average cost, $c_j(n)$, incurred during a decision interval. Let L_j be the average number of jobs in the system at time τ, given that there were j jobs at time 0 and n servers were hired. That average is obtained:

$$
L_j = \sum_{k=1}^{J} k p_{j,k}(\tau) . \tag{16}
$$

The average number of jobs present *during* the decision interval can be approximated by taking the mean of the queue sizes at the beginning and end of the interval, i.e. $(j + L_j)/2$. Hence, the total cost incurred during the interval is given by:

$$c_j(n) = \left[c_1 \frac{j + L_j}{2} + c_2 n \right] \tau. \qquad (17)$$

Using these expressions, the optimal policy can be computed as described in section 2.

N.B. One might wish to relax the assumptions that jobs arrive in a Poisson stream during a decision interval, and their lengths are distributed exponentially. Some generalizations using phase-type distributions could be treated numerically, but replacing the $M/M/n/J$ queue with a $GI/G/n/J$ one would require major approximations.

5   Heuristics and Experiments

When the computation of the optimal becomes expensive, it may be worth exploring policies that are sub-optimal, but offering good performance and ease of implementation.

A promising heuristic policy for any given model is the one which, at every decision epoch, minimizes the average cost incurred during the current decision interval. In other words, when the current state is j, take the action n^* such that:

$$c_j(n^*) = \min_n c_j(n), \qquad (18)$$

where $c_j(n)$ is the cost appropriate to the model. This short-term policy that looks only at the current state and does not care about the future. It will be called the 'greedy' heuristic, as this type of policies are commonly referred to.

The implementation of the greedy heuristic does not require any iterations; it is enough to evaluate the costs $c_j(n)$ for different values of n. Hence, the complexity of implementing the greedy heuristic is $O(JC)$, where C is the complexity of evaluating an individual cost. In practice, the greedy heuristic is orders of magnitude faster to find than the optimal policy.

The performance of the greedy heuristic will be compared with that of the optimal policy, for each of our models. In addition, an even simpler policy will be introduced to use as a benchmark. The latter abandons dynamic decision-making altogether and hires a fixed number of servers, n^*, regardless of the system state. This is, in fact, the policy often adopted in practice. To avoid saturating the queue, n^* should be chosen so that the average long-term server occupancy is less than 100%. For example, one could aim for an occupancy of 70%. In the case of batch arrivals, bearing in mind that the offered load is $\rho = \lambda b/\mu$, where b is the average batch size, the above condition implies:

$$n^* = \left\lceil \frac{\lambda b}{0.7\mu} \right\rceil. \qquad (19)$$

For the second model, the offered load is $\rho = \lambda/\mu$, so the allocation becomes:

$$n^* = \left\lceil \frac{\lambda}{0.7\mu} \right\rceil. \qquad (20)$$

That policy will be referred to as the 'fixed policy'.

Figure 1 illustrates and compares the behaviour of the three policies for the batch arrivals model, in the case where batch sizes are distributed geometrically with parameter α. That is, the probability that a batch contains j jobs is $\alpha(1 - \alpha)^{j-1}$. The average batch size is $b = 1/\alpha$. The offered load is increased by decreasing α, and the long-term average cost, g, is plotted against the average batch size. The average service time is $1/\mu = 1$, while the batch arrival rate is $\lambda = 0.1$. In this experiment, it was assumed that the unit holding cost and the unit server cost are equal: $c_1 = c_2 = 1$.

The bound on the number of jobs in the system was taken as $J = 100$. Under all three policies, the probability of reaching that bound is small. For example, when the average batch size is 50, the probability that a batch of size 100 will be submitted is about 0.1.

A notable feature of the figure is that the greedy heuristic is almost optimal over the entire range of offered loads. One would therefore be justified in using the heuristic in practice, knowing that its performance cannot be improved significantly. By contrast, the costs of the fixed policy are considerably higher. That remains the case if the 70% occupancy of the servers is replaced by 80% occupancy. Of course, the more the fixed policy over-provides servers unnecessarily, the poorer its performance would be. The non-monotone character of the graph for the fixed policy is due to the rounding-up operation in (19).

Next, we experiment with a batch size distribution that has been constructed to have a large coefficient of variation. More precisely, batches consist of a single job with probability 0.7, and B jobs with probability 0.3. The average batch size is $b = 0.7 + 0.3B$. The coefficient of variation grows roughly linearly with B. In figure 2, B is varied between 20 and 100, and the average achieved cost is plotted against b.

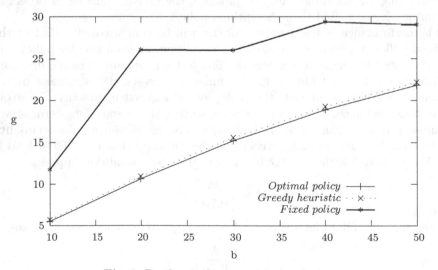

Fig. 1. Batch arivals: geometric batch sizes

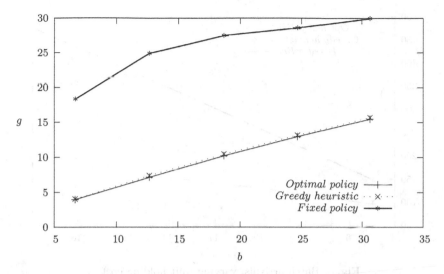

Fig. 2. Batch arrivals: skewed batch size distribution

It seems that large coefficients of variation do not prevent the greedy heuristic from performing well. Its costs are almost indistinguishable from those of the optimal policy. On the other hand, the fixed policy is, if anything, worse than before in comparison.

In the third experiment, the characteristics of the demand are held fixed, at $\lambda = 0.1$, $\mu = 1$, $\alpha = 0.04$ (i.e., average batch size of 25). Also, the unit server cost is fixed at $c_2 = 10$. What is varied is the unit holding cost, from $c_1 = 5$ to $c_1 = 20$. That is, the relative cost of keeping jobs in the system is varied from half to double the cost of a server.

The results are shown in figure 3, where the average long-term costs g achieved by the optimal policy, the greedy heuristic and the fixed policy are plotted against c_1.

Again, it is notable that the greedy heuristic achieves nearly optimal costs over the entire range of c_1 values. By contrast, the performance of the fixed policy is rather poor. Moreover, whereas the cost of the fixed policy grows linearly with c_1 (as can be expected), those of the optimal and greedy policies grow slower than linearly.

It is perhaps worth pointing out that, for all points in these three figures, the policy improvement algorithm took no more than 3 iterations to find the optimal policy.

The remaining experiments concern the model with fixed hire periods. In figure 4, the offered load is increased from $\rho = 10$ to $\rho = 18$ by varying the job arrival rate. The service rate is kept at $\mu = 1$, and the unit holding cost is half of the server cost: $c_1 = 0.5$, $c_2 = 1$. The bound on the number of jobs is $J = 50$. The hire period length is $\tau = 4$, meaning on average, between 40 and 72 jobs arrive during a decision period. The fixed policy is based on equation (20).

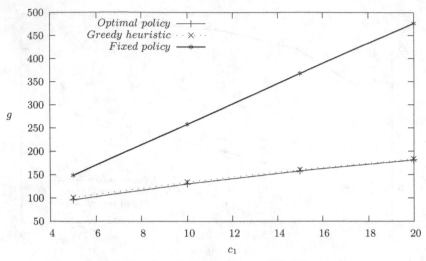

Fig. 3. Batch arrivals: varying unit holding cost

We observe that the difference between the worst policy (fixed) and the best one (optimal) is now much narrower. This is due to the fact that jobs continue to arrive throughout a decision period, and the rate of arrivals does not depend on the action taken. This reduces the advantages derived from making dynamic decisions. The costs achieved by the optimal policy are about 15% lower than those of the fixed policy. The greedy heuristic still performs quite well, but its costs are now about 10% higher than those of the optimal policy.

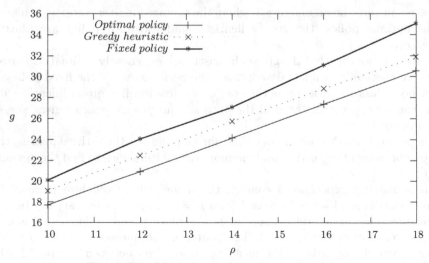

Fig. 4. Fixed hiring periods: varying offered load

In figure 5, the job arrival rate is kept fixed at $\lambda = 12$. The service rate, server cost and decision period length have the same values as before, $\mu = 1$, $c_2 = 1$, $\tau = 4$, while the unit holding cost is varied from half to twice the server cost: $0.5 \le c_1 \le 2$.

The average costs achieved by the three policies are quite close over the entire range of c_1 values. Moreover, it is notable that the higher the value of c_1 relative to c_2, the closer those costs are, i.e. the lower the benefit of dynamic decision-making. Indeed, one could have expected that when the dominant factor

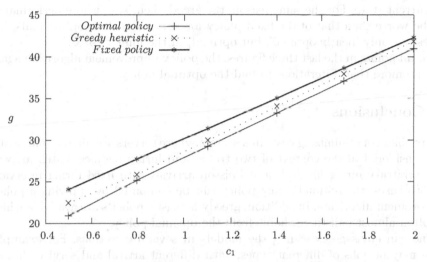

Fig. 5. Fixed hiring periods: varying unit holding cost

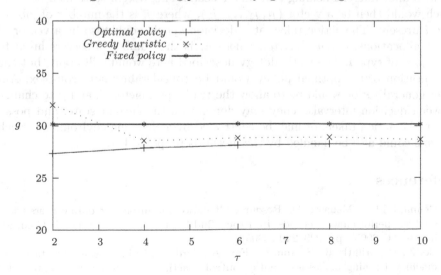

Fig. 6. Fixed hiring periods: varying hire period length

is customer performance, the most important part of the policy is to always maintain enough servers to cope with the load.

In the final experiment, traffic characteristics and unit costs are kept fixed ($\lambda = 12$, $\mu = 1$, $c_1 = c_2 = 1$), while the length of the decision interval is varied from $\tau = 2$ to $\tau = 10$. That is, the average number of arrivals during a decision interval varies from 24 to 120.

The fixed policy is independent of τ, so its graph is a horizontal line. The optimal and greedy policies also approach a horizontal asymptote. This is predictable, since the system tends to reach steady state during a large decision interval, and the distribution at the next decision epoch becomes independent of the current state. For the same reason, the greedy heuristic, whose performance can be worse than that of the fixed policy for very short decision intervals, becomes not only 'nearly optimal', but optimal, in the limit $\tau \to \infty$.

For all points in the last three figures, the policy improvement algorithm again took no more that 3 iterations to find the optimal policy.

6   Conclusions

The problem of minimizing costs in a system where servers are hired dynamically was considered in the context of two traffic and hiring regimes: batch arrivals with arbitrary hiring intervals and Poisson arrivals with fixed hiring intervals. In both cases, the optimal hiring policy can be computed by applying a policy improvement algorithm. In addition, greedy heuristic policies are available which are often almost indistinguishable from the optimal policy.

One can envisage extending the models in several directions. For example, there may be jobs of different types, with different arrival and service characteristics and different holding and server costs. The system state at a decision epoch would then be a vector $(j_1, j_2, \ldots, j_k)$, where j_i is the number of jobs of type i present. The action taken at a decision epoch would also be a vector of server allocations, $(n_1, n_2, \ldots, n_k)$, where n_i is the number of servers hired to serve jobs of type i. The methodology described here would still apply, but the computation of the optimal policy would be considerably more complex. Another generalization would be to allow the traffic parameters λ and μ to change between decision intervals. They may depend on the current state, and possibly on the action taken, or may be controlled by a changing environment. Such systems could also be handled by the methods proposed here.

References

1. Bennani, M.N., Menascé, D.: Resource allocation for autonomic data centers using analytic performance methods. In: Procs. 2nd IEEE Conf. on Autonomic Computing, ICAC 2005), pp. 229–240 (2005)
2. Bodík, P., Griffith, R., Sutton, C., Fox, A., Jordan, M., Patterson, D.: Statistical machine learning makes automatic control practical for internet datacenters. In: Conf. on Hot Topics in Cloud Computing, HotCloud 2009, Berkeley, CA, USA (2009)

3. Byun, E.-K., Kee, Y.-S., Kim, J.-S., Maeng, S.: Cost optimized provisioning of elastic resources for application workflows. Future Generation Computer Systems 27(8), 1011–1026 (2011), http://dx.doi.org/10.1016/j.future.2011.05.001
4. Byun, E.-K., Kee, Y.-S., Kim, J.-S., Deelman, E., Maeng, S.: BTS: Resource capacity estimate for time-targeted science workflows. Journal of Parallel and Distributed Computing 71(6), 848–862 (2011), doi:10.1016/j.jpdc.2011.01.008
5. Chandra, A., Gong, W., Shenoy, P.: Dynamic resourse allocation for shared data centers using online measurements. In: Procs. 11th ACM/IEEE Int. Workshop on Quality of Service (IWQoS), pp. 381–400 (2003)
6. Chaisiri, S., Lee, B.S., Niyato, D.: Optimization of resource provisioning cost in cloud computing. IEEE Transactions on Services Computing 5(2), 164–177 (2012)
7. Fox, B.L., Glynn, P.W.: Computing Poisson Probabilities. Management Science and Operations Research 31(4), 440–445 (1988)
8. Hiden, H., Woodman, S., Watson, P., Cala, J.: Developing cloud applications using the e-science central platform. Royal Soc. of London, Phil. Trans. A. (Mathematical, Physical and Engineering Science), 371 (2013)
9. Lampe, U., Siebenhaar, M., Hans, R., Schuller, D., Steinmetz, R.: Let the clouds compute: Cost-efficient workload distribution in infrastructure clouds. In: Vanmechelen, K., Altmann, J., Rana, O.F. (eds.) GECON 2012. LNCS, vol. 7714, pp. 91–101. Springer, Heidelberg (2012)
10. Mazzucco, M., Dyachuk, D., Dikaiakos, M.: Profit-aware server allocation for green internet services. In: IEEE Int. Symp. on Modeling, Analysis and Simulation of Computer and Telecommunication Systems (MASCOTS), pp. 277–284 (2010)
11. Mazzucco, M., Mitrani, I., Fisher, M., McKee, P.: Allocation and Admission Policies for Service Streams. In: Procs. MASCOTS 2008, Baltimore, pp. 155–162 (2008)
12. Mazzucco, M., Vasar, M., Dumas, M.: Squeezing out the cloud via profit-maximizing resource allocation policies. In: IEEE Int. Symp. on Modeling, Analysis and Simulation of Computer and Telecommunication Systems (MASCOTS), pp. 19–28 (2012)
13. Mitrani, I.: Managing Performance and Power Consumption in a Server Farm. Annals of Operations Research (2011), doi:10.1007/s10479-011-0932-1
14. Reibman, A., Trivedi, K.: Numerical transient analysis of Markov models. Computing and Operations Research 15(1), 19–36 (1988)
15. D. Thain, T. Tannenbaum and Miron Livny, "Distributed computing in practice: the Condor experience", Concurrency and Computation: Practice and Experience, 17 (2-4),323-356, doi:http://dx.doi.org/10.1002/cpe.v17:2/4
16. Tijms, H.C.: Stochastic Models. John Wiley and sons (1994)
17. Urgaonkar, R., Kozat, U.C., Igarashi, K., Neely, M.J.: Dynamic Resource Allocation and Power Management in Virtualized Data Centers. In: IEEE/IFIP NOMS 2010, Osaka, Japan (2010)

Performance Evaluation of NoSQL Databases

Andrea Gandini[1], Marco Gribaudo[1], William J. Knottenbelt[2], Rasha Osman[2],
and Pietro Piazzolla[1]

[1] Politecnico di Milano, via Ponzio 34/5, 20133 Milano, Italy
{andrea.gandini,marco.gribaudo,pietro.piazzolla}@polimi.it
[2] Department of Computing, Imperial College London, London SW7 2AZ, UK
{rosman,wjk}@imperial.ac.uk

Abstract. NoSQL databases have emerged as a backend to support Big
Data applications. NoSQL databases are characterized by horizontal scalability, schema-free data models, and easy cloud deployment. To avoid
overprovisioning, it is essential to be able to identify the correct number
of nodes required for a specific system before deployment. This paper
benchmarks and compares three of the most common NoSQL databases:
Cassandra, MongoDB and HBase. We deploy them on the Amazon EC2
cloud platform using different types of virtual machines and cluster sizes
to study the effect of different configurations. We then compare the behavior of these systems to high-level queueing network models. Our results show that the models are able to capture the main performance
characteristics of the studied databases and form the basis for a capacity
planning tool for service providers and service users.

1 Introduction

Recently NoSQL databases have become widely adopted on cloud platforms due
to their horizontal scalability, schema-free data model and the capability to manage large amounts of data. Huge amounts of data require systems that are able
not only to retrieve information in very short timescales, but also to scale at
the same rate as the data increases. The growing importance of Big Data applications [15] has driven the development of a wide variety of NoSQL databases,
e.g. Google's BigTable [5], Amazon's Dynamo [10], Facebook's Cassandra [16],
Oracle's NoSQL DB [20], MongoDB [18] and Apache's HBase [14].

One of the main features of NoSQL databases is horizontal scalability [4]; that
is, the capacity to scale in performance when the number of machines added to
an existing cluster increases. This capability potentially wastes resources due
to over-provisioning. Thus, being able to identify the correct number of nodes
required for a specific workload is important. Moreover, correctly adding or removing nodes from a distributed database is often a time-consuming operation
whose impact can be minimised by proper planning.

The purpose of this work is to benchmark three of the most common NoSQL
databases [9], namely *Cassandra*, *MongoDB* and *HBase* in order to provide insights about their behavior under various settings. We deploy them on the Amazon EC2 cloud platform using different types of virtual machines (in term of

A. Horváth and K. Wolter (Eds.): EPEW 2014, LNCS 8721, pp. 16–29, 2014.

CPUs, memory, I/O speed, etc.) and variable number of nodes, in order to study the effect of different configurations on the performance of these systems. Finally, we present two simple high-level queuing network models that are able to capture the main features of the considered databases. These models are able to provide insight into suitable cluster sizes for NoSQL applications and form the basis of a capacity planning tool for service providers.

To date there has been limited work in benchmarking the performance of NoSQL datastores. Rabl et al. [25] benchmark six NoSQL datastores, identifying their ability to support application performance management tools. The authors report response times and throughput for workloads that scale as the number of nodes increases in a configuration. The experiments are conducted on a fixed physical hardware architecture. By contrast, the present work benchmarks three popular datastores on a visualized architecture and explores their performance under different hardware, software and workload configurations. In [8], the authors evaluate and compare the performances of four different NoSQL systems (Cassandra and HBase among them) when used for RDF processing. Their work is mainly aimed at characterizing the differences between NoSQL systems and native triple stores.

In terms of modelling efforts, the performance community has concentrated on the modelling of traditional relational databases [23], using mainly queueing networks, e.g. [19, 12, 11, 21] and more recently queueing Petri nets, e.g. [22, 7]. In [3] Mean Field analysis is used to model the replication of resources in a NoSQL application, while [1] uses a multi-formalism approach to model queries in the Apache Hive data warehousing NoSQL solution. The authors in [24], use queueing Petri nets to study the replication performance of the Cassandra NoSQL datastore. We are unaware of previous work that attempts to depict the main characteristics of multiple NoSQL datastores in one model as presented in this paper.

The remainder of this paper is organized as follows. Section 2 explains the architecture and primary characteristics of our target NoSQL databases. Section 3 discusses the experimental setup, Section 4 presents benchmarking results, and Section 5 introduces our queueing models. Finally, Section 6 concludes and considers avenues for future work.

2   NoSQL Database Architecture

Here we give a brief introduction to the main characteristics of the NoSQL databases considered, highlighting those aspects that impact on performance:

Cassandra [13] belongs to the *wide-column store* NoSQL family [26] and provides an extended key-value store method built on a column-oriented structure. Its architecture, shown in Fig. 1a, is based on a *ring* topology, in which every node is identical to the others, guaranteeing that the system has no single point of failure. Each record inserted in the database has an associated hash value called *token*. The range of tokens is partitioned among the nodes to balance the ring. Cassandra allows replication among the nodes in the cluster by duplicating

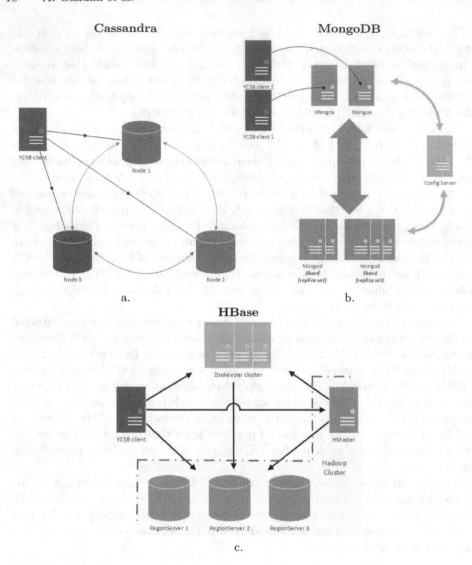

Fig. 1. The architectures of our target NoSQL Databases

data from a node to subsequent nodes on the ring. The number of replicas for each data item can be controlled by a parameter set by the user, called the *replication factor*. The *consistency level* defines the number of replicas that should respond to a data request. It is also possible to define which nodes communicate outside the ring. Such nodes are called *entry points*.

MongoDB [18] belongs to the *document-store* NoSQL family. As shown in Fig. 1b, the main components of the system are:

- `mongod`: data nodes for storing and retrieving data.
- `mongos`: the only instances able to communicate outside of the cluster.
- `config-server`: the containers of the metadata about the objects stored in the `mongod`. The metadata is used in case of a node failure. A cluster allows only one or three `config-server` instances.

Each running component constitutes one node in the MongoDB cluster.

Replication is achieved by means of *shards*. Every shard is a group of one or more nodes. The number of nodes in a shard determines the replication factor of the system: each member of the shard contains a copy of the data, creating a so-called *replica set*. Nodes belonging to the same shard have the same data. Within a shard only one node can be a master, able to execute write and read tasks; the others are considered *slaves* and can only perform read operations.

HBase [14] is based on the Hadoop map-reduce framework and Hadoop Distributed File System (HDFS). It belongs to the *wide-column store* NoSQL family. As shown in Fig. 1c, HBase relies on two supporting applications:

- `Hadoop`: a distributed map-reduce framework that provides high-throughput access to application data and which manages replication.
- `Zookeeper`: provides a distributed configuration and synchronization service for large distributed systems.

HBase has two main types of nodes: the *master*, that accounts for the nodes that are alive and which provides communication services, the zookeeper cluster and clients; and the *region servers*, which distribute data using the notion of *regions*. Regions are initially allocated to a node and split when they become too large. Thus, HBase tends to accumulate data on some nodes with a non-uniform distribution of data, especially when the system is lightly loaded.

3   Experimental Setup

We used the *Yahoo! Cloud Serving Benchmark (YCSB)* [6], a workload generator developed by `Yahoo!`, as our benchmarking framework. YCSB provides means to stress-test multiple databases and compare them in a fair and consistent way. It operates in two phases: first, data is loaded onto the data nodes (the ones responsible for storing actual data, irrespective of the naming convention of each database architecture). In our experiments, the data was generated randomly and stored by the database in its specific data format. In the second phase, YCSB executes the actual tests, in which random key requests are sent to the data nodes. The requests are randomly mixed with 50% reads and 50% writes for each client thread. The number of client threads concurrently querying the database is either varied to simulate different workload levels, or fixed to a value that saturates the DB servers. For each execution, the benchmark measures the latency (in microseconds per operation) and throughput (in operations per second).

The servers used for the tests are provided by the Amazon EC2 cloud platform. Table 1 presents the virtual hardware specifications of the machines employed for the benchmarking experiments. All nodes of a database, independently of their role in the architecture, are run on a virtual machine with the same specification during a given test to allow for a fair comparison. To ensure that the

Table 1. Virtual Hardware Specifications

Instance Type	vCPU	Physical Processor	Memory(GiB)
m1.large	2	Intel Xeon Family	7.5
c1.xlarge	8	Intel Xeon Family	7
c3.2xlarge	8	Intel Xeon E5-2680 v2	15

benchmark application does not affect the performance of the database under test, it is hosted on a dedicated machine that is not part of the DB cluster. We will refer to this machine as the *ycsb client*. We used the m1.large virtual machine specification for the machines hosting the nodes of the database and the ycsb client for most of our experiments. The m1.large virtual machines were chosen since they offer a good trade-off between cost and available memory. In particular, it satisfies the minimum memory requirement of HBase which is at least 7 GiB. When the number of cores per server became the focus of the tests, we switched to c1.xlarge instances to have a larger number of cores available. During some of the tests, due to the high number of client threads required for the total number of nodes, the ycsb client became the bottleneck of the system. In those cases, we switched the ycsb client machine to a c3.2xlarge instance. Every machine runs the Ubuntu Server 12.04 operating system release provided by Amazon. In order to reduce the variability inherent to the cloud environment on the measured results, some tests were performed during particular time slots when we observed that the provisioned servers showed more stability.

Each test is performed using a specific DB configuration in terms of active cores per server, number of nodes and data replication. The ycsb tool is executed at least 20 times for each configuration before collecting the average values for response time and throughput. In addition to the performance indexes collected by the benchmark, we used several bash scripts to track the cpu utilization of every single node in the cluster. Testing the scalability of MongoDB was not as straightforward as for the other databases. For this DB we had to experimentally determine which of its configurations was able to scale more uniformly. For MongoDB both mongos and mongod instances run on every node, while the config-server (that has a negligible load) runs in addition to the other two services on a randomly chosen node. However, the ycsb tool is not able to set more than one mongos connection at a time; therefore we perform the tests on MongoDB by scaling the number of clients to be equal to the number of nodes. In addition, the number of client threads running on one node is scaled by an appropriate factor to maintain the same average load used for Cassandra and HBase.

4   Benchmarking Results

Our initial tests focused on infrastructure-dependent parameters that influence the performance of a NoSQL database. We began with investigating the effect of

node capacity, i.e. number of cores, on performance. The impact of the number
of nodes in a cluster and the number of threads (workload) on the client was
then studied. Finally, we examined the effect of the replication factor.

4.1 Number of Cores

The first set of experiments measure the database performance as the number
of cores on a single node is increased. The number of client threads was fixed
at 50 threads irrespective of the number of cores. These tests are performed on
a single machine of the c1.xlarge type for both Cassandra and MongoDB, in
which each was configured with one node only. For HBase, the *region server*
was hosted on a single machine of the c1.xlarge type, and the *master* and the
zookeeper were each hosted on a different m1.xlarge VM. In Fig. 2a, we show the
total throughput as function of the number of cores active on a single node. It is
evident that HBase is able to take advantage of the number of cores better than
the other databases, scaling almost linearly. MongoDB scales well initially but
then it shows a plateau at around 6 cores. Such a tendency seems absent from
Cassandra which, although significantly slower that HBase in absolute terms,
also scales quite linearly. Fig. 2b shows the mean update latency as function
of the number of cores. In this case the mean response time required for write
operations is more or less constant for both Cassandra and HBase, showing,
presumably that this task is mainly I/O bound on these databases. HBase shows
an extremely low update latency, benefiting from its client-side write buffer.
MongoDB shows an improvement in performance when the number of cores
increases from one to three, then stabilizes. This is due to the fact that the
data node must communicate updates with the config-server. For the read
operations in Fig. 2c, the behavior is more hyperbolic, as read operations are
more likely to benefit from parallelization. In this case, MongoDB has the lowest
latency; Cassandra shows the highest latency when running on a VM with a low
number of cores. HBase has a much lower performance for read requests than
for write requests since the former do not exploit a client-side buffer.

a.   b.   c.

Fig. 2. System throughput and latencies as a function of the number of cores

4.2   Number of Nodes

Distributed database performance is strongly affected by the size of the cluster and the workload it has to handle. To investigate the effect of the number of nodes on performance we simulate higher workload intensities by increasing the number of threads running simultaneously on the client for each request. For Cassandra and MongoDB, the number of entry points is scaled to match the number of nodes; for HBase the number of entry points is not configurable. For the single node case, we consider the performance of both on a single and a double core CPU. For more than one node, we only present results considering VMs with two cores.

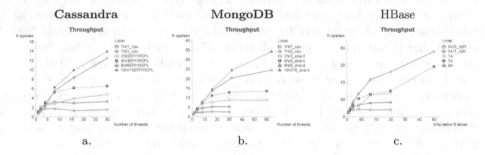

Fig. 3. Throughput as a function of the number of nodes

The throughput of Cassandra is considered in Fig. 3a. The throughput generally increases with an increase in the number of nodes. Fig. 3a shows the saturation point of the database, that is the point at which the database is fully utilized and throughput does not increase along with the number client threads. Configurations with 8 nodes and above do not saturate the system even at 30 threads. Similar results are shown for MongoDB in Fig. 3b and for HBase in Fig. 3c. Note that, in terms of absolute performance, HBase presents higher throughput than Cassandra and MongoDB. However, its scalability is less predictable. This is due to the non-uniform distribution of data among the nodes, in contrast to Cassandra and MongoDB.

Read and write latencies for the three DBs are reported in Fig. 4. Cassandra does not show a significant gain when the number of nodes jumps from eight to sixteen: this might be an indication of some scalability issues for very large configurations. MongoDB does not saturate with the considered workloads when distributed on four or eight nodes. HBase presents an almost linear behavior for all the considered configurations.

4.3   Replication

Replication of data in a DB cluster introduces yet another performance impacting factor. A replication factor equal to one, means that there is only one copy

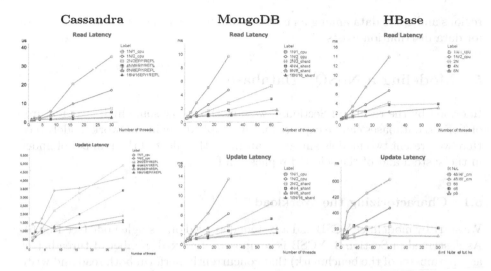

Fig. 4. Latency for the three databases with different number of nodes

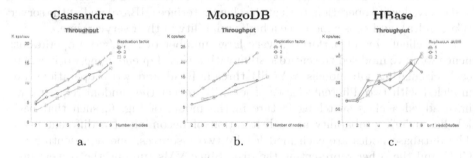

Fig. 5. Mean throughput for different replication factor

of the data item among all the nodes of the cluster. Higher replication factors means that, aside from the original data, copies are stored on different nodes.

Fig. 5a shows that increased replication in the cluster produces degradation of throughput. This degradation increases as more copies of the same data are requested in order to achieve consistency within the database. Fig. 5b shows the behavior of MongoDB when the replication factor is set to values higher than one. In this case, having replicas of data decreases the throughput, but differently from Cassandra, having more than one replica does not decrease performance. This is due to the fact that increase of replication in MongoDB consists in having more than one data node belonging to a shard, which basically distributes the nodes to more than one master. In Fig. 5c the effects of replication in HBase are shown. Because of the architecture of HBase, the behavior is not what was expected. There was no noted decrease in performance, nor in throughput, nor in response time. This is due to the fact that Hadoop is responsible for data allocation in a process transparent to the user. Moreover, HBase packs its data regions and splits data among its nodes randomly, hence no uniform architecture for data distribution exists.

5 Modeling a NoSQL Database

Informed by the results of Section 4, we now show how some high level features of NoSQL databases can be captured using simple queueing network models [17]. Here we present two models aimed at capturing the effect of the number of nodes in the system and of the selected replication factor.

5.1 Characterizing the Workload

We start by modelling YCSB and a system deployed on a single node (see Fig. 6). As described in Section 3, YCSB is composed of a fixed number of threads (set as a parameter of the benchmark) that concurrently perform both read and write requests with equal probability. As seen in Section 4, read and write present different performance characteristics. Moreover, write requests are subject to caching and delayed write operation, both at the client interface (HBase) and at the server (Cassandra), and thus experience a high variability in their service time.

Each client thread performs a very large number of read/write operations; hence we have modeled the entire system with a closed queueing network, where each circulating job models a YCSB thread. Read and write operations are modeled with two different classes of jobs. To reflect the random requests, we have added a class switching feature in the model of Fig. 6, such that each job has a 50% probability at each iteration of becoming of a different class. The database nodes are each modeled by two resources: one representing the CPU and the other representing the disk. Since VMs running the servers may have more than one core, the number of servers in the CPU station is set to the actual number of cores, which in this case is 2 virtual cores. The resource representing the disks are single-server with FIFO service discipline to represent the serial access that characterizes storage components. Clients are modeled by a queueing station whose number of servers is set to the number of cores of the VMs where the YCSB client are executed. This is done to reflect the resource contention of the YCSB threads. The network latency is explicitly included in the model as a infinite-server delay station as it directly affects read and write response times.

The service time of read requests is not affected by caching; therefore, read requests are executed by all the resources when submitted by the client (i.e., the network, the server CPU and the server disk). Write requests can be cached either at the client side (do not enter the network after being produced in the client, e.g., the *write buffer* in HBase), or at the server side (requests are not immediately written to the disk, as in Cassandra). The write behavior is modeled by class-dependent routing that connects the client to itself immediately after the class switching (caching at the client), and by the route that immediately returns to the client after the server cpu (caching at the server). We then determined

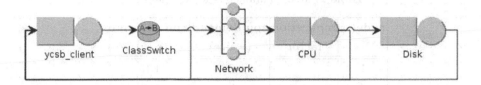

Fig. 6. Queueing network model representing a single node architecture

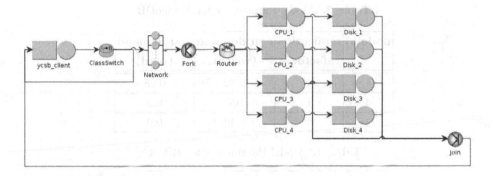

Fig. 7. Queueing network model representing a multi-node architecture

the service demands to use in the models. Network delay was measured using the *ping* command on the client machine, and estimated to be $0.45375ms$. The read and write service times for the disks were determined by benchmarking the I/O operations of the VMs and determined to be approximately $170ms$ (read) and $200ms$ (write). The CPU requirements and the caching probability are the parameters that reflect the behaviour of the considered NoSQL databases.

Tables 2, 3 and 4 show in the first column the estimated CPU demands for the read and write classes for Cassandra, MongoDB and HBase, respectively. For HBase, since it buffers write requests before sending them to the server, Table 4 has an extra column called $P(flush)$. It represents the probability that the generated request fills the buffer and that its content is sent to the server by routing it through the route that connects the class-switch node to the network delay. With probability $1 - P(flush)$, requests are not sent to the servers, so the corresponding job is immediately routed back to the client. For the other DBs, since they do not have client-side buffering, $P(flush) = 1$. To include server side caching of write commands, the probability of accessing the disk has been set to 10% on the DB: i.e. 90% of the write requests are served by the CPU only, and immediately return to the client.

The models have been analyzed using the JMT (Java Modeling Tool) [2]. The mean total throughput and mean read and write throughput for the single node configuration are shown in Fig. 8 for Cassandra, MongoDB and HBase. The `JSimGraph` component was used to solve the models by discrete event simulation. Confidence intervals were set to 95%; the resulting confidence intervals were too

Table 2. Model Parameters for Cassandra

node(s)	replication factor	cpu demand (write)	cpu demand (read)
1	-	0.075	1.15
4	1	0.8	1.25
4	4	0.15	0.55

Table 3. Model Parameters for MongoDB

node(s)	replication factor	cpu demand (write)	cpu demand (read)
1	-	0.42	0.28
4	1	0.7	0.3
4	4	0.42	0.9

Table 4. Model Parameters for HBase

node(s)	cpu demand (write)	cpu demand (read)	P(flush)
1	0.25	0.43	0.03
4	0.2	0.87	0.01

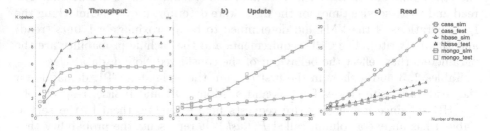

Fig. 8. Mean throughput and latency for one node for Cassandra, MongoDB and HBase

tight to appear on the graphs. Table 5 shows the relative error in the first line. From the results, the model captures the system mean throughput and the mean response times of the read and of the write operations for both Cassandra and MongoDB. For HBase, the model has a large error for the throughput. HBase is in fact the most complex of the three DBs, so a simple queuing network cannot accurately capture all its features and produce acceptable results.

5.2   Configurations with Multiple Nodes

We then model the configurations with more than one node with the queueing network model, as shown in Fig. 7. In this case we have several groups of two

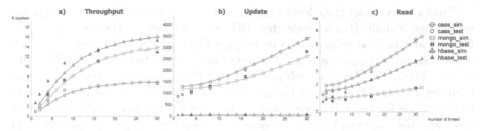

Fig. 9. Mean throughput and latency for four nodes and replication factor set to one for Cassandra, MongoDB and HBase

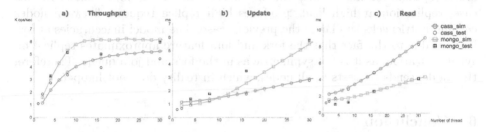

Fig. 10. Mean throughput and latency for four nodes and replication factor set to four for Cassandra, MongoDB and HBase

Table 5. Relative errors of the proposed models: throughput (**th**), write latency (**wr**), read latency (**re**)

node(s) repl.	Cassandra (th/wr/re)	MongoDB (th/wr/re)	HBase (th/wr/re)
1 / 1	4% / 14% / 8.4%	12.2% / 16.9% / 21.9%	24.4% / 10.7% / 9%
4 / 1	15.1% / 19.7% / 23.4%	7.5% / 6.3% / 16.3%	19.6% / 33% / 30.5
4 / 4	19.3% / 23.8% / 22.4%	8.7% / 23.4% / 2.4%	N.A.

queueing stations per node, representing respectively the CPU and the disk. A router models the choice of the node that will serve the requests: for Cassandra and HBase it implements a random selection policy. For MongoDB, it behaves differently for the read and for the write class. Read requests are randomly sent to the nodes. Write requests are sent to a single node (the topmost node of the model) to represent single-master replication of this specific database. We assume that network latency is the same for all the nodes: in this way we can model the network delay with a single delay station inserted before the router. Since the presence of more nodes introduces extra overhead in the computations necessary to serve the requests, we have to estimate a different service demand for each of the considered DB, as shown in the second line of Tables 2, 3 and 4. Fig. 9 compares the model results with the measurement of Section 4, and the second line of Table 5 reports the relative errors. The model performs quite well for high loads, but it presents larger errors for a small number of client threads.

Finally we consider the effect of replication in both Cassandra and MongoDB: we do not include HBase since that DB does not have a specific replication feature, and instead relies on the underlying Hadoop infrastructure. We model replication with the *fork and join* feature of queueing networks (nodes *Fork* and *Join* in Figure 7). We fork each job based on the number of replicas, and join them again before continuing to the client. In this case, we have to estimate the demands for the CPU operations to account for the overheads required to consider this additional feature. These corresponding values are presented in the last line of Tables 2 and 3. Results are compared with measurements in Fig. 10 and relative errors are reported in the last line of Table 5. The results are acceptable for high loads, but are inaccurate for the lower loads. In this case, replication at high loads produces high replica requests between nodes causing bottlenecks similar to the previous case. The model inaccuracies at low loads is due to the fact that the fork and join feature approximates replication synchronization, as it is not symmetric as in the fork and join queue. Therefore, the model sends requests to all nodes, which in reality does not happen.

6   Conclusion

This paper has presented benchmarks and models for three of the most common NoSQL databases: Cassandra, MongoDB and HBase. We deployed them on the Amazon EC2 cloud platform using different types of virtual machines and cluster sizes to study the effect of different configurations and to characterize the performance behavior of the databases. Using a high-level queueing network model we represented these characteristics. Our results showed that the models are able to capture much of the main performance characteristics of the studied databases at high workloads. Further investigation into modelling complex replication is required to accurately reflect the performance of replicated data. Future work includes benchmarking other NoSQL databases and providing a generic modelling framework.

References

[1] Barbierato, E., Gribaudo, M., Iacono, M.: Performance evaluation of nosql big-data applications using multi-formalism models. Future Generation Computer Systems (2013) (to appear) (available online)
[2] Bertoli, M., Casale, G., Serazzi, G.: JMT: Performance engineering tools for system modeling. SIGMETRICS Perform. Eval. Rev. 36(4), 10–15 (2009)
[3] Castiglione, A., Gribaudo, M., Iacono, M., Palmieri, F.: Exploiting mean field analysis to model performances of big data architectures. Future Generation Computer Systems (2013) (article in press); cited by (since 1996)
[4] Cattell, R.: Scalable sql and nosql data stores. SIGMOD Rec. 39(4), 12–27 (2011)
[5] Chang, F., Dean, J., Ghemawat, S., Hsieh, W.C., Wallach, D.A., Burrows, M., Chandra, T., Fikes, A., Gruber, R.E.: Bigtable: A distributed storage system for structured data. ACM Trans. Comput. Syst. 26(2), 4:1–4:26 (2008)

[6] Cooper, B.F., Silberstein, A., Tam, E., Ramakrishnan, R., Sears, R.: Benchmarking cloud serving systems with ycsb. In: Proceedings of the 1st ACM Symposium on Cloud Computing, SoCC 2010, pp. 143–154. ACM, New York (2010)

[7] Coulden, D., Osman, R., Knottenbelt, W.J.: Performance modelling of database contention using queueing petri nets. In: ICPE, pp. 331–334 (2013)

[8] Cudré-Mauroux, P., et al.: NoSQL databases for RDF: An empirical evaluation. In: Alani, H., et al. (eds.) ISWC 2013, Part II. LNCS, vol. 8219, pp. 310–325. Springer, Heidelberg (2013)

[9] Db engines. Db-engines ranking of database management systems (March 2014) (accessed: March 04, 2014)

[10] De Candia, G., Hastorun, D., Jampani, M., Kakulapati, G., Lakshman, A., Pilchin, A., Sivasubramanian, S., Vosshall, P., Vogels, W.: Dynamo: Amazon's highly available key-value store. SIGOPS Oper. Syst. Rev. 41(6), 205–220 (2007)

[11] Di Sanzo, P., Palmieri, R., Ciciani, B., Quaglia, F., Romano, P.: Analytical modeling of lock-based concurrency control with arbitrary transaction data access patterns. In: WOSP/SIPEW 2010, pp. 69–78. ACM, New York (2010)

[12] Elnikety, S., Dropsho, S., Cecchet, E., Zwaenepoel, W.: Predicting replicated database scalability from standalone database profiling. In: EuroSys 2009, pp. 303–316. ACM, New York (2009)

[13] Apache Software Foundation. Cassandra, http://cassandra.apache.org/ (accessed: March 04, 2014)

[14] Apache Software Foundation. Hbase project, https://hbase.apache.org/ (accessed: March 04, 2014)

[15] Labrinidis, A., Jagadish, H.V.: Challenges and opportunities with big data. Proc. VLDB Endow. 5(12), 2032–2033 (2012)

[16] Lakshman, A., Malik, P.: Cassandra: A decentralized structured storage system. SIGOPS Oper. Syst. Rev. 44(2), 35–40 (2010)

[17] Lazowska, E.D., Zahorjan, J., Graham, G.S., Sevcik, K.C.: Quantitative System Performance. Prentice-Hall (1984)

[18] MongoDB, Inc. Mongodb, http://www.mongodb.org/ (accessed: March 04, 2014)

[19] Nicola, M., Jarke, M.: Performance modeling of distributed and replicated databases. IEEE Trans. on Knowl. and Data Eng. 12(4), 645–672 (2000)

[20] Oracle. Oracle nosql database. An oracle white paper. white paper (September 2011)

[21] Osman, R., Awan, I., Woodward, M.E.: Queped: Revisiting queueing networks for the performance evaluation of database designs. Simulation Modelling Practice and Theory 19(1), 251–270 (2011)

[22] Osman, R., Coulden, D., Knottenbelt, W.J.: Performance modelling of concurrency control schemes for relational databases. In: Dudin, A., De Turck, K. (eds.) ASMTA 2013. LNCS, vol. 7984, pp. 337–351. Springer, Heidelberg (2013)

[23] Osman, R., Knottenbelt, W.J.: Database system performance evaluation models: A survey. Perform. Eval. 69(10), 471–493 (2012)

[24] Osman, R., Piazzolla, P.: Modelling replication in nosql datastores. In: QEST (2014)

[25] Rabl, T., Sadoghi, M., Jacobsen, H.-A., Gómez-Villamor, S., Muntés-Mulero, V., Mankowskii, S.: Solving big data challenges for enterprise application performance management. PVLDB 5(12), 1724–1735 (2012)

[26] Weber, S.: Nosql databases. University of Applied Sciences HTW Chur, Switzerland

A Systematic Approach for Composing General Middleware Completions to Performance Models

Adnan Faisal, Dorina Petriu, and Murray Woodside

Carleton University, Dept. of Systems and Computer Eng., Ottawa K1S5B6, Canada
{faisal,petriu,cmw}@sce.carleton.ca

Abstract. A software design often does not describe the software infrastructure it will need to run, but a performance analysis must account for its effects. "Performance completions" represent the infrastructure and must be incorporated in the application performance model. This paper considers completions for middleware. It proposes a unified framework for describing all kinds of middleware in the Layered Queuing Network (LQN) model, based on a generic template and elaborations for middleware features. The template is applied to several common request-reply middleware systems. A process is given for building a new middleware completion model and for incorporating it into a LQN model.

Keywords: Performance analysis, middleware, role-based modeling, layered queuing network, aspect-oriented modeling, performance completion.

1 Introduction

Application models for software products describe the functional properties of the application being developed. In distributed systems, application software runs on a platform of third-party software (e.g., middleware, operating systems) and networks, which are usually not modeled by the system designer. For performance analysis this infrastructure needs to be added to the application software model. This issue was identified by Woodside et al. in [1] and the authors called these added performance model elements *performance completions*. A key completion for a distributed application is its middleware (See **Fig. 1**). As described in [2], middleware is software that masks the heterogeneity of the underlying networks, hardware, operating systems and programming language by providing a higher level of abstraction for communication.

There may be options for middleware which can significantly impact the performance, as described in [3, 4, 5]. In [3], the authors showed that the transferred message sizes depend on the middleware and also on its features (e.g., security). SOAP messages were observed to be on average ~4.3 times larger than RMI JRMP messages. In the same work Web services were found to be ~9.6 times slower than RMI for simple data types (e.g., boolean, integer etc.) and strings, and ~14% slower for instantiation. These examples underline the importance of modeling the particular middleware used. If the performance analyst can estimate the performance effect of various middleware products, the system designer can make platform decisions that meet both functional and Quality of Service (QoS) requirements.

A. Horváth and K. Wolter (Eds.): EPEW 2014, LNCS 8721, pp. 30–44, 2014.
© Springer International Publishing Switzerland 2014

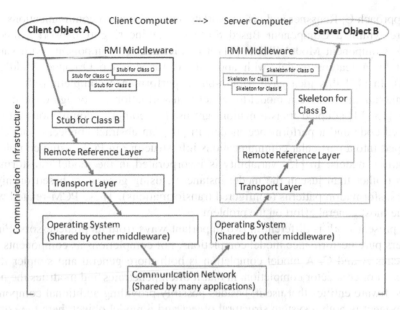

Fig. 1. Roles [17] that participate in a typical distributed method invocation are shown in rectangles. The communication infrastructure includes middleware (here RMI), operating system and network elements.

This paper proposes a systematic approach for generating middleware sub-models as performance completions and composing them with performance models of software. The performance models can come from any model-creation process, but here they are assumed to be layered queuing network (LQN) models [6]. The proposed systematic approach reduces the time and effort required to evaluate the performance of an application under different middleware and different deployments. The approach is scalable too, because the middleware and their features are always composed locally in any model.

Three lines of research which have addressed parts of this problem will be identified as A, B, and C. In A, an approach and tool for composing subsystems into LQN models was described by Wu et al. [7, 8]. Subsystem models may have numerical parameters, but this is the only form of variability. While this capability may be used to model some kinds of middleware, it has a limited ability to adapt to the context of each interaction.

In approach B, Verdickt et al. [9, 10] modeled CORBA middleware (by hand) at the UML level, automatically combined it with a UML model of the application, and finally derived LQN performance models (by hand). That is, they solved only the automatic composition problem, and did that at the UML level. While this has the advantage it can address middleware issues within the software design, it has the disadvantage that the performance model must come from UML (our process can exploit LQN models from any source). Approach B did not consider middleware in general, only CORBA.

In approach C, Reussner and his co-workers have addressed communications infrastructure within a Component Based Software Engineering framework called the Palladio Component Model (PCM) [11]. Completions are components with variants defined by a feature model, with applications to Message Oriented Middleware (MOM). In [12] this idea gave pattern-based performance completions for MOM, with annotations to drive a model-to-model transformation to configure the completion. In [13, 14] coupled transformations are used to compose completions into both generated code and a performance model. In [16], an abstract connector model was developed into a particular communications infrastructure in PCM using completions and feature models. In [15] variability is incorporated in the model transformation process (rather than just in the model instances), using generators based on higher-order transformation patterns (configured transformations). These PCM-related works form the most general effort on the problem.

The present work is different in two important ways from its predecessors. First, it considers pure performance model completions, not completions as components as in approaches A and C. A model completion is both more general and simpler than a component or connector completion, in that it often penetrates and modifies the model of the software entities that use it, besides possibly providing additional components. A component is both a system structural object and a model object; here the completion is only considered in the performance model and its implementation is not a complicating concern. The second difference is that (unlike as done in B or C) the completion is defined and composed at the performance model level, not in the software specification. This is often simpler because the performance model is at a more abstract level than a design model defined in PCM or UML.

These differences give two advantages. First, the resulting performance model is smaller and simpler. Second, design details may be confidential or simply unavailable, and for such cases a performance model may be constructed as a "black box" and its middleware then composed using the methods proposed here.

The main contribution of this paper is a general systematic process to represent the performance impact of most kinds of middleware, including optional features. This is achieved by identifying and grouping the different roles that take part in various middleware products into two categories: mandatory roles and optional roles (Section 3). A base middleware sub-model (Section 4) is proposed that can represent the architecture of many middleware products, which are defined as variations on it, using feature models and realizations (Section 5). A kind of aspect composition for LQN (which has not previously been defined) is used to compose the feature realizations into the base sub-model, and the middleware sub-models into the primary LQN model (Sections 5 and 6). The method is demonstrated by an example in Section 7.

2   Layered Queuing Network (LQN) Performance Models

The LQN performance models [6] used here offer the advantage that they directly represent the software components and their deployment, treat software entities as resources, and capture inter-component communications, and resource interactions between layers of the application. They behave like ordinary queuing networks where that is appropriate.

Fig. 2 shows two LQN models, without and with network latency for a web service application. Each concurrent entity (called a *task*) is represented by a bold rectangle showing its name and a parameter for its thread-pool multiplicity (e.g. {$N}, {inf}). It has attached rectangles that represent its operations, called *entries* and labeled with the host (CPU) demand for one invocation of the entry (e.g. [$web], [20 ms] etc.).

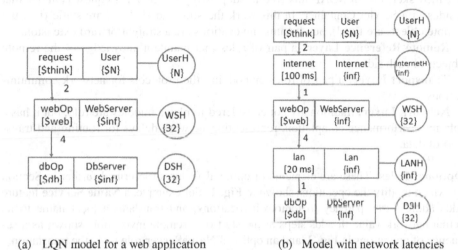

(a) LQN model for a web application   (b) Model with network latencies

Fig. 2. LQN models of a generic 3-tier web application

Each task has a host processor drawn as an oval, with a multiplicity (e.g., {32}, {$N} etc.) which can represent multiple cores. A *call* from one entry to another is represented by an arrow labeled with the average number of calls.

An infinite server provides a pure time delay and is modeled as an infinite task running on an infinite host (using the multiplicity {inf}). Infinite servers represent network latencies in **Fig. 2(b)**.

3   Roles in Middleware

This paper considers middleware products which support remote service invocations through request-reply interactions and implement some of the roles in **Fig. 1**. This section describes these roles along with the components/processes of three different kinds of middleware (RMI, CORBA and SOAP) [2] that realize them. Note that roles do not correspond directly to software modules or components, one of which may realize multiple roles. There are *Mandatory roles* and *Optional roles*, as follows.

Mandatory Roles (always present):
    Client and Servant: the originator and responder of a remote invocation.

Stub: a client-side proxy that represents the remote object at the client. It hides the details of the remote object reference from the client object, it marshalls/unmarshalls arguments and it sends/receives messages to/from the client object. Both RMI and CORBA define stubs. In SOAP the stub role is played by the SOAP Client API.

Skeleton: a server-side proxy that plays the same role for the servant as the stub plays for the client. A servant has one skeleton for each remote client. The component that plays the server-side proxy role in most middleware including RMI and CORBA is called skeleton. In SOAP this role is adopted by the SOAP Request Handler that resides in a Servlet Container. In this work the stub and skeleton are static (i.e., the remote objects are fixed), but dynamic invocations are a straightforward extension.

Remote Reference Layer: a pair of roles that maintain knowledge of the remote objects at each side.

Transport Layer: a pair of roles embodying the protocols for network communications.

Network Layer: traditionally not considered to play a middleware role, but it has a role in a performance completion, representing network delays for routing and transport of data.

Optional Roles: These are elements of optional features (described further in Section 5), which modify the operations shown in **Fig. 1**. For example, a **Name Service** feature adds address transparency to remote invocations, and translates a local name to its actual network value. It adds steps at the start of a remote invocation, shown as a sequence of operations in **Fig. 3**, and an optional **NameServer** role to do the translation.

Other examples of optional roles are an **Encryption agent** to implement an encryption feature and a **ServiceManager** to implement the features of a web application server such as JBoss. Every request going from the skeleton to the servant must pass through the **ServiceManager** (in EJB the beans act as client and servant).

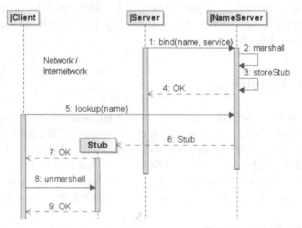

Fig. 3. A simplified Name Service interaction in a middleware

4   The Base Middleware LQN Sub-model

The *base LQN sub-model* for middleware (*BaseSM* for brevity) in **Fig. 4** has host, component (task) and operation (entry) roles (following concepts of role-based modeling as in [17]). It provides concurrent components for the client-side and server-side operations |cOP and |sOp, which combine the stub/skeleton, remote reference and transport layer roles in **Fig. 1**. These mandatory roles construct the essential parts of most middleware, including RMI, CORBA and SOAP. Their execution demand is defined by expressions with a fixed part ($cFix per request) and a part proportional to message size with multiplier $cVar. $msgSize, which is a parameter of each application call, is the sum of the call and reply message sizes. The sum of call and reply processing is included in the operation demands.

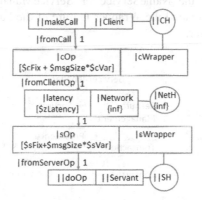

Fig. 4. BaseSM, the base middleware sub-model in LQN

The tasks in BaseSM are called *role-tasks*; they have *role-entries* which make *role-calls* and run on hosts we call *role-hosts*. Role names are preceded by a single bar (for a role that must be explicitly bound to some entity in the application model, possibly one created for the purpose) or by double bars (for a role that is bound implicitly, so the modeler need not specify it). The role-entry ||makeCall is implicitly bound to the operation that makes the remote invocation, and the role-entry ||doOp is implicitly bound to the remote operation. The role-tasks ||Client, ||Servant and role-hosts ||CH, ||SH are defined implicitly as the task and host for the two role-entries. The |Network role-task is a place-holder for the network delay.

When a feature is added to the base sub-model it replaces a role-call or a role-entry, which are called *join-calls* and *join-entries* (because they function like join-points in aspect weaving) respectively. To identify such a call, we specify the pair (callingEntry, calledEntry). For example, in **Fig. 4**, |fromCall can be specified by (||makeCall, |cOp).

Although BaseSM represents middleware, it can also model a style without explicit middleware, such as REST (which uses HTTP without marshalling/unmarshalling). The wrappers then would include only the HTTP and TCP/UDP overhead.

5   Middleware Features

Most middleware has additional features, which may be optional (e.g. encryption) or intrinsic to the middleware (e.g. a location server for CORBA). Features are organized in a *feature model* and defined in LQN notation by *feature realizations*.

5.1   Feature Models to Represent Variability

As a foundation, BaseSM is defined as a feature called Middleware, with a Wrapper feature realized by the role-tasks |ClientWrapper and |ServantWrapper. In the feature model of **Fig. 5(a)**, a directed arc labelled "°" indicates an optional feature that can be nested within its parent feature, whereas "•" indicates a mandatory feature. Thus Middlware may include the NameService or ServiceManager features. **Fig. 5(b)** shows features that may extend the Wrapper feature of the base sub-model, and sub-features of ServiceManager.

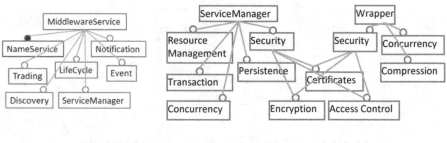

(a)   Middleware features   (b)   ServiceManager and Wrapper features

Fig. 5. Feature model for optional features

5.2   Library of Feature Realizations

The features are specified in a feature library. A feature has four attributes:
- *compositionType* = one of   InFlowStruct , OutOfFlowStruct, Value; InFlow-Struct and OutOfFlowStruct are called Structure-type Features.
- *joinList* = a list of join-calls (for InFlowStruct types) or join-entries (otherwise);
- *realization* = a LQN feature submodel (for structural types) or an expression for an added CPU demand value in a join-entry (for the Value type);
- *parameterList* = parameters of the realization, which may be expressions.

The feature realization is composed at each join-call or join-entry, by binding the roles in the realization with roles in BaseSM for structural types, or by modifying role-entry demand values, for the Value type. Each feature realization begins with a dummy role-task ||Requester with role-entry ||makeCall, and ends with dummy role-task ||Replier with role-entry ||doOp.

The compositionType describes how the feature is composed:

- InFlowStruct features replace each join-call by a copy of the submodel (with role-entry feature.||makeCall replaced by joinCall.callingEntry, and role-entry feature.||doOp replaced by joinCall.calledEntry);
- OutOfFlowStruct features add one or more calls to each join-entry;
- Value-type features add the amount feature.value to the demand of each identified join-entry. For example the Compression feature might add a CPU demand to the ClientWrapper task defined by a linear function of the form:

```
Compression.value =
        baseSM.$msgSize * Compression.$compressionFactor
```

A feature may have two alternative specifications, one for a Value type and one for a structure type. **Fig. 6(a)**, **6(b)** and **6(c)**   show realizations for three structure-modifying features: Encryption, NameService and ServiceManager. The first two are of type InFlowStruct and are composed into the message flow, while the last is of type OutOfFlowStruct and adds a call to the calling role-entry.

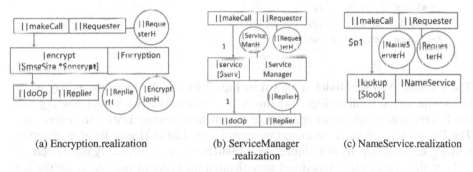

| (a) Encryption.realization | (b) ServiceManager .realization | (c) NameService.realization |

Fig. 6. Feature realizations in LQN

5.3 Obtaining the Specialized Middleware LQN Sub-model

Given a set of features, their realizations are composed with the basic sub-model as directed by a Feature Composition Descriptor. It specifies the starting model (BaseSM), a list of features to be composed, and the resulting specialized submodel.

Consider a ServiceManager with encryption (call it EJB+E). To add encryption and a service manager as features, and call the result MWSubModel, the Feature Composition Descriptor would be:

```
ComposeFeatures (BaseSM -> MWSubModel)
    feature Encryption
    feature ServiceManager
end
```

For bindings, the composition is directed by the joinList for each feature, which identifies all the joinCalls/joinEntries in BaseSM where it is to be composed. It is bound as follows:

For this feature type	This roleEntry	Is bound to this BaseSM.roleEntry
InFlowStruct	feature.‖makeCall	joinCall.callingEntry
InFlowStruct	feature.‖doOp	joinCall.calledEntry
OutOfFlowStruct	feature.‖makeCall	joinEntry

When a feature role-entry is bound to a role-entry of BaseSM, it is merged with and takes on the identity of the latter, and its task and host are bound to the task and host of the BaseSM entity. The calls and model-entities between the two entries ‖makeCall and ‖doOp for each feature are added to the base sub-model.

The added entities of the feature can be bound differently to BaseSM entities by optional binding overrides, using statements of form:

> (**bind** feature-entity baseModel-entity)

For example the EJB+E feature composition descriptor given above introduces a new role-task |Encryption and a new host |EncryptionH. But, the following descriptor causes the |Encryption task to be deployed instead on its caller's host:

```
ComposeFeatures (baseSM -> MWSubModel)
  feature Encryption
    (bind |EncryptionH ||CallerH)
  feature ServiceManager
end
```

The resulting MWSubModel is shown in **Fig. 7(b)**, which shows the newly added tasks with dotted borders. **Fig. 7(a)** shows the intermediate result after binding only the Encryption, which (with appropriate parameters) models RMI with encryption. The Encryption feature is composed twice as its joinList is {fromClientOp, fromServerOp}. Call multiplicity is assumed to be 1 where it is not specifically mentioned.

Note that, though in a functional specification the order of operations by the features is important, for a performance model it is sufficient to capture the total workload and its concurrency (attachment to tasks) and host loading (deployment). Thus having the encryption called by the sWrapper (rather than executing before the wrapper) is backwards in execution order but gives a correct performance model.

A high-level algorithm for composing realization fragments is:

```
FEATURE-COMPOSITION(fcd:featureCompositionDescriptor)
BEGIN  FOR every feature feature in array fcd
    IF feature.compositionType is InFlowStruct
      GET its join-calls array feature.joinList
      FOR every element joinCall in feature.joinList
        REPLACE joinCall with feature
      ENDFOR
      BIND hosts of feature
    ELSEIF feature.compositionType is OutOfFlowStruct
      GET its join-entry array feature.joinList
      FOR every element joinEntry in feature.joinList
        ADD feature
      ENDFOR
```

```
    BIND hosts of feature
ELSE   /* feature.compositionType is Value*/
    UPDATE service demands
ENDIF  ENDFOR  END
```

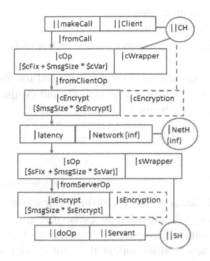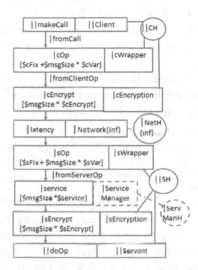

(a) Encryption is Composed with BaseSM first, giving RMI with Encryption (RMI+E)

(b) then Service Manager is added to give the EJB+E specialized sub-model

Fig. 7. Specialization of the base sub-model

As indicated in this example, standard types of middleware are modeled by adding their features to BaseSM. For example, RMI and REST have a basic version which is the same as the base sub-model (with appropriate parameter values); they can also have additional features.

6 Obtaining the Final Composed LQN Model

The same approach can be used to compose the appropriate middleware submodels with the application LQN model. Each call in the application that uses middleware becomes a join-call for the specialized submodel for that middleware, which is composed with the call in the same way as an InFlowStruct feature. The choice of middleware for each call is specified by a set of records of the form (call-specification MWSubModel). To simplify cases where many calls use the same middleware, the following rules apply:

(1) Sets of application calls can be specified by a generalized call specification:

generalized-call-specification = (set of calling entries, set of called entries)

which defines the set of all calls between entries in the first set, and entries in the second. The following special designations for sets of entries are used:

- AllEntries, for all entries in the application model,
- EntriesT(task), for all entries of the given task, and
- EntriesH(host), for all entries of tasks deployed on the given host.

(2) Sets are defined hierarchically, so that later specifications override earlier ones.

For example, suppose middleware with the specialized sub-model m1 is used for the entire application, except for using m2 for calls from task t1 to task t2, and (within that group) m3 for calls from entry e1 to e2. The specification is:

```
MiddlewareSpec
  (AllEntries, AllEntries) m1
  (EntriesT(t1), EntriesT(t2)) m2
  (e1, e2) m3
end
```

The process for binding a particular MWSubModel at a joinCall of the applicationn is the same as for the feature binding in Section 5.3. Its role-entries ∥makeCall and ∥doOp are bound to joinCall.callingEntry and joinCall.calledEntry (respectively), and the entities between them are instantiated as new entities unless they are explicitly bound to existing entities. Just as for features, tasks and hosts in the MWSubModel may be bound to entities in the application model by a **bind** directive following the join-call specification. The following example specifies completion of the application model of Fig. 2(b) with the RMI sub-model of Fig. 4 over the whole application, except for RMI+E (RMI with encryption, as in Fig. 7(a)) for calls between tasks User and WebServer. The encryption hosts are bound to the hosts of the client and servant.

```
MiddlewareSpec
  (*, *) RMI
  (EntriesT(User), EntriesT(WebServer)) RMIPlusEnc
  (bind |EncryptionCH UserH)
  (bind |EncryptionSH WSH)
end
```

The MiddlewareSpec above is converted into a Middleware Composition Description (analogous to the Feature Composition Description) with a join-call list for each specialized submodel, and an algorithm outlined as follows is applied:

```
MIDDLEWARE-COMPOSITION(mcd:MiddlewareCompositionDescriptor)
BEGIN FOR every MWSubModel m in mcd
    GET all JoinCalls where m is to be applied
    FOR every JoinCall c
      ADD tasks of m at c
      ADD AND BIND calls and hosts of the added tasks
    ENDFOR  ENDFOR  END
```

The resulting model is called the *final model* and is shown in **Fig. 8**.

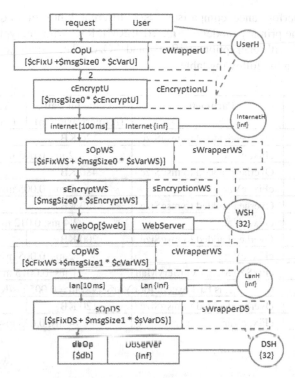

Fig. 8. RMI middleware in the application model of Figure 2(b), with encryption of user messages. All unmentioned call multiplicities are 1.

7   Experimental Results

This section describes an example of a 3-tier web application for the management of a medium-size hospital. The application designers must determine how many users it can support while keeping the mean response time below 2 seconds, and must make a decision regarding what middleware to use. The LQN of **Fig. 2(b)** shows the application architecture and the high-level LQN performance model, and this table gives the application model parameters:

Application Model parameter	Value
User external delay (think time)	3000 ms
Message size for call (request, webOp)	200 KB
webOp CPU demand	20 ms
Message size for call (webOp, dbOp)	300 KB
dbOp CPU demand	14 ms
Internet latency	60 ms
Lan latency	10 ms
Host (WSH and DSH) multiplicity	32

To make performance comparisons, six different kinds of middleware were composed with the primary model of **Fig. 2(b)**: RMI, RMI with encryption (RMI+E), REST, EJB, EJB with encryption (EJB+E) and SOAP. The middleware parameters were assigned as in the following table:

Middleware	Parameter	Value for call (request, webOp)	Value for call (webOP, dbOp)
RMI	Overhead added to msgSize	120 KB	200KB
	cFix, cVar	2 ms, 0.005 ms/KB	
RMI+E	Overhead added to msgSize	480 KB	800 KB
	cFix, cVar	2 ms, 0.005 ms/KB	
REST	Overhead added to msgSize	120 KB	200 KB
	cFix, cVar	2.4 ms, 0.012 ms/KB	
EJB	Overhead added to msgSize	200 KB	800 KB
	cFix, cVar	2 ms, 0.005 ms/KB	
	cFixServMan, cVarServMan	1 ms. 0.0015 ms/KB	
EJB+E	…as EJB except for cVarServMan = 0.005 ms/KB		
SOAP	Overhead added to msgSize	120 KB	200 KB
	cFix, cVar	4 ms, 0.023 ms/KB	

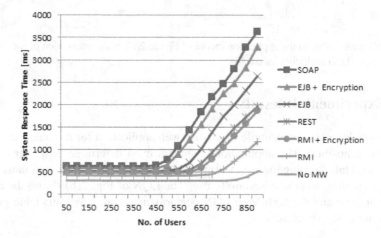

Fig. 9. System Response Time [ms] for various middleware compositions

The performance models were solved for a user think time of 3000 ms and a number of users ranging from 50 to 900, with the results in **Fig. 9** and **Fig. 10**. From **Fig. 9**, the system response time is small (about half a second) for all cases, up to about 400 users. Above this number the curves turn up at a point we will call the point of contention, which is different for different middleware.   SOAP and EJB+E saturate first, due to the large XML messages of SOAP and to the overheads caused by the ServiceManager and encryption in EJB+E. At the opposite extreme "no middleware" enters contention last, preceded by RMI, RMI+E and REST. RMI is fastest among all the middleware but unlike the others, is not language-independent. REST and (RMI+E) are similar because the marshalling plus encryption overhead in RMI have a similar effect to the exchange of 'resources' as large messages in REST. The 'No MW' line shows the 'ideal' case with only network latency as in **Fig. 2(b)**, to compare with the cases with middleware overhead and additional messages. **Fig. 10** shows the corresponding throughputs, saturating at values ranging from below 500 to 900 requests/s.

To keep the response time below 2 seconds, the number of users for SOAP, EJB+E, EJB, REST, RMI+E must be kept below 700, 750, 800, 900 and 950 respectively. For all these cases it was observed that the DataBase host is the bottleneck, and the critical middleware overhead is in the features executed on that host.

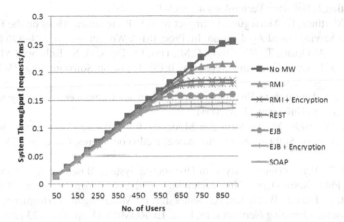

Fig. 10. System Throughput [requests/ms] for various middleware compositions

8   Conclusion

This paper presents a systematic approach for adding existing or to-be-created middleware completions in a LQN performance model of an application. The generic base middleware sub-model encapsulates any kind of functionality which is interposed in the message flow. For specific middleware it is customized both in structure and in value. Customization can represent any infrastructure which transparently conveys messages. Customization incorporates the features offered by the middleware, as additional components and modified performance parameters, within a specialized middleware sub-model that is finally composed with the application model. The composed models are correct by construction, as long as the specifications given by the user were correct.

The authors are currently extending the library of middleware to include group communication and implementing an automated process on the Eclipse platform.

Acknowledgements. This research was funded by the Natural Sciences and Engineering Research Council of Canada through its Strategic Network SAVI (Smart Applications on Virtual Infrastructure) and its Discovery Grant programs.

References

1. Woodside, M., Petriu, D.B., Siddiqui, K.H.: Performance-related Completions for Software Specifications. In: Proc. 24th Int. Conf. on Software Engineering, Orlando (May 2002)
2. Coulouris, G., Dollimore, J., Kindberg, T., Blair, G.: Distributed Systems Concepts and Design, 5th edn. Pearson Education Inc. (2012)
3. Juric, M.B., Rozman, I., Brumen, B., Colnaric, M., Hericko, M.: Comparison of performance of Web services, WS-Security, RMI, and RMI-SSL. Journal of Systems and Software (79), 689–700 (2006)
4. Juric, M.B., Rozman, I., Hericko, M.: Performance comparison of CORBA and RMI. Information & Software Technology 42, 915–933 (2000)
5. Gómez-Martínez, E., Merseguer, J.: Impact of SOAP Implementations on the Performance of a Web Service-Based Application. In: Proc. ISPA Workshops, pp. 884–896 (2006)
6. Franks, G., Al-Omari, T., Woodside, C.M., Das, O., Derisavi, S.: Enhanced Modeling and Solution of Layered Queueing Networks. IEEE Trans. on Software Eng. 35(2), 148–161 (2009)
7. Wu, X.: An Approach to Predicting Performance for Component-based Systems, MASc thesis, Carleton University (July 2003)
8. Wu, X., Woodside, M.: Performance Modeling for Software Components. In: Proc. 4th Int. Workshop on Software and Performance, Redwood Shores, Calif., pp. 290–301 (January 2004)
9. Verdickt, T.: Performance Analysis of Distributed Systems Based on Architectural System Models, PhD Thesis, Dept. of Inf. Technology, Universiteit Gent, Belgium (2007)
10. Verdickt, T., Dhoedt, B., Gielen, F., Demeester, P.: Modelling the performance of CORBA using Layered Queueing Networks. In: Proc. EUROMICRO, pp. 117–123 (2003)
11. Becker, S., Koziolek, H., Reussner, R.: Model-based Performance Prediction with the Palladio Component Model. In: Proc. 6th Int. Workshop on Software and Performance, pp. 56–67 (2007)
12. Happe, J., Friedrich, H., Becker, S.: A pattern-based performance completion for Message-oriented Middleware. In: Proc. 7th Int. Workshop on Software and Performance, pp. 165–176 (2008)
13. Happe, J., Becker, S., Rathfelder, C., Friedrich, H., Reussner, R.H.: Parametric Performance Completions for Model-Driven Performance Prediction. Performance Evaluation 16(8), 694–717 (2010)
14. Becker, S.: Coupled model transformations. In: Proc. 7th Int. Workshop on Software and Performance, pp. 165–176 (2008)
15. Kapova, L.: Configurable Software Performance Completions through Higher-Order Model Transformations, PhD Thesis, Karlsruhe Institute of Technology (2011)
16. Strittmatter, M., Happe, L.: Compositional performance abstractions of software connectors. In: Proc. 3rd International Conf. on Performance Engineering, pp. 275–278 (2013)
17. Kim, D.-K., France, R.B., Ghosh, S., Song, E.: Using Role-Based Modeling Language (RBML) to Characterize Model Families. In: Proc. ICECCS, pp. 107–116 (2002)
18. Alhaj, M.: Automatic Derivation of Performance Models in the Context of Model-Driven SOA, Doctoral dissertation, Carleton University, Ottawa, Canada (2013)

Vacation and Polling Models with Retrials

Onno Boxma and Jacques Resing

EURANDOM and Department of Mathematics and Computer Science
Eindhoven University of Technology
P.O. Box 513, 5600 MB Eindhoven, The Netherlands
{o.j.boxma,j.a.c.resing}@tue.nl

Abstract. We study a vacation-type queueing model, and a single-server multi-queue polling model, with the special feature of retrials. Just before the server arrives at a station there is some deterministic glue period. Customers (both new arrivals and retrials) arriving at the station during this glue period will be served during the visit of the server. Customers arriving in any other period leave immediately and will retry after an exponentially distributed time. Our main focus is on queue length analysis, both at embedded time points (beginnings of glue periods, visit periods and switch- or vacation periods) and at arbitrary time points.

Keywords: Vacation queue, polling model, retrials.

1 Introduction

Queueing systems with retrials are characterized by the fact that arriving customers, who find the server busy, do not wait in an ordinary queue. Instead of that they go into an orbit, retrying to obtain service after a random amount of time. These systems have received considerable attention in the literature, see e.g. the book by Falin and Templeton [7], and the more recent book by Artalejo and Gomez-Corral [3].

Polling systems are queueing models in which a single server, alternatingly, visits a number of queues in some prescribed order. Polling systems, too, have been extensively studied in the literature. For example, various different service disciplines (rules which describe the server's behaviour while visiting a queue) and both models with and without switchover times have been considered. We refer to Takagi [20,21] and Vishnevskii and Semenova [22] for some literature reviews and to Boon, van der Mei and Winands [5], Levy and Sidi [12] and Takagi [18] for overviews of the applicability of polling systems.

In this paper, motivated by questions regarding the performance modelling of optical networks, we consider vacation and polling systems with retrials. Despite the enormous amount of literature on both types of models, there are hardly any papers having both the features of retrials of customers and of a single server polling a number of queues. In fact, the authors are only aware of a sequence of papers by Langaris [9,10,11] on this topic. In all these papers the author

A. Horváth and K. Wolter (Eds.): EPEW 2014, LNCS 8721, pp. 45–58, 2014.

determines the mean number of retrial customers in the different stations. In [9] the author studies a model in which the server, upon polling a station, stays there for an exponential period of time and if a customer asks for service before this time expires, the customer is served and a new exponential stay period at the station begins. In [10] the author studies a model with two types of customers: primary customers and secondary customers. Primary customers are all customers present in the station at the instant the server polls the station. Secondary customers are customers who arrive during the sojourn time of the server in the station. The server, upon polling a station, first serves all the primary customers present and after that stays an exponential period of time to wait for and serve secondary customers. Finally, in [11] the author considers a model with Markovian routing and stations that could be either of the type of [9] or of the type of [10].

In this paper we consider a polling station with retrials and so-called glue periods. Just before the server arrives at a station there is some deterministic glue period. Customers (both new arrivals and retrials) arriving at the station during this glue period "stick" and will be served during the visit of the server. Customers arriving in any other period leave immediately and will retry after an exponentially distributed time.

The study of queueing systems with retrials and glue periods is motivated by questions regarding the performance modelling and analysis of optical networks. Performance analysis of optical networks is a challenging topic (see e.g. Maier [13] and Rogiest [17]). In a telecommunication network, packets must be routed from source to destination, passing through a series of links and nodes. In copper-based transmission links, packets from different sources are time-multiplexed. This is often modeled by a single server polling system. Optical fibre offers some big advantages for communication w.r.t. copper cables: huge bandwidth, ultra-low losses, and an extra dimension – the wavelength of light. However, in an optical routing node, opposite to electronics, it is difficult to store photons, and hence buffering in optical routers can only be very limited. Buffering in these networks is typically realized by sending optical packets into fibre delay loops, i.e., letting them circulate in a fibre loop and extracting them after a certain number of circulations. This feature can be modelled by retrial queues. Recent experiments with 'slow light', where light is slowed down by significantly increasing the refractive index in waveguides, have up to now shown very modest buffering times [8]. It should be noted that with the very high speeds achievable in fibre, packet durations are very short, so that small buffering times may already allow sufficient storage of small packets. We represent the effect of slowing down light by introducing a glue period at a queue just before the server arrives.

The paper is organized as follows. In Section 2 we consider the case of a single queue with vacations and retrials; arrivals and retrials only "stick" during a glue period. We study this case separately because (i) it is of interest in its own right, (ii) it allows us to explain the analytic approach as well as the probabilistic meaning of the main components in considerable detail, (iii) it makes the analysis of the multi-queue case more accessible, and (iv) results for the one-queue case may serve as a first-order approximation for the behaviour of a particular queue in the N-queue case, switchover periods now also representing glue and visit periods at other queues. In Section 3 the two-queue case is analyzed. We do not have the space to treat the N-queue case in this paper, but the analysis in Sections 2 and 3 lays the groundwork for analyzing the N-queue case. Section 4 presents some conclusions and suggestions for future research.

2   Queue Length Analysis for the Single-Queue Case

2.1   Model Description

In this section we consider a single queue Q in isolation. Customers arrive at Q according to a Poisson process with rate λ. The service times of successive customers are independent, identically distributed (i.i.d.) random variables (r.v.), with distribution $B(\cdot)$ and Laplace-Stieltjes transform (LST) $\tilde{B}(\cdot)$. A generic service time is denoted by B. After a visit period of the server at Q it takes a vacation. Successive vacation lengths are i.i.d. r.v., with S a generic vacation length, with distribution $S(\cdot)$ and LST $\tilde{S}(\cdot)$. We make all the usual independence assumptions about interarrival times, service times and vacation lengths at the queues. After the server's vacation, a *glue* period of deterministic (i.e., constant) length begins. Its significance stems from the following assumption. Customers who arrive at Q do not receive service immediately. When customers arrive at Q during a glue period G, they stick, joining the queue of Q. When they arrive in any other period, they immediately leave and retry after a retrial interval which is independent of everything else, and which is exponentially distributed with rate ν. The glue period is immediately followed by a visit period of the server at Q.

The service discipline at Q is gated: During the visit period at Q, the server serves all "glued" customers in that queue, i.e., all customers waiting at the end of the glue period (but none of those in orbit, and neither any new arrivals).

We are interested in the steady-state behaviour of this vacation model with retrials. We hence make the assumption that $\rho := \lambda \mathbb{E}B < 1$; it may be verified that this is indeed the condition for this steady-state behaviour to exist.

Some more notation:

G_n denotes the nth glue period of Q.

V_n denotes the nth visit period of Q (immediately following the nth glue period).

S_n denotes the nth vacation of the server (immediately following the nth visit period).

X_n denotes the number of customers in the system (hence in orbit) at the start of G_n.

Y_n denotes the number of customers in the system at the start of V_n. Notice that here we should distinguish between those who are queueing and those who are in orbit: We write $Y_n = Y_n^{(q)} + Y_n^{(o)}$, where q denotes queueing and o denotes in orbit.

Finally,

Z_n denotes the number of customers in the system (hence in orbit) at the start of S_n.

2.2   Queue Length Analysis at Embedded Time Points

In this subsection we study the steady-state distributions of the numbers of customers at the beginning of (i) glue periods, (ii) visit periods, and (iii) vacation periods. Denote by X a r.v. with as distribution the limiting distribution of X_n. Y and Z are similarly defined, and $Y = Y^{(q)} + Y^{(o)}$, the steady-state numbers of customers in queue and in orbit at the beginning of a visit period (which coincides with the end of a glue period). In the sequel we shall introduce several generating functions, throughout assuming that their parameter $|z| \leq 1$. For conciseness of notation, let $\beta(z) := \tilde{B}(\lambda(1 - z))$ and $\sigma(z) := \tilde{S}(\lambda(1 - z))$. Then it is easily seen that

$$\mathbb{E}[z^X] = \sigma(z)\mathbb{E}[z^Z], \tag{2.1}$$

since X equals Z plus the new arrivals during the vacation;

$$\mathbb{E}[z^Z] = \mathbb{E}[\beta(z)^{Y^{(q)}} z^{Y^{(o)}}], \tag{2.2}$$

since Z equals $Y^{(o)}$ plus the new arrivals during the $Y^{(q)}$ services; and

$$\mathbb{E}[z_q^{Y^{(q)}} z_o^{Y^{(o)}}] = e^{-\lambda(1-z_q)G}\mathbb{E}[\{(1 - e^{-\nu G})z_q + e^{-\nu G}z_o\}^X]. \tag{2.3}$$

The last equation follows since $Y^{(q)}$ is the sum of new arrivals during G and retrials who return during G; each of the X customers which were in orbit at the beginning of the glue period have a probability $1 - e^{-\nu G}$ of returning before the end of that glue period.

Combining Equations (2.1)-(2.3), and introducing

$$f(z) := (1 - e^{-\nu G})\beta(z) + e^{-\nu G}z, \tag{2.4}$$

we obtain the following functional equation for $\mathbb{E}[z^X]$:

$$\mathbb{E}[z^X] = \sigma(z)e^{-\lambda(1-\beta(z))G}\mathbb{E}[f(z)^X].$$

Introducing $K(z) := \sigma(z)e^{-\lambda(1-\beta(z))G}$ and $X(z) := \mathbb{E}[z^X]$, we have:

$$X(z) = K(z)X(f(z)). \tag{2.5}$$

This is a functional equation that naturally occurs in the study of queueing models which have a branching-type structure; see, e.g., [6] and [16]. Typically, one may view customers who newly arrive into the system during a service as children of the served customer ("branching"), and customers who newly arrive into the system during a vacation or glue period as immigrants. Such a functional equation may be solved by iteration, giving rise to an infinite product – where the jth term in the product typically corresponds to customers who descend from an ancestor of j generations before. In this particular case we have after n iterations:

$$X(z) = \prod_{j=0}^{n} K(f^{(j)}(z))X(f^{(n+1)}(z)), \tag{2.6}$$

where $f^{(0)}(z) := z$ and $f^{(j)}(z) := f(f^{(j-1)}(z))$, $j = 1, 2, \ldots$. Below we show that this product converges for $n \to \infty$ iff $\rho < 1$, thus proving the following theorem:

Theorem 1. *If $\rho < 1$ then the generating function $X(z) = \mathbb{E}[z^X]$ is given by*

$$X(z) = \prod_{j=0}^{\infty} K(f^{(j)}(z)). \tag{2.7}$$

Proof. Equation (2.5) is an equation for a branching process with immigration, where the number of immigrants has generating function $K(z)$ and the number of children in the branching process has generating function $f(z)$. Clearly, $K'(1) = \lambda \mathbb{E}S + \lambda \rho G < \infty$ and $f'(1) = e^{-\nu G} + \left(1 - e^{-\nu G}\right)\rho < 1$, if $\rho < 1$. The result of the theorem now follows directly from the theory of branching processes with immigration (see e.g., Theorem 1 on page 263 in Athreya and Ney [4]).

Having obtained an expression for $\mathbb{E}[z^X]$ in Theorem 1, expressions for $\mathbb{E}[z^Z]$ and $\mathbb{E}[z_q^{Y^{(q)}} z_o^{Y^{(o)}}]$ immediately follow from (2.2) and (2.3). Moments of X may be obtained from Theorem 1, but it is also straightforward to obtain $\mathbb{E}X$ from Equations (2.1)-(2.3):

$$\mathbb{E}X = \lambda \mathbb{E}S + \mathbb{E}Z, \tag{2.8}$$

$$\mathbb{E}Z = \rho \mathbb{E}Y^{(q)} + \mathbb{E}Y^{(o)}, \tag{2.9}$$

$$\mathbb{E}Y^{(q)} = \lambda G + (1 - e^{-\nu G})\mathbb{E}X, \tag{2.10}$$

$$\mathbb{E}Y^{(o)} = e^{-\nu G}\mathbb{E}X, \tag{2.11}$$

yielding

$$\mathbb{E}X = \frac{\lambda \mathbb{E}S + \lambda \rho G}{(1-\rho)(1-e^{-\nu G})}. \tag{2.12}$$

Hence

$$\mathbb{E}Y^{(q)} = \lambda G + (1 - e^{-\nu G})\frac{\lambda \mathbb{E}S + \lambda \rho G}{(1-\rho)(1-e^{-\nu G})} = \frac{\lambda \mathbb{E}S + \lambda G}{1-\rho}, \tag{2.13}$$

$$\mathbb{E}Y^{(o)} = e^{-\nu G}\frac{\lambda \mathbb{E}S + \lambda \rho G}{(1-\rho)(1-e^{-\nu G})}, \tag{2.14}$$

$$\mathbb{E}Z = \frac{\lambda \rho G + \lambda \mathbb{E}S[\rho(1 - e^{-\nu G}) + e^{-\nu G}]}{(1-\rho)(1-e^{-\nu G})}. \tag{2.15}$$

Notice that the denominators of the above expressions equal $1 - f'(1)$. Also notice that it makes sense that the denominators contain both the factor $1 - \rho$ and the probability $1 - e^{-\nu G}$ that a retrial returns during a glue period.

In a similar way as the first moments of X, $Y^{(q)}$, $Y^{(o)}$ and Z have been obtained, we can also obtain their second moment. Here we only mention $\mathbb{E}X^2$:

$$\mathbb{E}X^2 = \frac{K''(1)}{(1-\rho)(1-e^{-\nu G})(1 + \rho(1 - e^{-\nu G}) + e^{-\nu G})} \tag{2.16}$$
$$+ \frac{K'(1)[1 - (\rho(1 - e^{-\nu G}) + e^{-\nu G})^2 + 2K'(1)(\rho(1 - e^{-\nu G}) + e^{-\nu G}) + (1 - e^{-\nu G})\lambda^2 \mathbb{E}B^2]}{(1-\rho)^2(1-e^{-\nu G})^2(1 + \rho(1 - e^{-\nu G}) + e^{-\nu G})},$$

where $K'(1) = \lambda \mathbb{E}S + \lambda \rho G$ and $K''(1) = \lambda^2 \mathbb{E}S^2 + 2\rho \lambda^2 G \mathbb{E}S + \lambda^3 G \mathbb{E}B^2 + (\lambda G \rho)^2$.

Remark 1. Special cases of the above analysis are, e.g.:
(i) Vacations of length zero. Simply take $\sigma(z) \equiv 1$ and $\mathbb{E}S = 0$ in the above formulas.
(ii) $\nu = \infty$. Retrials now always return during a glue period. We then have $f(z) = \beta(z)$, which leads to minor simplifications.

Remark 2. It seems difficult to handle the case of non-constant glue periods, as it seems to lead to a process with complicated dependencies. If G takes a few distinct values $G_1, \ldots, G_N$ with different probabilities, then one might still be able to obtain a kind of multinomial generalization of the infinite product featuring in Theorem 1. One would then have several functions $f_i(z) := (1 - e^{-\nu G_i})\beta(z) + e^{-\nu G_i}z$, and all possible combinations of iterations $f_i(f_h(f_k(\ldots(z))))$ arising in functions $K_i(z) := \sigma(z)e^{-\lambda(1-\beta(z))G_i}$, $i = 1, 2, \ldots, N$. By way of approximation, one might stop the iterations after a certain number of terms, the number depending on the speed of convergence (hence on $1 - \rho$ and on $1 - e^{-\nu G_i}$).

2.3   Queue Length Analysis at Arbitrary Time Points

Having found the generating functions of the number of customers at the beginning of (i) glue periods ($\mathbb{E}[z^X]$), (ii) visit periods ($\mathbb{E}[z_q^{Y^{(q)}} z_o^{Y^{(o)}}]$), and (iii) vacation periods ($\mathbb{E}[z^Z]$), we can also obtain the generating function of the number of customers at arbitrary time points.

Theorem 2. *If $\rho < 1$, we have the following results:*

a) *The joint generating function, $R^{va}(z_q, z_o)$, of the number of customers in the queue and in the orbit at an arbitrary time point in a vacation period is given by*

$$R^{va}(z_q, z_o) = \mathbb{E}[z_o^Z] \cdot \frac{1 - \tilde{S}(\lambda(1 - z_o))}{\lambda(1 - z_o)\mathbb{E}S}. \tag{2.17}$$

b) *The joint generating function, $R^{gl}(z_q, z_o)$, of the number of customers in the queue and in the orbit at an arbitrary time point in a glue period is given by*

$$R^{gl}(z_q, z_o) = \int_{t=0}^{G} e^{-\lambda(1-z_q)t}\mathbb{E}[\{(1 - e^{-\nu t})z_q + e^{-\nu t}z_o\}^X]\frac{dt}{G}. \tag{2.18}$$

c) *The joint generating function, $R^{vi}(z_q, z_o)$, of the number of customers in the queue and in the orbit at an arbitrary time point in a visit period is given by*

$$R^{vi}(z_q, z_o) = \frac{z_q\left[\mathbb{E}[z_q^{Y^{(q)}} z_o^{Y^{(o)}}] - \mathbb{E}[\tilde{B}(\lambda(1 - z_o))^{Y^{(q)}} z_o^{Y^{(o)}}]\right]}{\mathbb{E}[Y^{(q)}]\left(z_q - \tilde{B}(\lambda(1 - z_o))\right)} \cdot \frac{1 - \tilde{B}(\lambda(1 - z_o))}{\lambda(1 - z_o)\mathbb{E}B}. \tag{2.19}$$

d) *The joint generating function, $R(z_q, z_o)$, of the number of customers in the queue and in the orbit at an arbitrary time point is given by*

$$R(z_q, z_o) = \rho R^{vi}(z_q, z_o) + (1 - \rho)\frac{G}{G+\mathbb{E}S}R^{gl}(z_q, z_o) + (1 - \rho)\frac{\mathbb{E}S}{G+\mathbb{E}S}R^{va}(z_q, z_o). \tag{2.20}$$

Proof.

a) Follows from the fact that during vacation periods the number of customers
   in the queue is 0 and the fact that the number of customers at an arbitrary
   time point in the orbit is the sum of two independent terms: The number
   of customers at the beginning of the vacation period and the number that
   arrived during the past part of the vacation period. The generating function
   of the latter is given by
   $$\frac{1 - \tilde{S}(\lambda(1 - z_o))}{\lambda(1 - z_o)\mathbb{E}S}.$$

b) Follows from the fact that if the past part of the glue period is equal to t,
   the generating function of the number of new arrivals in the queue during
   this period is equal to $e^{-\lambda(1-z_q)t}$ and each customer present in the orbit at
   the beginning of the glue period is, independent of the others, still in orbit
   with probability $e^{-\nu t}$ and has moved to the queue with probability $1 - e^{-\nu t}$.

c) During an arbitrary point in time in a visit period the number of customers
   in the system consists of two parts:
   - the number of customers in the system at the beginning of the service
     time of the customer currently in service, leading to the term
     $$\frac{z_q \left(\mathbb{E}[z_q^{Y^{(q)}} z_o^{Y^{(o)}}] - \mathbb{E}[\tilde{B}(\lambda(1 - z_o))^{Y^{(q)}} z_o^{Y^{(o)}}] \right)}{\mathbb{E}[Y^{(q)}] \left(z_q - \tilde{B}(\lambda(1 - z_o)) \right)};$$

   (see Takagi [19], formula (5.14) on page 206, for a similar formula in the
   ordinary $M/G/1$ vacation queue with gated service but without retrials).
   - the number of customers that arrived during the past part of the service
     of the customer currently in service, leading to the term
     $$\frac{1 - \tilde{B}(\lambda(1 - z_o))}{\lambda(1 - z_o)\mathbb{E}B}.$$

d) Follows from the fact that the fraction of time the server is visiting Q is
   equal to ρ, and if the server is not visiting Q, with probability $\mathbb{E}S/(G+\mathbb{E}S)$
   the server is on vacation and with probability $G/(G+\mathbb{E}S)$ the system is in
   a glue phase.

From Theorem 2, we now can obtain the steady-state mean number of customers in the system at arbitrary time points in vacation periods ($\mathbb{E}[R_{va}]$), in
glue periods ($\mathbb{E}[R_{gl}]$), in visit periods ($\mathbb{E}[R_{vi}]$) and in arbitrary periods ($\mathbb{E}[R]$).
These are given by

$$\mathbb{E}[R_{va}] = \mathbb{E}[Z] + \lambda\frac{\mathbb{E}[S^2]}{2\mathbb{E}[S]},$$

$$\mathbb{E}[R_{gl}] = \mathbb{E}[X] + \lambda\frac{G}{2},$$

$$\mathbb{E}[R_{vi}] = 1 + \lambda\frac{\mathbb{E}[B^2]}{2\mathbb{E}[B]} + \frac{\mathbb{E}[Y^{(q)}Y^{(o)}]}{\mathbb{E}[Y^{(q)}]} + \frac{(1+\rho)\mathbb{E}[Y^{(q)}(Y^{(q)}-1)]}{2\mathbb{E}[Y^{(q)}]},$$

$$\mathbb{E}[R] = \rho\mathbb{E}[R_{vi}] + (1 - \rho)\frac{G}{G+\mathbb{E}S}\mathbb{E}[R_{gl}] + (1 - \rho)\frac{\mathbb{E}S}{G+\mathbb{E}S}\mathbb{E}[R_{va}].$$

Remark that the quantities $\mathbb{E}[Y^{(q)}Y^{(o)}]$ and $\mathbb{E}[Y^{(q)}(Y^{(q)}-1)]$ can be obtained using (2.3):

$$\mathbb{E}[Y^{(q)}Y^{(o)}] = \lambda G e^{-\nu G}\mathbb{E}[X] + \left(1 - e^{-\nu G}\right)e^{-\nu G}\mathbb{E}[X(X-1)],$$
$$\mathbb{E}[Y^{(q)}(Y^{(q)}-1)] = (\lambda G)^2 + \left(1 - e^{-\nu G}\right)^2\mathbb{E}[X(X-1)] + 2\lambda G\left(1 - e^{-\nu G}\right)\mathbb{E}[X].$$

Finally, the mean sojourn time of an arbitrary customer now immediately follows from Little's formula. The results of this section can, e.g., be used to determine the value of G which minimizes the mean sojourn time of an arbitrary customer.

3   Queue Length Analysis for The Two-Queue Case

3.1   Model Description

In this section we consider a one-server polling model with two queues, Q_1 and Q_2. Customers arrive at Q_i according to a Poisson process with rate λ_i; they are called type-i customers, $i = 1, 2$. The service times at Q_i are i.i.d. r.v., with B_i a generic r.v., with distribution $B_i(\cdot)$ and LST $\tilde{B}_i(\cdot)$, $i = 1, 2$. After a visit of the server at Q_i, it switches to the other queue. Successive switchover times from Q_i to the other queue are i.i.d. r.v., with S_i a generic r.v., with distribution $S_i(\cdot)$ and LST $\tilde{S}_i(\cdot)$, $i = 1, 2$. We make all the usual independence assumptions about interarrival times, service times and switchover times at the queues. After a switch of the server to Q_i, there first is a deterministic (i.e., constant) glue period G_i, $i = 1, 2$, before the visit of the server at Q_i begins. As in the one-queue case, the significance of the glue period stems from the following assumption. Customers who arrive at Q_i do not receive service immediately. When customers arrive at Q_i during a glue period G_i, they stick, joining the queue of Q_i. When they arrive in any other period, they immediately leave and retry after a retrial interval which is independent of everything else, and which is exponentially distributed with rate ν_i, $i = 1, 2$.

The service discipline at both queues is gated: During the visit period at Q_i, the server serves all "glued" customers in that queue, i.e., all type-i customers waiting at the end of the glue period – but none of those in orbit, and neither any new arrivals.

We are interested in the steady-state behaviour of this polling model with retrials. We hence assume that the stability condition $\sum_{i=1}^2 \rho_i < 1$ holds, where $\rho_i := \lambda_i\mathbb{E}B_i$.

Some more notation:

G_{ni} denotes the nth glue period of Q_i.

V_{ni} denotes the nth visit period of Q_i.

S_{ni} denotes the nth switch period out of Q_i, $i = 1, 2$.

$(X_{n1}^{(i)}, X_{n2}^{(i)})$ denotes the vector of numbers of customers of type 1 and of type 2 in the system (hence in orbit) at the start of G_{ni}, $i = 1, 2$.

$(Y_{n1}^{(i)}, Y_{n2}^{(i)})$ denotes the vector of numbers of customers of type 1 and of type 2 in the system at the start of V_{ni}, $i = 1, 2$. We distinguish between those who are queueing in Q_i and those who are in orbit for Q_i: We write $Y_{n1}^{(1)} = Y_{n1}^{(1q)} + Y_{n1}^{(1o)}$ and $Y_{n2}^{(2)} = Y_{n2}^{(2q)} + Y_{n2}^{(2o)}$, where q denotes queueing and o denotes in orbit. Finally, $(Z_{n1}^{(i)}, Z_{n2}^{(i)})$ denotes the vector of numbers of customers of type 1 and of type 2 in the system (hence in orbit) at the start of S_{ni}, $i = 1, 2$.

3.2   Queue Length Analysis

In this section we study the steady-state joint distribution of the numbers of customers in the system at beginnings of glue periods. This will also immediately yield the steady-state joint distributions of the numbers of customers in the system at the beginnings of visit periods and of switch periods. We follow a similar generating function approach as in the one-queue case, throughout making the following assumption regarding the parameters of the generating functions: $|z_i| \leq 1$, $|z_{iq}| \leq 1$, $|z_{io}| \leq 1$. Observe that the generating function of the vector of numbers of arrivals at Q_1 and Q_2 during a type-i service time B_i is $\beta_i(z_1, z_2) := \tilde{B}_i(\lambda_1(1 - z_1) + \lambda_2(1 - z_2))$. Similarly, the generating function of the vector of numbers of arrivals at Q_1 and Q_2 during a type-i switchover time S_i is $\sigma_i(z_1, z_2) := \tilde{S}_i(\lambda_1(1 - z_1) + \lambda_2(1 - z_2))$. We can successively express (in terms of generating functions) $(X_{n1}^{(2)}, X_{n2}^{(2)})$ into $(Z_{n1}^{(1)}, Z_{n2}^{(1)})$, $(Z_{n1}^{(1)}, Z_{n2}^{(1)})$ into $(Y_{n1}^{(1q)}, Y_{n1}^{(1o)}, Y_{n2}^{(1)})$, and $(Y_{n1}^{(1q)}, Y_{n1}^{(1o)}, Y_{n2}^{(1)})$ into $(X_{n1}^{(1)}, X_{n2}^{(1)})$; etc. Denote by $(X_1^{(i)}, X_2^{(i)})$ the vector with as distribution the limiting distribution of $(X_{n1}^{(i)}, X_{n2}^{(i)})$, $i = 1, 2$, and similarly introduce $(Z_1^{(i)}, Z_2^{(i)})$ and $(Y_1^{(i)}, Y_2^{(i)})$, with $Y_1^{(1)} = Y_1^{(1q)} + Y_1^{(o)}$ and with $Y_2^{(2)} = Y_2^{(2q)} + Y_2^{(o)}$, for $i = 1, 2$. We have:

$$\mathbb{E}[z_1^{X_1^{(2)}} z_2^{X_2^{(2)}}] = \sigma_1(z_1, z_2)\mathbb{E}[z_1^{Z_1^{(1)}} z_2^{Z_2^{(1)}}]. \tag{3.1}$$

$$\mathbb{E}[z_1^{Z_1^{(1)}} z_2^{Z_2^{(1)}} | Y_1^{(1q)} = h_{1q}, Y_1^{(1o)} = h_{1o}, Y_2^{(1)} = h_2] = z_1^{h_{1o}} z_2^{h_2} \beta_1^{h_{1q}}(z_1, z_2), \tag{3.2}$$

yielding

$$\mathbb{E}[z_1^{Z_1^{(1)}} z_2^{Z_2^{(1)}}] = \mathbb{E}[\beta_1(z_1, z_2)^{Y_1^{(1q)}} z_1^{Y_1^{(1o)}} z_2^{Y_2^{(1)}}]. \tag{3.3}$$

Furthermore,

$$\mathbb{E}[z_{1q}^{Y_1^{(1q)}} z_{1o}^{Y_1^{(1o)}} z_2^{Y_2^{(1)}} | X_1^{(1)} = i_1, X_2^{(1)} = i_2]$$
$$= z_2^{i_2} e^{-\lambda_2(1-z_2)G_1} e^{-\lambda_1(1-z_{1q})G_1} [(1 - e^{-\nu_1 G_1})z_{1q} + e^{-\nu_1 G_1} z_{1o}]^{i_1}, \tag{3.4}$$

yielding

$$\mathbb{E}[z_{1q}^{Y_1^{(1q)}} z_{1o}^{Y_1^{(1o)}} z_2^{Y_2^{(1)}}] = e^{-\lambda_2(1-z_2)G_1} e^{-\lambda_1(1-z_{1q})G_1}$$
$$\times \mathbb{E}[[(1 - e^{-\nu_1 G_1})z_{1q} + e^{-\nu_1 G_1} z_{1o}]^{X_1^{(1)}} z_2^{X_2^{(1)}}]. \tag{3.5}$$

It follows from (3.1), (3.3) and (3.5), with

$$f_1(z_1, z_2) := (1 - e^{-\nu_1 G_1})\beta_1(z_1, z_2) + e^{-\nu_1 G_1} z_1, \tag{3.6}$$

that

$$\mathbb{E}[z_1^{X_1^{(2)}} z_2^{X_2^{(2)}}] = \sigma_1(z_1, z_2) e^{-\lambda_1(1-\beta_1(z_1,z_2))G_1 - \lambda_2(1-z_2)G_1}$$
$$\times \mathbb{E}[f_1(z_1, z_2)^{X_1^{(1)}} z_2^{X_2^{(1)}}]. \tag{3.7}$$

Similarly we have, with

$$f_2(z_1, z_2) := (1 - e^{-\nu_2 G_2})\beta_2(z_1, z_2) + e^{-\nu_2 G_2} z_2, \tag{3.8}$$

that

$$\mathbb{E}[z_1^{X_1^{(1)}} z_2^{X_2^{(1)}}] = \sigma_2(z_1, z_2) e^{-\lambda_1(1-z_1)G_2 - \lambda_2(1-\beta_2(z_1,z_2))G_2}$$
$$\times \mathbb{E}[z_1^{X_1^{(2)}} f_2(z_1, z_2)^{X_2^{(2)}}]. \tag{3.9}$$

It follows from (3.7) and (3.9) that

$$\mathbb{E}[z_1^{X_1^{(1)}} z_2^{X_2^{(1)}}] = \sigma_1(z_1, f_2(z_1, z_2))\sigma_2(z_1, z_2) e^{-\lambda_1(1-z_1)G_2 - \lambda_2(1-\beta_2(z_1,z_2))G_2}$$
$$\times e^{-\lambda_1(1-\beta_1(z_1,f_2(z_1,z_2)))G_1 - \lambda_2(1-f_2(z_1,z_2))G_1}$$
$$\times \mathbb{E}[f_1(z_1, f_2(z_1, z_2))^{X_1^{(1)}} f_2(z_1, z_2)^{X_2^{(1)}}]. \tag{3.10}$$

We can rewrite this, with

$$h_1(z_1, z_2) := f_1(z_1, f_2(z_1, z_2)), \quad h_2(z_1, z_2) := f_2(z_1, z_2), \tag{3.11}$$

and

$$X(z_1, z_2) := \mathbb{E}[z_1^{X_1^{(1)}} z_2^{X_2^{(1)}}], \tag{3.12}$$

and with an obvious definition of $K(\cdot, \cdot)$, as

$$X(z_1, z_2) = K(z_1, z_2) X(h_1(z_1, z_2), h_2(z_1, z_2)). \tag{3.13}$$

Define

$$h_i^{(0)}(z_1, z_2) := z_i, \quad h_i^{(n)}(z_1, z_2) := h_i(h_1^{(n-1)}(z_1, z_2), h_2^{(n-1)}(z_1, z_2)), \quad i = 1, 2. \tag{3.14}$$

Theorem 3. *If $\rho_1 + \rho_2 < 1$, then the generating function $X(z_1, z_2)$ is given by*

$$X(z_1, z_2) = \prod_{m=0}^{\infty} K(h_1^{(m)}(z_1, z_2), h_2^{(m)}(z_1, z_2)). \tag{3.15}$$

Proof. Equation (3.15) follows from (3.13) by iteration. We still need to prove that the infinite product converges if $\rho_1 + \rho_2 < 1$. Equation (3.13) is an equation for a multi-type branching process with immigration, where the number of immigrants of different types has generating function $K(z_1, z_2)$ and the number of children of different types of a type 1 individual in the branching process has generating function $h_1(z_1, z_2)$ and the number of children of different types of a type 2 individual in the branching process has generating function $h_2(z_1, z_2)$. An important role in the analysis of such a process is played by the mean matrix M of the branching process,

$$M = \begin{pmatrix} m_{11} & m_{12} \\ m_{21} & m_{22} \end{pmatrix}, \tag{3.16}$$

where m_{ij} represents the mean number of children of type j of a type i individual. In our case, the elements of the matrix M are given by

$$m_{11} = e^{-\nu_1 G_1} + \left(1 - e^{-\nu_1 G_1}\right)\rho_1 + \left(1 - e^{-\nu_1 G_1}\right)\left(1 - e^{-\nu_2 G_2}\right)\rho_1\rho_2, \tag{3.17}$$

$$m_{12} = \left(1 - e^{-\nu_1 G_1}\right)\lambda_2 \mathbb{E}B_1 \left(e^{-\nu_2 G_2} + \left(1 - e^{-\nu_2 G_2}\right)\rho_2\right), \tag{3.18}$$

$$m_{21} = \left(1 - e^{-\nu_2 G_2}\right)\lambda_1 \mathbb{E}B_2, \tag{3.19}$$

$$m_{22} = e^{-\nu_2 G_2} + \left(1 - e^{-\nu_2 G_2}\right)\rho_2. \tag{3.20}$$

For example, formula (3.18) can be explained as follows. A type 1 customer present in the system at the beginning of a glue period of Q_1 is served with probability $1 - e^{-\nu_1 G_1}$. If he is served, on average $\lambda_2 \mathbb{E}B_1$ type 2 customers will arrive during his service time. During the visit period of Q_2 each of these customers is not served with probability $e^{-\nu_2 G_2}$ or served with probability $1 - e^{-\nu_2 G_2}$, in which case on average ρ_2 type 2 customers will arrive during this service time.

The theory of multitype branching processes with immigration (see Quine [15] and Resing [16]) now states that if (i) the expected total number of immigrants in a generation is finite and (ii) the maximal eigenvalue λ_{max} of the mean matrix M satisfies $\lambda_{max} < 1$, then the generating function of the steady state distribution of the process is given by (3.15). To complete the proof of Theorem 3, we shall now verify (i) and (ii).

Ad (i): The expected total number of immigrants in a generation is

$$(\lambda_1 + \lambda_2)\mathbb{E}S_2 + \lambda_1 G_2 + \lambda_2 G_2(\lambda_1 + \lambda_2)\mathbb{E}B_2$$
$$+\lambda_1 \mathbb{E}S_1 + \lambda_2 \mathbb{E}S_1 \left(e^{-\nu_2 G_2} + \left(1 - e^{-\nu_2 G_2}\right)(\lambda_1 + \lambda_2)\mathbb{E}B_2\right)$$
$$+\lambda_1 G_1 \left(\lambda_1 \mathbb{E}B_1 + \lambda_2 \mathbb{E}B_1 \left(e^{-\nu_2 G_2} + \left(1 - e^{-\nu_2 G_2}\right)(\lambda_1 + \lambda_2)\mathbb{E}B_2\right)\right)$$
$$+\lambda_2 G_1 \left(e^{-\nu_2 G_2} + \left(1 - e^{-\nu_2 G_2}\right)(\lambda_1 + \lambda_2)\mathbb{E}B_2\right), \tag{3.21}$$

and hence indeed finite. Here, the term $(\lambda_1 + \lambda_2)\mathbb{E}S_2$ corresponds to the customers arriving during the switch period out of Q_2. The term $\lambda_1 G_2$ corresponds to the type 1 customers arriving during the glue period of Q_2. The term $\lambda_2 G_2(\lambda_1 + \lambda_2)\mathbb{E}B_2$ corresponds to the type 2 customers arriving during the glue period of Q_2. These customers are served during the visit period of Q_2 and during their service time other customers will arrive. The term $\lambda_1 \mathbb{E}S_1$ corresponds to the type 1 customers arriving during the switch period out of Q_1. The term $\lambda_2 \mathbb{E}S_1 \left(e^{-\nu_2 G_2} + \left(1 - e^{-\nu_2 G_2}\right)(\lambda_1 + \lambda_2)\mathbb{E}B_2\right)$ corresponds to the type

2 customers arriving during the switch period out of Q_1. These customers are served during the visit period of Q_2 with probability $1 - e^{-\nu_2 G_2}$ (in which case again other customers will arrive during their service time) and with probability $e^{-\nu_2 G_2}$ they are not served during this visit period. The last two terms in (3.21) correspond to the type 1 and type 2 customers arriving during the glue period of Q_1.

Ad (ii): Define the matrix

$$
H = \begin{pmatrix} e^{-\nu_1 G_1} + \left(1 - e^{-\nu_1 G_1}\right) \rho_1 & \left(1 - e^{-\nu_1 G_1}\right) \lambda_2 \mathbb{E} B_1 \\ \left(1 - e^{-\nu_2 G_2}\right) \lambda_1 \mathbb{E} B_2 & e^{-\nu_2 G_2} + \left(1 - e^{-\nu_2 G_2}\right) \rho_2 \end{pmatrix},
\tag{3.22}
$$

where the elements h_{ij} of the matrix H represent the mean number of type j customers that replace a type i customer during a visit period of Q_i (either new arrivals if the customer is served, or the customer itself if it is not served). We have that

$$
H \begin{pmatrix} \mathbb{E} B_1 \\ \mathbb{E} B_2 \end{pmatrix} = \begin{pmatrix} \left[e^{-\nu_1 G_1} + \left(1 - e^{-\nu_1 G_1}\right) (\rho_1 + \rho_2) \right] \mathbb{E} B_1 \\ \left[e^{-\nu_2 G_2} + \left(1 - e^{-\nu_2 G_2}\right) (\rho_1 + \rho_2) \right] \mathbb{E} B_2 \end{pmatrix} < \begin{pmatrix} \mathbb{E} B_1 \\ \mathbb{E} B_2 \end{pmatrix}
\tag{3.23}
$$

if and only if $\rho_1 + \rho_2 < 1$. Using this result and following the same line of proof as in Section 5 of Resing [16], we can show that the stability condition $\rho_1 + \rho_2 < 1$ implies that also the maximal eigenvalue λ_{max} of the mean matrix M satisfies $\lambda_{max} < 1$. This concludes the proof.

4   Conclusions and Suggestions for Future Research

In this paper we have studied vacation queues and two-queue polling models with the gated service discipline and with retrials. Motivated by optical communications, we have introduced a glue period just before a server visit; during such a glue period, new customers and retrials "stick" instead of immediately going into orbit. For both the vacation queue and the two-queue polling model, we have derived steady-state queue length distributions at an arbitrary epoch and at various specific epochs. This was accomplished by establishing a relation to branching processes. We have thus laid the groundwork for the performance analysis of an N-queue polling model with retrials.

In future studies, we shall not only turn to that N-queue model; we also would like to consider other service disciplines. Furthermore, the following model variants seem to fall within our framework: (i) customers may *not* retry with a certain probability; (ii) the arrival rates may be different for visit, vacation and glue periods; (iii) one might allow that new arrivals during a glue period are already served during that glue period.

We would also like to explore the possibility to study the heavy traffic behavior of these models via the relation to branching processes, cf. [14].

Finally, we would like to point out an important advantage of optical fibre: the wavelength of light. A fibre-based network node may thus route incoming packets not only by switching in the time-domain, but also by wavelength division multiplexing. In queueing terms, this gives rise to *multiserver* polling models, each server representing a wavelength. We refer to [1] for the stability analysis of multiserver polling models, and to [2] for a mean field approximation of large passive optical networks. It would be very interesting to study multiserver polling models with the additional features of retrials and glue periods.

Acknowledgment. The authors gratefully acknowledge fruitful discussions with Kevin ten Braak and Tuan Phung-Duc about retrial queues and with Ton Koonen about optical networks. The research is supported by the IAP program BESTCOM, funded by the Belgian government.

References

1. Antunes, N., Fricker, C., Roberts, J.: Stability of multi-server polling system with server limits. Queueing Systems 68, 229–235 (2011)
2. Antunes, N., Fricker, C., Robert, P., Roberts, J.: Traffic capacity of large WDM Passive Optical Networks. In: Proceedings 22nd International Teletraffic Congress, ITC 22, Amsterdam (September 2010)
3. Artalejo, J.R., Gomez-Corral, A.: Retrial Queueing Systems: A Computational Approach. Springer, Berlin (2008)
4. Athreya, K.B., Ney, P.E.: Branching Processes. Springer, Berlin (1972)
5. Boon, M.A.A., van der Mei, R.D., Winands, E.M.M.: Applications of polling systems. SORMS 16, 67–82 (2011)
6. Boxma, O.J., Cohen, J.W.: The $M/G/1$ queue with permanent customers. IEEE J. Sel. Areas in Commun. 9, 179–184 (1991)
7. Falin, G.I., Templeton, J.G.C.: Retrial Queues. Chapman and Hall, London (1997)
8. Koonen, A.M.J.: Personal communication (2014)
9. Langaris, C.: A polling model with retrial of customers. Journal of the Operations Research Society of Japan 40, 489–507 (1997)
10. Langaris, C.: Gated polling models with customers in orbit. Mathematical and Computer Modelling 30, 171–187 (1999)
11. Langaris, C.: Markovian polling system with mixed service disciplines and retrial customers. Top 7, 305–322 (1999)
12. Levy, H., Sidi, M.: Polling models: Applications, modeling and optimization. IEEE Trans. Commun. 38, 1750–1760 (1990)
13. Maier, M.: Optical Switching Networks. Cambridge University Press, Cambridge (2008)
14. Olsen, T.L., van der Mei, R.D.: Periodic polling systems in heavy traffic: Distribution of the delay. Journal of Applied Probability 40, 305–326 (2003)
15. Quine, M.P.: The multitype Galton-Watson process with immigration. Journal of Applied Probability 7, 411–422 (1970)
16. Resing, J.A.C.: Polling systems and multitype branching processes. Queueing Systems 13, 409–426 (1993)
17. Rogiest, W.: Stochastic Modeling of Optical Buffers. Ph.D. Thesis, Ghent University, Ghent, Belgium (2008)

18. Takagi, H.: Application of polling models to computer networks. Comput. Netw. ISDN Syst. 22, 193–211 (1991)
19. Takagi, H.: Queueing Analysis: A Foundation of Performance Evaluation. Vacation and Priority Systems, vol. 1. Elsevier Science Publishers, Amsterdam (1991)
20. Takagi, H.: Queueing analysis of polling models: progress in 1990-1994. In: Dshalalow, J.H. (ed.) Frontiers in Queueing: Models, Methods and Problems, pp. 119–146. CRC Press, Boca Raton (1997)
21. Takagi, H.: Analysis and application of polling models. In: Reiser, M., Haring, G., Lindemann, C. (eds.) Dagstuhl Seminar 1997. LNCS, vol. 1769, pp. 424–442. Springer, Heidelberg (2000)
22. Vishnevskii, V.M., Semenova, O.V.: Mathematical methods to study the polling systems. Autom. Remote Control 67, 173–220 (2006)

Fluid Vacation Model with Markov Modulated Load and Exhaustive Discipline

Zsolt Saffer[1] and Miklós Telek[1,2]

[1] Budapest University of Technology and Economics, Hungary
[2] MTA-BME Information systems research group, Hungary
{saffer,telek}@webspn.hit.bme.hu

Abstract. In this paper we analyze a fluid vacation model with exhaustive discipline, in which the fluid source is modulated by a background continuous-time Markov chain and the fluid is removed by constant rate during the service period. Due to the continuous nature of the fluid the state space of the model becomes continuous, which is the major novelty and challenge of the analysis. We adapt the descendant set approach used in polling models to the fluid vacation model. We provide steady-state vector Laplace Transform and mean of the fluid level at arbitrary epoch.

1 Introduction

Fluid vacation model is an extension of the classical vacation model (see in [2], [7]), in which fluid takes the role of the customer of the classical model. Due to the the continuous nature of the fluid, the flow in and the removal of fluid are characterized by rates. Hence the state space becomes continuous, which is a challenge in the analysis comparing to that of the discrete state space of the classical vacation model. This requires different analysis techniques.

In this paper we investigate a fluid vacation model with exhaustive service when the fluid source is modulated by a background Markov chain. The main idea of the analysis is the extension of the descendant set approach (see in [1]) to the continuous fluid model context. This together with the transient analysis of the input fluid flow enable to describe the evolution of the joint fluid level and the state of the background Markov chain between the vacation end and vacation start epochs - on the Laplace transform (LT) level. The resulted relations are called as governing equations. From them we determine the steady-state probability vector of the background Markov chain at the vacation start epochs. In the course of the analysis we derive a relation for the steady-state vector LT and vector mean of the fluid level at arbitrary epoch in terms of the previously mentioned steady-state probability vector. We also derive the steady-state LT of the service time, which is the counterpart of the busy period analysis in the classical vacation queue.

The rest of the paper is organized as follows. In section 2 we present the fluid vacation model and the concept of embedding matrix LTs, which is needed to the extension of the descendant set approach to fluid model. In section 3

A. Horváth and K. Wolter (Eds.): EPEW 2014, LNCS 8721, pp. 59–73, 2014.

we establish the governing equations of the model. The derivation of steady-state results follows in section 4. A section with numerical example is omitted here by space limitation.

2   Model and Notation

2.1   Model Description

We consider a fluid vacation model with Markov modulated load and exhaustive discipline. The model has an infinite fluid buffer.

The input fluid flow of the buffer is determined by a modulating CTMC ($\Omega(t)$ for $t \geq 0$) with state space $\mathcal{S} = \{1, \ldots, L\}$ and generator $\mathbf{Q}$. When this Markov chain is in state j ($\Omega(t) = j$) then fluid flows to the buffer at rate r_j for $j \in \{1, \ldots, L\}$. We define the diagonal matrix $\mathbf{R} = diag(r_1, \ldots, r_L)$. During the service period the server removes fluid from the fluid buffer at finite rate $d > 0$. Consequently, when the overall Markov chain is in state j ($\Omega(t) = j$) then the fluid level of the buffer during the service period changes at rate $r_j - d$, otherwise during the vacation periods it changes at rate r_j, because there is no service. In the vacation model the length of the service period is determined by the applied discipline. In this work we consider the exhaustive discipline. Under exhaustive discipline the fluid is removed during the service period until the buffer becomes empty. Each time the buffer becomes empty the server takes a vacation period. During vacation periods there is no service thus the fluid level of the buffer is increasing by the actual flowing rates. The consecutive vacation times are independent and identically distributed (i.i.d.). The random variable of the vacation time, its probability distribution function (pdf), its Laplace transform (LT) and its mean are denoted by $\tilde{\sigma}$, $\sigma(t) = \frac{d}{dt} Pr(\tilde{\sigma} < t)$ and $\sigma^*(s) = E(e^{-s\tilde{\sigma}})$, $\sigma = E(\tilde{\sigma})$, respectively. We define the cycle time (or simple cycle) as the time between just after the starts of two consecutive service periods.

We set the following assumptions on the fluid vacation model:

- **A.1** The generator matrix $\mathbf{Q}$ of the modulating CTMC is irreducible.
- **A.2** The fluid rates are positive and finite, i.e. $r_j > 0$ for $j \in \{1, \ldots, L\}$.
- **A.3** The fluid is removed from the buffer according to the FCFS discipline.
- **A.4** The fluid level decreases during the service period, i.e., $r_j < d$ for $j \in \{1, \ldots, L\}$.

Let $\boldsymbol{\pi}$ be the stationary probability vector of the modulating Markov chain. Due to assumption **A.1** the equations

$$\boldsymbol{\pi}\mathbf{Q} = 0, \quad \boldsymbol{\pi}\mathbf{e} = 1. \tag{1}$$

uniquely determine $\boldsymbol{\pi}$, where $\mathbf{e}$ is the $L \times 1$ column vector of ones. The stationary fluid flow rate, λ, and and the utilization ρ, is given as

$$\lambda = \boldsymbol{\pi}\mathbf{R}\mathbf{e}, \quad \rho = \frac{\lambda}{d}, \tag{2}$$

respectively. The necessary condition of the stability of the fluid vacation model is that mean fluid arrival rate $\lambda = \pi \mathbf{Re}$ is less than d, which is equivalent with $\rho < 1$.

If the amount of fluid served during a service period were limited, like e.g. in case of a model with time-limited discipline, then further restriction would be needed for the sufficiency. However the model with the exhaustive discipline does not have any load-independent limitation for a service period, therefore the above necessary condition is also a sufficient one for the stability of the system.

For the i,j-th element of the matrix $\mathbf{Z}$ the notations $\mathbf{Z}_{ij}$ or $[\mathbf{Z}]_{ij}$ are used. Similarly $\mathbf{z}_j$ and $[\mathbf{z}]_j$ denote the j-th element of vector $\mathbf{z}$. When $\mathbf{X}^*(s)$, $Re(s) \geq 0$ is a matrix LT, $\mathbf{X}^{(k)}$ denotes its k-th ($k \geq 1$) derivative at $s = 0$, i.e., $\mathbf{X}^{(k)} = \frac{d^k}{ds^k}\mathbf{X}^*(s)|_{s=0}$ and $\mathbf{X}^{(0)}$ denotes its value at $s = 0$, i.e., $\mathbf{X}^{(0)} = \mathbf{X}^*(0)$. Similar notations are applied for vector LT $\mathbf{x}^*(s)$ and scalar LT $x^*(s)$.

2.2 Embedded Matrix LTs

Let $\mathbf{Z}$ be an $L \times L$ rate matrix which has the following properties:

- the diagonal elements are negative ($\mathbf{Z}_{i,i} < 0$) and the other elements are non-negative ($\mathbf{Z}_{i,j} \geq 0$, for $i \neq j$),
- the row sums are zero.

For $Re(v) \geq 0$ let

$$\mathbf{H}(v) = \mathbf{T}v - \mathbf{Z}, \qquad (3)$$

be a linear $L \times L$ matrix function of the complex variable v, where $\mathbf{Z}$ is a rate matrix and $\mathbf{T}$ is diagonal and its diagonal elements are positive, i.e. $[\mathbf{T}]_{j,j} > 0$ for $j \in \{1, \ldots, L\}$. That is $\mathbf{Z}$ and $\mathbf{T}$ are real. The matrix function $-\mathbf{H}(v)$ has the following properties:

- **P.1** it is analytic for $Re(v) \geq 0$,
- **P.2** it is a rate matrix when $v = 0$,
- **P.3** the real part of its diagonal elements are negative for $Re(v) \geq 0$, i.e. $(Re(-\mathbf{H}_{j,j}(v)) < 0)$,
- **P.4** it is a diagonal dominant matrix for $Re(v) \geq 0$, i.e., $|Re(-\mathbf{H}_{j,j}(v))| \geq \sum_{k,k \neq j} |-\mathbf{H}_{j,k}(v))|$.

We define the operator $\mathcal{O}()$ on a complex variable v and on a linear matrix function $\mathbf{G}(v) = \mathbf{G}_1 v + \mathbf{G}_2$ as the operator performing the substitution $v \to \mathbf{H}(v)$. That is $\mathcal{O}(v) = \mathbf{H}(v) = \mathbf{T}v - \mathbf{Z}$ and $\mathcal{O}(\mathbf{G}(v)) = \mathbf{G}_1\mathbf{H}(v) + \mathbf{G}_2 = \mathbf{G}_1\mathbf{T}v - \mathbf{G}_1\mathbf{Z} + \mathbf{G}_2$, which are linear matrix functions as well. The order of non-commuting matrices are kept according to this definition. The multifold operator $\mathcal{O}^k(\bullet)$ is defined recursively as $\mathcal{O}^k(\bullet) = \mathcal{O}(\mathcal{O}^{k-1}(\bullet))$, $k \geq 2$, where $\mathcal{O}^1(\bullet) = \mathcal{O}(\bullet)$ by definition. Additionally, we introduce $\mathcal{O}^0(\bullet) = \bullet$. Starting from (3), the $L \times L$ linear matrix function $\mathcal{O}^k(v)$ can be expressed recursively as

$$\mathcal{O}^k(v) = \mathbf{T}^k v - \sum_{i=0}^{k-1} \mathbf{T}^i \mathbf{Z}, \qquad (4)$$

The matrix function $-\mathcal{O}^k(v)$ has the following properties:

◇ $-\mathcal{O}^k(v)$ is analytic for $Re(v) \geq 0$ (due to **P.1** of $\mathbf{H}(v)$),
◇ $-\mathcal{O}^k(v)\big|_{v=0}$ is also a rate matrix (**P.2**), since multiplying rate matrix $\mathbf{Z}$ any times by positive diagonal matrices from left results in a rate matrix and the sum of $L \times L$ rate matrices is also an $L \times L$ rate matrix,
◇ $-\mathcal{O}^k(v)$ has also the properties **P.3** and **P.4**.

It follows from the recursive definition of the multifold operator $\mathcal{O}^k()$ that the matrix LT $\int_{x=0}^{\infty} p(x)e^{-\mathcal{O}^k(v)x}dx$ is created by consecutive embedding of the matrix $\mathbf{H}(v)$ in the previous matrix LT and therefore we call this matrix LT as *embedded matrix LT*. If $p(x) \geq 0$ for $x \geq 0$ and v is the complex argument of the LT $\int_{x=0}^{\infty} p(x)e^{-vx}dx$ then

$$\int_{x=0}^{\infty} p(x)e^{-\mathcal{O}(v)x}dx = \int_{x=0}^{\infty} p(x)e^{-\mathbf{H}(v)x}dx \tag{5}$$

is an $L \times L$ matrix LT. According to the Gerschgorin Circle Theorem [3] each eigenvalue of $-\mathbf{H}(v)$ is in one of the disks $\{z : |z - (-\mathbf{H}_{j,j}(v))| \leq \sum_{k \neq j} | - \mathbf{H}_{j,k}(v)|\}$ (i.e. disks in complex z-plane with center at $(-\mathbf{H}_{j,j}(v))$ and radius $\sum_{k \neq j} | - \mathbf{H}_{j,k}(v)|$), for $\forall j \in \{1, \ldots, L\}$. This together with properties **P.3** and **P.4** imply that the eigenvalues of $-\mathbf{H}(v)$ have negative or zero real part for $Re(v) \geq 0$. The matrix function $e^{-\mathbf{H}(v)x}$ can be written in Lagrange matrix polynomial form as

$$(e^x)^{-\mathbf{H}(v)} = \sum_{k=1}^{K} (e^x)^{\gamma_k} L_k(-\mathbf{H}(v)), \tag{6}$$

where K is the number of different eigenvalues of matrix $(-\mathbf{H}(v))$, γ_k and $L_k(v)$, for $k = 1, \ldots, K$ denotes the different eigenvalues of matrix $(-\mathbf{H}(v))$ and the finite Lagrange-polynomials belonging to the roots of the minimal-polynomial of matrix $(-\mathbf{H}(v))$, respectively. Applying (6) in (5) and rearranging it yields

$$\int_{x=0}^{\infty} p(x)e^{-\mathcal{O}(v)x}dx = \sum_{k=1}^{K} L_k(-\mathbf{H}(v)) \int_{x=0}^{\infty} p(x)e^{\gamma_k x}dx$$

Recall that the eigenvalues γ_k have negative or zero real part for $Re(v) \geq 0$. Consequently (5) is finite when the LT $\int_{x=0}^{\infty} p(x)e^{-vx}dx$ is finite for $Re(v) \geq 0$.

The same argument holds also for the matrix LT $\int_{x=0}^{\infty} p(x)e^{-\mathcal{O}^k(v)x}dx$, since the utilized properties **P.3** and **P.4** of $(-\mathbf{H}(v))$ hold also for $-\mathcal{O}^k(v)$. We remark here that the order of matrix and scalar $\mathbf{T}v$ in the definition of $\mathbf{H}(v)$ is crucial in order to ensure the validity of the properties **P.2** and **P.4** for the matrix function $\mathcal{O}^k(v)$.

3   The Governing Equations of the System

3.1   Transient Analysis of the Arriving Fluid

In this section we consider the accumulated fluid during time $t \geq 0$. More precisely we derive the matrix LT of the fluid flowing into the buffer as a function of time, where the rows and columns of the matrix LT represent the initial and the final states of the modulating Markov chain.

Let $Y(t) \in \mathbb{R}^+$ be the accumulated fluid arrived at the buffer until time t, $\mathbf{A}(t, y)$ be the transition density matrix composed by elements $\mathbf{A}_{j,k}(t, y) = \frac{\partial}{\partial y} Pr(\Omega(t) = k, Y(t) < y | \Omega(0) = j, Y(0) = 0)$ and its Laplace transform be $\mathbf{A}^*(t, v) = \int_{y=0}^{\infty} \mathbf{A}(t, y) e^{-vy} dy$.

Proposition 1. *The matrix LT of the fluid generated by the Markov modulated fluid source in interval $(0, t]$ can be expressed as*

$$\mathbf{A}^*(t, v) = e^{-t(\mathbf{R}v - \mathbf{Q})}. \tag{7}$$

The proof of the proposition is provided in [6].

3.2   The Descendant Fluid

We extend the concept of descendant set (see in Borst and Boxma [1]) to fluid model and describe the exhaustive service period as consecutive gated service intervals without vacations. We define the 1-st descendant fluid level of the given fluid amount as the fluid flowing into the buffer during the service of the given fluid amount. This is similar to the descendant set of a customer in the regular vacation model, which consists of the group of customers arrived during the service of the original customer. Similarly we define the k-th descendant fluid level of the given fluid amount recursively as the fluid accumulated during the service of the $k - 1$-th descendant fluid level. This is the same as the fluid level after k cycles in a gated system without vacation initiated by the given fluid amount. By definition the 0-th descendant fluid level of a given fluid amount equals to itself. The k-th descendant period is defined as the removal time of the k-th descendant fluid for $k \geq 0$. Moreover we consider the evolution of the fluid level jointly with the evolution of the state of the modulating Markov chain. This joint evolution is described by the help of matrix LT formalism. When $\mathbf{g}^*(v)$ is the vector LT of a given initial fluid density then the pdf and the vector LT of its k-th descendant fluid level, for $k \geq 1$, are denoted by $\mathbf{g}^{<k>}(x)$ and $\mathbf{g}^{*<k>}(v)$. Furthermore $\mathbf{g}^{<0>}(x) = \mathbf{g}(x)$ and $\mathbf{g}^{*<0>}(v) = \mathbf{g}^*(v)$. Let

$$\mathcal{O}(v) = \mathbf{H}(v) = \frac{\mathbf{R}v - \mathbf{Q}}{d}, \tag{8}$$

that is $\mathbf{T} = \frac{\mathbf{R}}{d}$ and $\mathbf{Z} = \frac{\mathbf{Q}}{d}$ in (3).

Furthermore we introduce a notation for the LT with respect to the $L \times L$ matrix function $\mathbf{H}(v)$ as follows

$$\mathbf{g}^*(\mathbf{H}(v)) = \int_{x=0}^{\infty} \mathbf{g}(x) e^{-\mathbf{H}(v)x} dx, \tag{9}$$

where $\mathbf{g}()$ is an $1 \times L$ vector function.

Proposition 2. *Starting from the initial fluid amount whose vector LT is* $\mathbf{g}^*(v)$ *the vector LT of the k-th ($k \geq 0$) descendant fluid can be expressed as*

$$\mathbf{g}^{*<k>}(v) = \mathbf{g}^*(\mathcal{O}^k(v)) . \tag{10}$$

Proof. The k-th descendant fluid is defined as the fluid accumulated during the service of the $k-1$-th descendant fluid for $k \geq 1$. The fluid density vector of the $k-1$-th descendant fluid is $\mathbf{g}^{<k-1>}(\xi)$. When the $k-1$-th descendant fluid is ξ, then its service duration is $\frac{\xi}{d}$, from which we can express $[\mathbf{g}^{<k>}(x)]_k$ as

$$[\mathbf{g}^{<k>}(x)]_k = \sum_{j=1}^{L} \int_{\xi=0}^{\infty} [\mathbf{g}^{<k-1>}(\xi)]_j \mathbf{A}_{jk}(\frac{\xi}{d}, x) d\xi , \tag{11}$$

whose vector-matrix form is

$$\mathbf{g}^{<k>}(x) = \int_{\xi=0}^{\infty} \mathbf{g}^{<k-1>}(\xi) \mathbf{A}(\frac{\xi}{d}, x) d\xi .$$

Applying (7) the LT of $\mathbf{g}^{<k>}(x)$ with respect to x is

$$\mathbf{g}^{*<k>}(v) = \int_{\xi=0}^{\infty} \mathbf{g}^{<k-1>}(\xi) \mathbf{A}^*(\frac{\xi}{d}, v) d\xi = \int_{\xi=0}^{\infty} \mathbf{g}^{<k-1>}(\xi) e^{-\frac{\xi}{d}(\mathbf{R}v - \mathbf{Q})} d\xi . \tag{12}$$

Utilizing that the right hand side of (12) is a matrix LT according to (9) and using (8) we have

$$\mathbf{g}^{*<k>}(v) = \mathbf{g}^{*<k-1>}\left(\frac{\mathbf{R}v - \mathbf{Q}}{d}\right) = \mathbf{g}^{*<k-1>}(\mathbf{H}(v)) = \mathbf{g}^{*<k-1>}(\mathcal{O}(v)). \tag{13}$$

Using the definition $\mathbf{g}^{*<0>}(v) = \mathbf{g}^*(v)$ for $k = 1$ we get

$$\mathbf{g}^{*<1>}(v) = \mathbf{g}^*(\mathcal{O}(v)). \tag{14}$$

Applying (14) recursively in (13) gives the proposition for $k \geq 2$. For $k = 0$ the proposition follows from the definitions $\mathbf{g}^{*<0>}(v) = \mathbf{g}^*(v)$ and $\mathcal{O}^0(v) = v$. $\square$

Proposition 3. *If the diagonal elements of* $\frac{\mathbf{R}}{d}$ *are less than one then* $\lim_{k\to\infty} \mathcal{O}^k(v)$ *exists, finite, independent of v and it is*

$$\mathcal{O}^\infty(v) = \lim_{k\to\infty} \mathcal{O}^k(v) = \left(\frac{\mathbf{R}}{d} - \mathbf{I}\right)^{-1} \frac{\mathbf{Q}}{d}. \tag{15}$$

Proof. Applying $\mathbf{T} = \frac{\mathbf{R}}{d}$ and $\mathbf{Z} = \frac{\mathbf{Q}}{d}$ in (4) gives

$$\mathcal{O}^k(v) = \left(\frac{\mathbf{R}}{d}\right)^k v - \sum_{i=0}^{k-1} \left(\frac{\mathbf{R}}{d}\right)^i \frac{\mathbf{Q}}{d}. \tag{16}$$

When the diagonal elements of $\frac{\mathbf{R}}{d}$ are less than one then $\lim_{k\to\infty} \left(\frac{\mathbf{R}}{d}\right)^k = \mathbf{0}$ and $\lim_{k\to\infty} \sum_{i=0}^{k-1} \left(\frac{\mathbf{R}}{d}\right)^i = (\mathbf{I} - \frac{\mathbf{R}}{d})^{-1}$. Applying them in (16) results in the proposition. $\square$

Proposition 4. *Starting from the initial fluid amount whose vector LT is* $\mathbf{g}^*(v)$ *the limiting vector LT of the k-th descendant fluid as k tends to infinity can be expressed as*

$$\lim_{k\to\infty}\mathbf{g}^{*<k>}(v)=\mathbf{g}^*(\mathcal{O}^\infty(v))=\mathbf{g}^*\left(\left(\frac{\mathbf{R}}{d}-\mathbf{I}\right)^{-1}\frac{\mathbf{Q}}{d}\right). \tag{17}$$

Proof. Applying (10) and the operator limit (15) we have

$$\lim_{k\to\infty}\mathbf{g}^{*<k>}(v)=\mathbf{g}^*(\lim_{k\to\infty}\mathcal{O}^k(v))=\mathbf{g}^*(\mathcal{O}^\infty(v))=\mathbf{g}^*\left(\left(\frac{\mathbf{R}}{d}-\mathbf{I}\right)^{-1}\frac{\mathbf{Q}}{d}\right). \qquad \Box$$

3.3   The Governing Equations of the System at Vacation Start and End Epochs

Let $X(t)\in\mathbb{R}^+$ denote the fluid level in the buffer at time t and $t^f(\ell)$ for $\ell\geq 1$ be the time at the end of the vacation in the ℓ-th cycle. We define the $1\times L$ row vector $\mathbf{f}(\ell,x)$ by its elements as

$$[\mathbf{f}(\ell,x)]_j=\frac{d}{dx}Pr(\Omega(t^f(\ell))=j,X(t^f(\ell))<x),\ j\in\Omega,$$

and its LT as $\mathbf{f}^*(\ell,v)=\int_{x=0}^\infty\mathbf{f}(\ell,x)e^{-vx}dx$. We also define the steady-state vector LT of the fluid level at end of vacation, the $1\times L$ row vector $\mathbf{f}^*(v)$ as $\mathbf{f}^*(v)=\lim_{\ell\to\infty}\mathbf{f}^*(\ell,v)$. Analogously let $t^m(\ell)$ be the time at the start of vacation in the ℓ-th cycle. The $1\times L$ row vector $\mathbf{m}(\ell,x)$ is defined by its elements as

$$[\mathbf{m}(\ell,x)]_j=\frac{d}{dx}Pr(\Omega(t^m(\ell))=j,X(t^m(\ell))<x),\ j\in\Omega,$$

and its LT and embedded steady-state vector are $\mathbf{m}^*(\ell,v)=\int_{x=0}^\infty\mathbf{m}(\ell,x)e^{-vx}dx$ and $\mathbf{m}^*(v)=\lim_{\ell\to\infty}\mathbf{m}^*(\ell,v)$.

Due to the exhaustive discipline the buffer is idle at the start of the vacation period. This implies that $\mathbf{m}^*(\ell,v)$ is independent of v, which is the phase distribution of the background Markov chain at the beginning of the ℓth vacation. Thus we introduce $\mathbf{m}(\ell)=\mathbf{m}^*(\ell,v)$.

Theorem 1. *In the fluid vacation model with exhaustive discipline the vector LTs of the fluid level at the end of the ℓth vacation, $\mathbf{f}^*(\ell,v)$, $\ell\geq 0$, and at the start of the ℓth vacation, $\mathbf{m}(\ell)$, $\ell\geq 1$, satisfy*

$$\mathbf{m}(\ell)=\mathbf{f}^*(\ell-1,\mathcal{O}^\infty(v))=\mathbf{f}^*\left(\ell-1,\left(\frac{\mathbf{R}}{d}-\mathbf{I}\right)^{-1}\frac{\mathbf{Q}}{d}\right), \tag{18}$$

$$\mathbf{f}^*(\ell,v)=\mathbf{m}(\ell)\sigma^*(\mathbf{R}v-\mathbf{Q}). \tag{19}$$

Proof. The k-th descendant fluid as $\lim_{k\to\infty}$ gives the fluid at the end of the exhaustive service period. Hence starting a service period with initial joint fluid level and phase distribution $\mathbf{g}^*(v)$ at the end of the service period the joint fluid level and phase distribution is $\lim_{k\to\infty}\mathbf{g}^{*<k>}(v)$. Proposition 4 states that $\mathbf{g}^*(\mathcal{O}^\infty(v)) = \mathbf{g}^*\left(\left(\frac{\mathbf{R}}{d}-\mathbf{I}\right)^{-1}\frac{\mathbf{Q}}{d}\right)$. (18) comes from the fact that the joint fluid level and phase distribution at the end of the $\ell-1$th vacation period is $\mathbf{f}^*(\ell-1, v)$.

In the exhaustive system the fluid level at the end of the ℓth vacation equals the fluid flowed into the buffer during the ℓth vacation, since the buffer is idle at the start of the ℓth vacation. Taking into account also the state of the modulating Markov chain we have

$$[\mathbf{f}(\ell, x)]_k = \sum_{j=1}^{L} \int_{t=0}^{\infty} [\mathbf{m}(\ell)]_j \mathbf{A}_{jk}(t, x)\sigma(t)dt. \tag{20}$$

Rearranging (20) to matrix form we get

$$\mathbf{f}^*(\ell, v) = \int_{t=0}^{\infty} \mathbf{m}(\ell)\mathbf{A}^*(t, v)\sigma(t)dt = \mathbf{m}(\ell) \int_{t=0}^{\infty} e^{-t(\mathbf{R}v-\mathbf{Q})}\sigma(t)dt, \tag{21}$$

where the explicit form of $\mathbf{A}^*(t, v)$ is taken from (7). Rearranging (21) results in (19).   □

In the rest of the paper we avoid the scalar versions of the equations like (11) and (20) and directly write their matrix versions.

4   The Steady-State Behavior of the Fluid Vacation Model

The main goal of this section is to compute the time stationary distribution of the fluid vector in transform domain. To this end we first provide the stationary distribution in service start and end epochs in Sec. 4.1, collect some subsequently used general properties of fluid vacation models in Sec. 4.2, compute the service time distribution in Sec. 4.3 and finally evaluate the time stationary behavior in Sec. 4.4.

4.1   Steady-State Behavior at Start and End of Vacation

We define $\mathbf{m}$ as $\mathbf{m} = \lim_{\ell\to\infty}\mathbf{m}(\ell)$ and $\mathbf{e}$ as the column vector of ones.

Theorem 2. *The stationary behavior of the stable fluid vacation model with exhaustive discipline is characterized by $\mathbf{m}$ and $\mathbf{f}^*(v)$ where $\mathbf{m}$ is the solution of the linear system*

$$\mathbf{m} = \mathbf{m}\sigma^*\left(\left(\frac{\mathbf{R}}{d}-\mathbf{I}\right)^{-1}\mathbf{Q}\right), \tag{22}$$

$$\mathbf{m}\mathbf{e} = 1,$$

and

$$\mathbf{f}^*(v) = \mathbf{m}\sigma^*(\mathbf{R}v - \mathbf{Q}). \tag{23}$$

Proof. Applying (19) in (18) gives

$$\mathbf{m}(\ell) = \mathbf{m}(\ell - 1)\sigma^*(\mathbf{R}\mathcal{O}^\infty(v) - \mathbf{Q}). \tag{24}$$

Taking the limit $\ell \to \infty$ in (24) and rearranging it leads to

$$\mathbf{m} = \mathbf{m}\sigma^*(d\frac{\mathbf{R}\mathcal{O}^\infty(v) - \mathbf{Q}}{d}) = \mathbf{m}\sigma^*(d\mathcal{O}(\mathcal{O}^\infty(v))) = \mathbf{m}\sigma^*(d\mathcal{O}^\infty(v)). \tag{25}$$

Applying (15) in (25) results in the first equation of (22). The normalizing condition of the system of linear equations comes from the fact that $\mathbf{m}$ is the phase distribution of the background Markov chain of the fluid source at service completion. Finally (23) comes by taking the limit $\ell \to \infty$ in (19). $\square$

4.2   Equilibrium Relationships

Let $S(\ell)$ be the service time in the ℓ-th cycle, $C(\ell)$ be the cycle time between two consecutive service starts in the ℓ-th cycle, $Z(\ell)$ be the amount of fluid served in the ℓ-th cycle, and $Z_i(\ell)$ be the amount of fluid served in the i-th descendant period of the ℓ-th service period. The related $lim_{\ell\to\infty}$ stationary quantities are S, C, Z, and Z_i, their LTs are $s^*(v)$, $c^*(v)$, $z^*(v)$, $z_i^*(v)$, and their means are s, c, z, z_i, respectively. The definitions imply that $Sd = Z$, $C = S + \tilde{\sigma}$ and $Z = \sum_{i=0}^{\infty} Z_i$ hold as well as these relations hold for their respective means.

Let $Y(t)$ be the accumulated fluid flowed into the buffer in interval $(0, t]$ and a be the mean amount of fluid, which flows into the buffer during a cycle in steady state. That is

$$a = \lim_{k\to\infty} \frac{\sum_{\ell=1}^{k} E[Y(t^f(\ell + 1)) - Y(t^f(\ell))]}{k},$$

whose right hand side can be rearranged as

$$a = \lim_{k\to\infty} \frac{E[\sum_{\ell=1}^{k} Y(t^f(\ell + 1)) - Y(t^f(\ell))]}{E[\sum_{\ell=1}^{k} C(\ell)]} \lim_{k\to\infty} \frac{E[\sum_{\ell=1}^{k} C(\ell)]}{k} = \lambda c. \tag{26}$$

Corollary 1. *In the stable fluid vacation model the steady-state mean cycle time can be expressed as*

$$c = \frac{\sigma}{1 - \rho}. \tag{27}$$

Proof. In the stable fluid vacation model the amount of fluid flowing into the buffer during a cycle equals the amount of fluid removed during the service period, that is $a = sd$. From this and $a = \lambda c$ we get $s = \frac{\lambda}{d}c = \rho c$ and $c = \sigma + s = \sigma + \rho c$, which gives the statement. $\square$

4.3   The Steady-State Distribution of the Service Time

Theorem 3. *The amount of fluid served in a service period and the length of the service period satisfy*

$$z^*(v) = \mathbf{f}^* \left(\left(\mathbf{I} - \frac{\mathbf{R}}{d} \right)^{-1} \left(v\mathbf{I} - \frac{\mathbf{Q}}{d} \right) \right) \mathbf{e}, \tag{28}$$

$$s^*(v) = z^* \left(v/d \right). \tag{29}$$

Proof. Let $f_{Z_0,Z_1,\ldots,Z_k}(x_0, x_1, \ldots, x_k)$ denote the joint density of $Z_0, Z_1, \ldots, Z_k$. Furthermore let the matrix $\mathbf{A}(\frac{x_i}{d}, x_{i+1})$ stand for the state dependent density of the fluid arrived to the buffer during the $\frac{x_i}{d}$ long i-th descendant period for $i = 0, \ldots, k-1$.

The fluid served in the i-th descendant period is the fluid flowed into the buffer during the $i-1$-th descendant period for $i = 1, \ldots, k$. Using it and taking into account also the evolution of the modulating Markov state the joint density $f_{Z_0,Z_1,\ldots,Z_k}(x_0, x_1, \ldots, x_k)$ can be given as

$$f_{Z_0,Z_1,\ldots,Z_k}(x_0, x_1, \ldots, x_k) = \mathbf{f}(x_0)\mathbf{A}(\frac{x_0}{d}, x_1) \ldots \mathbf{A}(\frac{x_{k-1}}{d}, x_k)\mathbf{e}. \tag{30}$$

By the help of (30) the mean $E(e^{-v\sum_{i=0}^{k} Z_i})$ can be expressed as

$$E(e^{-v\sum_{i=0}^{k} Z_i}) =$$

$$\int_{x_0} \int_{x_1} \ldots \int_{x_k} E(e^{-v\sum_{i=0}^{k} Z_i} | Z_0 = x_0, Z_1 = x_1, \ldots, Z_k = x_k)$$

$$\times f_{Z_0,Z_1,\ldots,Z_k}(x_0, x_1, \ldots, x_k)dx_k \ldots dx_1 dx_0 =$$

$$\int_{x_0} \mathbf{f}(x_0) \int_{x_1} \mathbf{A}(\frac{x_0}{d}, x_1) \ldots \underbrace{\int_{x_k} \mathbf{A}(\frac{x_{k-1}}{d}, x_k)e^{-vx_k}dx_k}_{\mathbf{A}^*(\frac{x_{k-1}}{d}, v)} \ldots e^{-vx_1}dx_1 e^{-vx_0}dx_0\mathbf{e} =$$

$$\int_{x_0} \mathbf{f}(x_0) \int_{x_1} \mathbf{A}(\frac{x_0}{d}, x_1) \ldots \underbrace{\int_{x_{k-1}} \mathbf{A}(\frac{x_{k-2}}{d}, x_{k-1}) \underbrace{e^{-x_{k-1}\frac{\mathbf{R}v-\mathbf{Q}}{d}} e^{-vx_{k-1}}}_{e^{-x_{k-1}(\mathcal{O}(v)+\mathbf{I}v)}} dx_{k-1}}_{\mathbf{A}^*(\frac{x_{k-2}}{d}, \mathcal{O}(v)+\mathbf{I}v)}$$

$$\ldots e^{-vx_1}dx_1 e^{-vx_0}dx_0\mathbf{e} =$$

$$\int_{x_0} \mathbf{f}(x_0) \int_{x_1} \mathbf{A}(\frac{x_0}{d}, x_1) \ldots \underbrace{\int_{x_{k-2}} \mathbf{A}(\frac{x_{k-3}}{d}, x_{k-2})e^{-x_{k-2}\frac{\mathbf{R}(\mathcal{O}(v)+\mathbf{I}v)-\mathbf{Q}}{d}} e^{-vx_{k-2}}dx_{k-2}}_{\mathbf{A}^*(\frac{x_{k-3}}{d}, \mathcal{O}^2(v)+\frac{\mathbf{R}}{d}v+\mathbf{I}v)}$$

$$\ldots e^{-vx_1}dx_1 e^{-vx_0}dx_0\mathbf{e} =$$

$$\ldots = \mathbf{f}^* \left(\mathcal{O}^k(v) + \sum_{i=0}^{k-1} \left(\frac{\mathbf{R}}{d} \right)^i v \right) \mathbf{e}.$$

This together with (15) yields

$$z^*(v) = E(e^{-v\sum_{i=0}^{\infty} Z_i}) = \mathbf{f}^* \left(\mathcal{O}^{\infty}(v) + \sum_{i=0}^{\infty} \left(\frac{\mathbf{R}}{d} \right)^i v \right) \mathbf{e} =$$

$$= \mathbf{f}^* \left(\left(\frac{\mathbf{R}}{d} - \mathbf{I} \right)^{-1} \frac{\mathbf{Q}}{d} + \left(\mathbf{I} - \frac{\mathbf{R}}{d} \right)^{-1} v \right) \mathbf{e},$$

from which rearrangement results in the first statement of the theorem. For the service time distribution we have

$$s^*(v) = E(e^{-vS}) = E(e^{-vZ/d}) = E(e^{-(v/d)Z}) = z^*(v/d). \qquad \square$$

4.4   The Steady-State Vector LT of the Fluid Level

The steady-state joint distribution of the fluid level and the state of the modulating Markov chain at an arbitrary epoch is defined by the $1 \times L$ row vector $\mathbf{q}(x)$ whose j-th element is

$$[\mathbf{q}(x)]_j = \lim_{t \to \infty} \frac{d}{dx} Pr(\Omega(t) = j, X(t) < x), \ j \in \Omega.$$

The LT of $\mathbf{q}(x)$ is $\mathbf{q}^*(v) = \int_{x=0}^{\infty} \mathbf{q}(x) e^{-vx} dx$.

Let d_k be the start time of the kth descendant service period for $k \geq 0$, where $d_0 = 0$. The steady-state joint density of the fluid level and the state of the modulating Markov chain at an arbitrary epoch in the kth ($k > 0$) descendant service period, the $1 \times L$ row vector $\mathbf{q}_k(x)$ is defined by its j-th element as

$$[\mathbf{q}_k(x)]_j = \lim_{t \to \infty} \frac{d}{dx} Pr(\Omega(t) = j, X(t) < x \mid t \in (d_k, d_{k+1})), \ j \in \Omega.$$

and corresponding LT is $\mathbf{q}_k^*(v)$.

Let $1_{(con)}$ denote the indicator of condition "con". Furthermore let $\mathbf{e}_j = (0, \ldots, 0, 1, 0, \ldots, 0)$ be the $1 \times L$ vector with 1 at the j-th position. We define the $1 \times L$ indicator vector $\mathbf{1}_{(\Omega(t))}$ as $\mathbf{1}_{(\Omega(t))} = \sum_{j=1}^{L} 1_{(\Omega(t)=j)} \mathbf{e}_j$.

Proposition 5. *For $k \geq 0$ the LT of the mean fluid level for the kth descendant interval, $E[\int_{t=d_k}^{d_{k+1}} e^{-X(t)v} \mathbf{1}_{(\Omega(t))} dt]$, satisfies*

$$E[\int_{t=d_k}^{d_{k+1}} e^{-X(t)v} \mathbf{1}_{(\Omega(t))} dt] ((\mathbf{R} - d\mathbf{I})v - \mathbf{Q}) = \mathbf{f}^*(\mathcal{O}^k(v)) - \mathbf{f}^*(\mathcal{O}^{k+1}(v)). \quad (31)$$

Proof. If the fluid level at the beginning of the k-th descendant period is x_k then the fluid level after time t in the k-th descendant period is $x_k - td + A(t)$ where $A(t)$ denotes the amount of fluid arrived in $(0, t)$ in the k-th descendant period. The LT of this quantity is $E(e^{-v(x_k - td + A(t))}) = E(e^{-vA(t)}) e^{-v(x_k - td)}$, where the first term is the LT of $A(t)$. Considering the state dependency of fluid level at the beginning of the k-th descendant period and the fluid arrival process for $k \geq 0$ we have

$$E[\int_{t=d_k}^{d_{k+1}} e^{-X(t)v} \mathbf{1}_{(\Omega(t))} dt] =$$

$$\int_{x_0} \mathbf{f}(x_0) \int_{x_1} \mathbf{A}(\frac{x_0}{d}, x_1) \ldots \int_{x_k} \mathbf{A}(\frac{x_{k-1}}{d}, x_k) \int_{t=0}^{x_k/d} \mathbf{A}^*(t, v) e^{-(x_k - td)v} \, dt \, dx_k \ldots dx_0 =$$

$$\int_{x_0} \mathbf{f}(x_0) \int_{x_1} \mathbf{A}(\frac{x_0}{d}, x_1) \ldots \int_{x_k} \mathbf{A}(\frac{x_{k-1}}{d}, x_k) \underbrace{\int_{t=0}^{x_k/d} e^{-t((\mathbf{R}-d\mathbf{I})v-\mathbf{Q})} \, dt}_{} e^{-x_k v} \, dx_k \ldots dx_0.$$

The underbraced integral can be evaluated by means of the following relation

$$\int_{t=0}^{x} e^{-t\mathbf{Z}} dt \mathbf{Z} = \mathbf{I} - e^{-x\mathbf{Z}}, \tag{32}$$

which leads to

$$E[\int_{t=d_k}^{d_{k+1}} e^{-X(t)v} \mathbf{1}_{(\Omega(t))} dt] ((\mathbf{R} - d\mathbf{I})v - \mathbf{Q}) =$$

$$\int_{x_0} \mathbf{f}(x_0) \int_{x_1} \mathbf{A}(\frac{x_0}{d}, x_1) \ldots \int_{x_k} \mathbf{A}(\frac{x_{k-1}}{d}, x_k) \left(\mathbf{I} - e^{-\frac{x_k}{d}(\mathbf{R}-d\mathbf{I})v - \mathbf{Q}} \right) e^{-x_k v} \, dx_k \ldots dx_0 =$$

$$\int_{x_0} \mathbf{f}(x_0) \int_{x_1} \mathbf{A}(\frac{x_0}{d}, x_1) \ldots \int_{x_{k-1}} \mathbf{A}(\frac{x_{k-2}}{d}, x_{k-1})$$

$$\left(\int_{x_k} \mathbf{A}(\frac{x_{k-1}}{d}, x_k) e^{-x_k v} dx_k - \int_{x_k} \mathbf{A}(\frac{x_{k-1}}{d}, x_k) e^{-\frac{x_k}{d}(\mathbf{R}v - \mathbf{Q})} dx_k \right) dx_{k-1} \ldots dx_0 =$$

$$\int_{x_0} \mathbf{f}(x_0) \int_{x_1} \mathbf{A}(\frac{x_0}{d}, x_1) \ldots \int_{x_{k-1}} \mathbf{A}(\frac{x_{k-2}}{d}, x_{k-1})$$

$$\left(\mathbf{A}^*(\frac{x_{k-1}}{d}, v) - \mathbf{A}^*(\frac{x_{k-1}}{d}, \mathcal{O}(v)) \right) dx_{k-1} \ldots dx_0 =$$

$$\int_{x_0} \mathbf{f}(x_0) \int_{x_1} \mathbf{A}(\frac{x_0}{d}, x_1) \ldots \int_{x_{k-1}} \mathbf{A}(\frac{x_{k-2}}{d}, x_{k-1})$$

$$\left(e^{-x_{k-1}\mathcal{O}(v)} - e^{-x_{k-1}\mathcal{O}^2(v)} \right) dx_{k-1} \ldots dx_0 =$$

$$\int_{x_0} \mathbf{f}(x_0) \int_{x_1} \mathbf{A}(\frac{x_0}{d}, x_1) \ldots \int_{x_{k-2}} \mathbf{A}(\frac{x_{k-3}}{d}, x_{k-2})$$

$$\left(\mathbf{A}^*(\frac{x_{k-2}}{d}, \mathcal{O}(v)) - \mathbf{A}^*(\frac{x_{k-2}}{d}, \mathcal{O}^2(v)) \right) dx_{k-2} \ldots dx_0 =$$

$$\ldots = \int_{x_0} \mathbf{f}(x_0) \left(\mathbf{A}^*(\frac{x_0}{d}, \mathcal{O}^{k-1}(v)) - \mathbf{A}^*(\frac{x_0}{d}, \mathcal{O}^k(v)) \right) dx_0 =$$

$$\mathbf{f}^*(\mathcal{O}^k(v)) - \mathbf{f}^*(\mathcal{O}^{k+1}(v)),$$

which completes the proof of the proposition.   □

Theorem 4. *In the stable fluid vacation model with exhaustive discipline the following relation holds for the steady-state vector LT of the fluid level at arbitrary epoch*

$$\mathbf{q}^*(v)(\mathbf{R}v - \mathbf{Q})\left((\mathbf{R} - d\mathbf{I})v - \mathbf{Q}\right) = \tfrac{vd}{c}\left(\mathbf{f}^*(v) - \mathbf{m}\right). \tag{33}$$

Proof. The fluid level at arbitrary epoch can be expressed by the help of the fluid level at the last service start on LT level by utilizing the transient behavior of the arrived fluid (relation (7)) and taking into account that it can fall either in service or vacation period as well as its position in the actual period. Thus it is enough to average over a cycle for determining the behavior at arbitrary epoch.

$$\mathbf{q}^*(v) = \frac{1}{c} E\left[\int_{t=0}^{C} e^{-X(t)v}\mathbf{1}_{(\Omega(t))}dt\right] \tag{34}$$

$$= \frac{1}{c}\left(\sum_{k=0}^{\infty} E\left[\int_{t=d_k}^{d_{k+1}} e^{-X(t)v}\mathbf{1}_{(\Omega(t))}dt\right] + E\left[\int_{t=S}^{C} e^{-X(t)v}\mathbf{1}_{(\Omega(t))}dt\right]\right).$$

From Proposition 5 we have

$$E[\int_{t=0}^{S} e^{-X(t)v}\mathbf{1}_{(\Omega(t))}dt]\left((\mathbf{R} - d\mathbf{I})v - \mathbf{Q}\right) =$$

$$\sum_{k=0}^{\infty} E[\int_{t=d_k}^{d_{k+1}} e^{-X(t)v}\mathbf{1}_{(\Omega(t))}dt]\left((\mathbf{R} - d\mathbf{I})v - \mathbf{Q}\right) =$$

$$\sum_{k=0}^{\infty} \mathbf{f}^*(\mathcal{O}^k(v)) - \mathbf{f}^*(\mathcal{O}^{k+1}(v)) = \mathbf{f}^*(v) - \mathbf{f}^*(\mathcal{O}^{\infty}(v)) = \mathbf{f}^*(v) - \mathbf{m}, \tag{35}$$

where we used the stationary version of (18) in the last step. For the evaluation of $E\left[\int_{t=S}^{C} e^{-X(t)v}\mathbf{1}_{(\Omega(t))}dt\right]$ it is enough to originate the fluid level from the start of the vacation instead of the last service start. This is because (34) expresses $\mathbf{q}^*(v)$ as sum of expected values. Relying on this and using again (32) we get for the vacation period

$$E\left[\int_{t=S}^{C} e^{-X(t)v}\mathbf{1}_{(\Omega(t))}dt\right](\mathbf{R}v - \mathbf{Q}) = \int_{t=0}^{\infty}\int_{x=0}^{t} \mathbf{m}\mathbf{A}^*(x,v)dx\sigma(t)dt\,(\mathbf{R}v - \mathbf{Q}) =$$

$$\mathbf{m}\int_{t=0}^{\infty}\int_{x=0}^{t} e^{-x(\mathbf{R}v - \mathbf{Q})}dx\sigma(t)dt\,(\mathbf{R}v - \mathbf{Q}) = \mathbf{m}\int_{t=0}^{\infty}\left(\mathbf{I} - e^{-t(\mathbf{R}v - \mathbf{Q})}\right)\sigma(t)dt =$$

$$\mathbf{m} - \mathbf{m}\sigma^*(\mathbf{R}v - \mathbf{Q}) = \mathbf{m} - \mathbf{f}^*(v), \tag{36}$$

where we used (23) in the last step.

Multiplying both sides of (34) by $(\mathbf{R}v - \mathbf{Q})\left((\mathbf{R} - d\mathbf{I})v - \mathbf{Q}\right)$ and substituting (35) and (36) we get statement of the theorem. $\qquad\square$

5   Moments of the Stationary Performance Measures

The goal of this section is to obtain computable moments expressions based on the transform domain expressions of the previous section.

Lemma 1. *In the stable fluid vacation model with exhaustive discipline the steady-state vector mean of the fluid level at arbitrary epoch is*

$$\mathbf{q}^{(1)} = \mathbf{q}^{(1)}\mathbf{e}\boldsymbol{\pi} + \left(\boldsymbol{\pi}\mathbf{R}\mathbf{Q} + \tfrac{d}{c}(\mathbf{f}^{(0)} - \mathbf{m})\right)\left(\mathbf{Q}^2 + \mathbf{e}\boldsymbol{\pi}\right)^{-1} , \tag{37}$$

where

$$\mathbf{q}^{(1)}\mathbf{e} = \tfrac{1}{\lambda}\left(\tfrac{1}{2}\mathbf{t}^{(2)}\mathbf{e} - \left(\boldsymbol{\pi}\mathbf{R} - \mathbf{t}^{(1)}\right)(\mathbf{Q} + \mathbf{e}\boldsymbol{\pi})^{-1}\mathbf{R}\mathbf{e}\right), \tag{38}$$

$$\mathbf{t}^{(1)}\mathbf{e} = \tfrac{1}{\lambda-d}\left(\tfrac{1}{2}\mathbf{r}^{(2)}\mathbf{e} + \mathbf{r}^{(1)}(\mathbf{Q}+\mathbf{e}\boldsymbol{\pi})^{-1}(\mathbf{R}-d\mathbf{I})\mathbf{e}\right),$$

$$\mathbf{t}^{(1)} = \mathbf{t}^{(1)}\mathbf{e}\boldsymbol{\pi} - \mathbf{r}^{(1)}(\mathbf{Q} + \mathbf{e}\boldsymbol{\pi})^{-1},$$

$$\mathbf{t}^{(2)}\mathbf{e} = \tfrac{1}{\lambda-d}\left(\tfrac{1}{3}\mathbf{r}^{(3)}\mathbf{e} - \left(2\mathbf{t}^{(1)} - \mathbf{r}^{(2)}\right)(\mathbf{Q}+\mathbf{e}\boldsymbol{\pi})^{-1}(\mathbf{R}-d\mathbf{I})\mathbf{e}\right),$$

and

$$\mathbf{r}^{(1)} = \tfrac{d}{c}\mathbf{m}(\sigma^*(-\mathbf{Q}) - \mathbf{I}), \quad \mathbf{r}^{(n)} = \tfrac{nd}{c}\mathbf{m}\left.\frac{d^{n-1}\sigma^*(\mathbf{R}v - \mathbf{Q})}{dv^{n-1}}\right|_{v=0} \quad \forall n > 1. \tag{39}$$

Proof. The derivative of (33) at $v = 0$ gives

$$\mathbf{q}^{(1)}\mathbf{Q}^2 - \mathbf{q}^{(0)}\mathbf{R}\mathbf{Q} - \mathbf{q}^{(0)}\mathbf{Q}(\mathbf{R} - d\mathbf{I}) = \tfrac{d}{c}(\mathbf{f}^{(0)} - \mathbf{m}).$$

Since $(\mathbf{Q} + \mathbf{e}\boldsymbol{\pi})$ as well as $(\mathbf{Q}^2 + \mathbf{e}\boldsymbol{\pi})$ are nonsingular [4,5], using $\mathbf{q}^{(0)} = \boldsymbol{\pi}$, $\boldsymbol{\pi}\mathbf{Q} = \mathbf{0}$, $\boldsymbol{\pi}(\mathbf{Q}^2 + \mathbf{e}\boldsymbol{\pi})^{-1} = \boldsymbol{\pi}$ and adding and subtracting $\mathbf{q}^{(1)}\mathbf{e}\boldsymbol{\pi}$ we obtain (37), where the only remaining unknown is the scalar $\mathbf{q}^{(1)}\mathbf{e}$, because according to (23) $\mathbf{f}^{(0)} = \mathbf{m}\sigma^*(-\mathbf{Q})$.

Unfortunately the computation of $\mathbf{q}^{(1)}\mathbf{e}$ is not that straight forward. To derive it we adopt a two step method. According to the structure of (33) we introduce $\mathbf{t}^*(v) = \mathbf{q}^*(v)(\mathbf{R}v - \mathbf{Q})$ and $\mathbf{r}^*(v) = \mathbf{t}^*(v)((\mathbf{R} - d\mathbf{I})v - \mathbf{Q})$ and express the moments of $\mathbf{q}^*(v)$ by the moment of $\mathbf{t}^*(v)$ in the first step and express the moments of $\mathbf{t}^*(v)$ by the moment of $\mathbf{r}^*(v)$ in the second step. Considering $\mathbf{q}^{(0)} = \boldsymbol{\pi}$, the first two derivatives of $\mathbf{t}^*(v) = \mathbf{q}^*(v)(\mathbf{R}v - \mathbf{Q})$ at $v = 0$ are

$$\mathbf{t}^{(1)} = -\mathbf{q}^{(1)}\mathbf{Q} + \boldsymbol{\pi}\mathbf{R}, \tag{40}$$

$$\mathbf{t}^{(2)} = -\mathbf{q}^{(2)}\mathbf{Q} + 2\mathbf{q}^{(1)}\mathbf{R}. \tag{41}$$

Adding and subtracting $\mathbf{q}^{(1)}\mathbf{e}\boldsymbol{\pi}$ to (40) and using $\boldsymbol{\pi}(\mathbf{Q} + \mathbf{e}\boldsymbol{\pi})^{-1} = \boldsymbol{\pi}$ leads to

$$\mathbf{q}^{(1)} = \left(\mathbf{q}^{(1)}\mathbf{e}\right)\boldsymbol{\pi} + \left(\boldsymbol{\pi}\mathbf{R} - \mathbf{t}^{(1)}\right)(\mathbf{Q} + \mathbf{e}\boldsymbol{\pi})^{-1}. \tag{42}$$

Post-multiplying (41) by $\mathbf{e}$, post-multiplying (42) by $\mathbf{Re}$ using $\pi\mathbf{Re} = \lambda$ and substituting the $\mathbf{q}^{(1)}\mathbf{Re}$ term from one equation to the other we obtain (38), where the unknowns are $\mathbf{t}^{(2)}\mathbf{e}$ and $\mathbf{t}^{(1)}$. Similarly, the nth derivative of $\mathbf{r}^*(v) = \mathbf{t}^*(v)((\mathbf{R} - d\mathbf{I})v - \mathbf{Q})$ at $v = 0$ is

$$\mathbf{r}^{(n)} = -\mathbf{t}^{(n)}\mathbf{Q} + n\mathbf{t}^{(n-1)}(\mathbf{R} - d\mathbf{I}). \qquad (43)$$

Applying the same steps as in the transformation of (40) to (42) we obtain

$$\mathbf{t}^{(n)} = \mathbf{t}^{(n)}\mathbf{e}\pi + \left(n\mathbf{t}^{(n-1)}(\mathbf{R} - d\mathbf{I}) - \mathbf{r}^{(n)}\right)(\mathbf{Q} + \mathbf{e}\pi)^{-1}, \qquad (44)$$

and applying the same steps as in the derivation of (38) based on (41) and (42) we obtain

$$\mathbf{t}^{(n)}\mathbf{e} = \frac{1}{\lambda - d}\left(\frac{1}{n+1}\mathbf{r}^{(n+1)}\mathbf{e} - \left(n\mathbf{t}^{(n-1)} - \mathbf{r}^{(n)}\right)(\mathbf{Q} + \mathbf{e}\pi)^{-1}(\mathbf{R} - d\mathbf{I})\mathbf{e}\right). \qquad (45)$$

Considering that $\mathbf{t}^{(0)} = -\mathbf{q}^{(0)}\mathbf{Q} = \mathbf{0}$, (44) and (45) allows the consecutive computation of $\mathbf{t}^{(n)}\mathbf{e}$ and $\mathbf{t}^{(n)}$. Finally, using (33) and (23) the derivatives of $\mathbf{r}^*(v) = \frac{vd}{c}(\mathbf{f}^*(v) - \mathbf{m})$ at $v = 0$ gives (39). $\square$

6   Conclusion

Finally, the steady-state mean vector $\mathbf{q}^{(1)}$ can be computed by the following steps.

1. Calculation of the steady-state phase distribution of the background Markov chain at start of vacation, $\mathbf{m}$, form the system of linear equations (22).
2. Computation of π, λ, ρ and c by applying (1), (2) and (27), respectively.
3. Computation of the steady-state mean $\mathbf{q}^{(1)}$ by applying Lemma 1.

Acknowledgment. This work is supported by the OTKA K101150 project.

References

1. Borst, S.C., Boxma, O.J.: Polling models with and without switchover times. Operations Research 45, 536–543 (1997)
2. Doshi, B.T.: Queueing systems with vacations - a survey. Queueing Systems 1, 29–66 (1986)
3. Gerschgorin, S.: Über die Abgrenzung der Eigenwerte einer Matrix. Izv. Akad. Nauk. USSR Otd. Fiz.-Mat. Nauk 7, 749–754 (1931)
4. Lucantoni, D.L.: New results on the single server queue with a batch markovian arrival process. Stochastic Models 7, 1–46 (1991)
5. Neuts, M.F.: Structured stochastic matrices of M/G/1 type and their applications. Marcel Dekker, New York (1989)
6. Saffer, Z., Telek, M.: Fluid vacation model with markov modulated load and gated discipline. In: QTNA (August 2014)
7. Takagi, H.: Queueing Analysis - A Foundation of Performance Evaluation, Vacation and Prority Systems, vol. 1. North-Holland, New York (1991)

Use of a Levy Distribution for Modeling Best Case Execution Time Variation

Jonathan C. Beard and Roger D. Chamberlain*

Dept. of Computer Science and Engineering
Washington University in St. Louis, St. Louis, Missouri, USA
{jbeard,roger}@wustl.edu

Abstract. Minor variations in execution time can lead to out-sized effects on the behavior of an application as a whole. There are many sources of such variation within modern multi-core computer systems. For an otherwise deterministic application, we would expect the execution time variation to be non-existent (effectively zero). Unfortunately, this expectation is in error. For instance, variance in the realized execution time tends to increase as the number of processes per compute core increases. Recognizing that characterizing the exact variation or the maximal variation might be a futile task, we take a different approach, focusing instead on the best case variation. We propose a modified (truncated) Levy distribution to characterize this variation. Using empirical sampling we also derive a model to parametrize this distribution that doesn't require expensive distribution fitting, relying only on known parameters of the system. The distributional assumptions and parametrization model are evaluated on multi-core systems with the common Linux completely fair scheduler.

1 Introduction and Background

Understanding the performance of software systems is often accomplished with the help of stochastic queueing models. These models typically require knowledge of the distributions for inputs such as arrival rate and service rate for compute kernels within an application. Directly influencing the aforementioned values is the execution time distribution of each compute kernel. Complete knowledge of the distribution is generally futile for modern systems. Yet understanding it, however incompletely, is critical to selecting proper model formulations. When understanding complex phenomena, it is often the practice to find a useful bound. We contend that the minimal expected execution time variation of a system, or best case execution time variation (BCETV), is such a bound. By forecasting BCETV for a particular software and hardware combination, we hope to improve the a *priori* knowledge of a models' applicability. This paper introduces the use of a modified Levy distribution for characterizing the BCETV of short execution,

---

* This work was supported by Exegy, Inc. Washington Univ. and R. Chamberlain receive income based on the license of technology by the university to Exegy, Inc.

A. Horváth and K. Wolter (Eds.): EPEW 2014, LNCS 8721, pp. 74–88, 2014.
© Springer International Publishing Switzerland 2014

compute bound kernels. A closed form expression for the probability density function as well as it's first and second moments are derived. The distributional assumptions and model are evaluated via empirical evaluation.

Several references simply assume that the distribution of a series of execution times should be Gaussian [7]. Other works (e.g., [10]) have shown some examples of successive execution times that are not Gaussian with any high probability. Other phenomena such as worst case execution time have been modeled with the Gumbel distribution [5]. Empirically measured execution time noise for a minimal workload of "no-op" instructions (the difference between the nominal and measured execution times, plotted in Figure 1) exhibits a heavily skewed distribution. Simply assuming a Gaussian distribution (green line) overestimates the mass of one tail while underestimating the other. A Gumbel distribution (blue line) is arguably even worse. Some might posit that a Gamma distribution is a good fit, however the support exists only for $x \geq 0$ which fits neither reality or our use case as a noise model. A modified Levy distribution (red line, exact modifications to be discussed) is plotted against the same data, visually it is the best fit to the observed data.

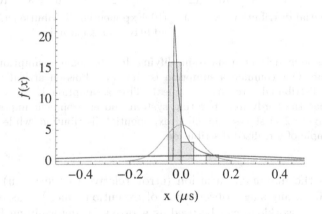

Fig. 1. Histogram of the discrete PDF for a simple "no-op" workload execution time absolute error (light blue bars) in μs plotted against the PDFs of a fitted Gaussian distribution (green line), a Gumbel distribution (blue line) and a modified Levy distribution (red line). Visually it is easy to see that the modified Levy distribution is the best fit for this data set.

Many performance models require details of the inner workings of the target processor [6]. When empirical evaluation is performed, often the results obtained are still uncertain. How well did the empirical evaluation sample the distribution of execution times? Even when detailed knowledge is assumed, or empirical evaluation is performed, there is still uncertainty in the values obtained. Causes of this execution time uncertainty can include cache behavior, interrupts, scheduling uncertainty as well as countless other factors. Distributional uncertainty can lead to poor stochastic model performance. Instead of focusing on the worst or even average case, our approach focuses on the best case and what this bound can do for the model decision making process. As an example, Figure 2(a) shows the distribution that a simple $M/M/k$ queueing model assumes for its inter-arrival distribution whereas Figure 2(b) might be closer to reality given a noisy system. One application of BCETV is to estimate how close a models' input assumptions will line up with reality assuming a best case variance. This could allow quick rejection of models whose assumptions are violated.

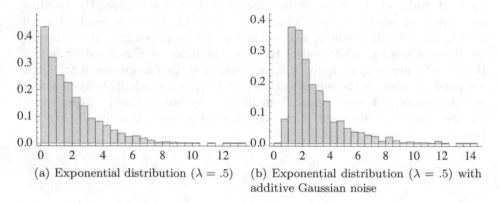

(a) Exponential distribution ($\lambda = .5$)   (b) Exponential distribution ($\lambda = .5$) with additive Gaussian noise

Fig. 2. Stochastic models often make simplifying distributional assumptions about the modeled system. One common assumption is that of a Poisson arrival process (i.e., exponentially distributed inter-arrival times). This assumption is often violated by the "noise" that the hardware, operating system and environment impose upon the application. Figure 2(a) shows a nominal exponential distribution, while Figure 2(b) shows an example of a realized distribution.

BCETV is the minimum variation (error relative to the mean) which can be expected from any single observation of execution time. We assert that the minimal "no-op" workload can be used as a proxy for determining BCETV for short execution, compute bound kernels. In principle, these workloads should be quite deterministic in execution time, but clearly are not. We will show that the distribution of BCETV experienced by these workloads represents a reasonable lower bound "noise" model for nominal execution time. Utilizing empirical data, the modified Levy distribution is revised in terms of system parameters (i.e., processes per core, nominal execution time). Evidence is provided that the modified Levy distribution is a good match for BCETV, especially as the number of processes per core grows.

2   Methodology

The motivation to use a Levy distribution to model the best case execution time variation (BCETV) came from empirical observation. Ultimately we must justify that decision by comparing model predictions to experimental observations.

To that end we start by describing the process through which these data are collected. This is followed by the description of the modified Levy distribution that we propose to use, and how to parametrize it.

2.1 Synthetic Workload

Our focus is the uncertainty in execution time of a running process due to factors other than the process itself. As such, we use an intentionally simple nominal workload so that the observed variation is due not to the application itself, but to other system related factors (e.g., operating system, hardware, etc.). Our nominal workload is the execution of a fixed number of null operations or "no-op" instructions. Aside from no instructions at all, we assume that a null operation is the least taxing instruction. It follows from this logic that a series of null operations should present the most consistent execution time out of any real executable instruction sequence.

One aspect under study is how changing the nominal workload time changes the observed variation in actual execution times. In order to produce a workload of "no-op" instructions that is calibrated to a specific nominal execution time we use sequences of instructions of various lengths which are timed and then used as input for regression to produce an equation for the number of instructions to use for each nominal execution time. Calibration timing is performed while the timed process is assigned to a single core and executing with no other processes.

In theory any duration of workload could be created using this method, however in practice the file sizes become prohibitively large proportional with the frequency of the processor and the desired running time (e.g., platorm A from Table 2 requires approximately 10 million "no-op" instructions for each second of execution time). Other approaches that reduce the file size could be used such as looping over a calibrated number of "no-op" instructions, however we've chosen to use the simpler aforementioned approach because it reduces the possibility of variation due to other factors, such as branching. Our method also assumes that cache pre-fetching will eliminate virtually all instruction cache misses which should then have no appreciable effect on the actual run time. One concern with huge numbers of instructions is that translation lookaside buffer (TLB) misses might increase the observed variation. With TLB misses we would expect an increase in the overall observed variation with longer duration executions with a random pattern (dependent on other processes operating on the same core, TLB algorithm, etc.). As we will show below, this is not the case; more variation is observed for short execution times.

2.2 Hardware, Software, and Data Collection

At the core of our efforts is empirical data collection. The distributional choice and subsequent verification depend upon it. To enable empirical data collection, a test harness was created that executes the synthetic "no-op" workloads while varying numbers of processes per core, nominal execution times, and execution platforms. As the synthetic workload processes are executing, the parameters in

Table 1 are collected. In order to reduce the possibility that results gleaned from this study might be an artifact of a particular hardware platform or operating system, two different platforms are used as shown in Table 2 (two of platform A and seven of platform B). All platforms support a version of the Linux completely fair scheduler [9] which will be exclusively used during data collection.

Table 1. Experimental Parameters

Parameter	Symbol
Nominal Execution Time	t_N
# Processes per Core	p
Voluntary Context Swaps	v
Non-Voluntary Context Swaps	nv
Actual Execution Time	t_A
Execution Time Noise $(t_A - t_N)$	Δ

Table 2. Hardware and Operating Systems

Label	Processor	Operating System
A	Intel E3 1220	Fedora 19, Linux Kernel v. 3.10.10
B	2 x AMD Opteron 2431	CentOS 5.9, Linux Kernel v. 3.0.27

Each data point collected consists of the dimensions outlined in Table 1. Nominal execution times vary from $0.25\mu s$ through $3.7ms$ with observations at an interval of $0.25\mu s$ throughout the range. The number of workload processes per core varies from 1 through 20 processes. Each sharing and nominal execution time pair is executed 1000 times to ensure a good distribution sample. The synthetic workloads are run on one of two of platform A or on one of seven of platform B from Table 2. In total 100+ million observations are made. Two factors limited the range of viable execution times: the lower bound on timer resolution (see below) and the memory needed to generate workloads of longer lengths (disk to store and physical memory to compile).

Generated data is divided into two sets. The first, a "training" set (of size 10^6) is segregated using uniform random sampling. The rest of the data is used for model evaluation and will be referred to as the "evaluation" set. We specifically want to judge the applicability of this noise model to multiple hardware types and operating systems using the same scheduler.

There has been much discussion about the best and most accurate way to time a section of code [3]. There are many methods including processor cycle counters and operating system "wall-clock" time. Given our reliance on empirical data for modeling and evaluation we feel it is important to cover how our timing measurements are made. In many cases, the use of a simple time stamp counter is effective assuming that the process will never migrate to another core. Another issue to consider is frequency scaling which can lead to wildly inaccurate timings when utilizing the processor cycle counter. To alleviate some of those concerns and provide a relatively universal timing interface we developed a system timer thread that utilizes the x86 time stamp counter instruction on a single reference core to update a user space timer. When a process or thread requests the current time, an in-lined function copies the current time struct which has two time references and it compares the two times. If they are the same then the calling code can be sure that the time has been fully updated and the function returns, if not the code loops until the values match. Frequency scaling is turned off for the time update thread.

This timing method has several advantages: (1) it is entirely in user space, (2) it is lock-free, and (3) it is monotonic even when the timed thread is shifted to a new core. Two concerns with this approach stem from the copying operation. How long does it take to copy the timer struct on a target system and what happens when there are multiple Non-uniform Memory Access (NUMA) nodes? To test the latter of these concerns a benchmark was constructed to ascertain how long a copy operation takes when the copy is from the same NUMA node as the calling process and when the timer thread and requester thread are on differing NUMA nodes. The results of this are shown in Figure 3 for platform B from Table 2. What we've found is that reading memory allocated on a NUMA node other than the one closest to the time requesting process the access times can vary somewhat. To eliminate this issue, all subsequent experiments only use a single NUMA node.

A common problem with highly accurate timing via software is determining what is ground truth. Short of an external atomic clock, there are only varying degrees of truth. In order to determine the precision and accuracy of our measurements, a standardized workload is created with a series of "no-op" instructions of varying lengths. Each "no-op" length is timed using either the

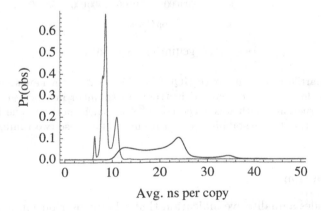

Fig. 3. Smooth histogram of 10^6 data points each representing timed averages of 500 copy operations, first on the same NUMA node (red line) and then across different NUMA nodes (blue line). The performance of a copy on the same NUMA node seems to be much more consistent.

x86 `rdtsc` instruction or the POSIX.1-2001 `clock_gettime()` function. Figure 4 shows the inter-quartile range (25th to 75th percentiles) difference of each timing measurement as a function of the length of the "no-op" instruction sequence, This plot informs us about the stability of the two timing methods. The system call to `clock_gettime()` is more stable than the `rdtsc` instruction, especially for these small workloads. A hypothesis as to why it is more stable is that the measurement of actual workload time is small relative to the time it takes to perform a system call. To test this theory the timing methods themselves are timed by executing five hundred of each method (either the `rdtsc` insn. or the `clock_gettime()` function) and using the average execution time of all five hundred to extrapolate the time to execute a single instruction. In this experiment the `rdtsc` instruction is used as the reference timer on platform A from Table 2. As expected (and shown in Figure 5) the system call to `clock_gettime()` takes almost 3× as long on average compared to the x86 `rdtsc` instruction. For this reason, we exclusively use the `rdtsc` instruction for all empirical timing measurements in this work.

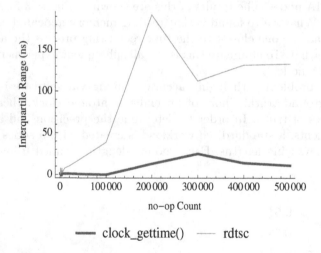

Fig. 4. Interquartile range difference (IQRD = $75^{th} - 25^{th}$) in nanoseconds for the times measured for each set of "no-op" instructions (number instructions listed on x-axis). Each instruction length was executed 10^6× for each method. The IQR gives a visual representation to the stability of measurements for these two timing methods.

2.3 Distribution

Figure 1 provides a qualitative indication that a Levy distribution makes a good choice for modeling the noise present in execution times of a nominally fixed minimal workload (the proxy for BCETV). Quantitatively Table 3 summarizes the p-values for each distribution (higher is better), the table shows the minimum, maximum and mean values. The Levy distribution is the only distribution with

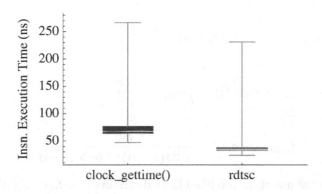

Fig. 5. Box and whisker plot showing a speed comparison of the `rdtsc` x86 assembly instruction compared to a `clock_gettime()` call to the Linux real-time clock. The `rdtsc` instruction's $25^{th} - 75^{th}$ percentiles are almost identical at the nanosecond scale. The `clock_gettime()` function overall takes much more time (approximately $3\times$).

Table 3. Summary of Anderson-Darling Goodness of Fit Test

Distribution	Min	10th	50th	90th	Max	Mean
Gaussian Distribution	0	1.17×10^{-15}	1.68×10^{-14}	3.0×10^{-5}	.719	.002
Levy Distribution	0	2.11×10^{-15}	2.15×10^{-14}	.038	.803	.025
Gumbel Distribution	0	8.93×10^{-16}	1.97×10^{-14}	6.39×10^{-06}	.357	.002
Cauchy Distribution	0	3.89×10^{-16}	1.34×10^{-14}	.002	.771	.009

greater than 10% of the data having a p-value $\geq .01$. Next, we will quantitatively describe the modified Levy distribution that we use.

Realized execution time is the sum of a nominal (mean) execution time and a noise term. If the nominal execution time is represented by a random variable N, and the noise is represented by a random variable V, then the realized execution time $R \sim N + V$. The goal of this work is to find a distribution to represent a lower bound for V which we term BCETV.

The Levy distribution [11] has a closed form probability density function (PDF, shown in Equation (1)), however in general it has no defined moments. Observations from the empirical data lead to a solution. Whereas the tail of the Levy distribution is infinite, the noise present within the real execution times has a limit. The limit, not surprisingly, is correlated with both the nominal length of execution and the number of processes assigned to a single compute core. This leads to the consideration of a variation of the Levy distribution that is truncated at a point represented by a new parameter Ω. The modified Levy distribution is defined using the truncation method of Equation (2) as Equation (3) where $F(\cdot)$ is the CDF of the PDF denoted by $f(\cdot)$. (Note: erfc(x) is the compliment of the Error Function, $1 - \text{erf}(x)$, and E_i is the exponential integral function [1].) In order to make the equations more concise, $w = \frac{\beta}{2(\alpha - \Omega)}$ and $z = \frac{\beta}{2(x - \alpha)}$.

$$f_L(x; \alpha, \beta) = \frac{e^{-z} (2z)^{3/2}}{\sqrt{2\pi}\beta} \tag{1}$$

$$f_{mL}(x; \alpha, \beta, \Omega) = \frac{f_L(x; \alpha, \beta)}{F_L(\Omega; \alpha, \beta) - F_L(-\infty; \alpha, \beta)} \tag{2}$$

$$f_{mL}(x; \alpha, \beta, \Omega) = \frac{\sqrt{\beta}e^{-z}}{\sqrt{2\pi}(x - \alpha)^{3/2} \left(\mathrm{erfc}\sqrt{-w}\right)} \tag{3}$$

Restricting the use of the modified Levy distribution, mL, to $x \leq \Omega$ and $x > \alpha$ leads to a closed form expression of the mean as shown in Equation (4). Lastly, a variance is also defined as Equation (5).

$$\mu_{mL}[\alpha, \beta, \Omega] = \frac{\beta\Gamma\left(-\frac{1}{2}, -w\right)}{2\sqrt{\pi}\mathrm{erfc}\left(\sqrt{-w}\right)} + \alpha \tag{4}$$

$$\sigma_{mL}^2[\alpha, \beta, \Omega]$$
$$=$$
$$\frac{(\alpha - \Omega)^2 \left(\frac{E_{\frac{5}{2}}(-w)\left(\sqrt{2\pi}\beta^{3/2}\mathrm{erfc}(\sqrt{-w})+3(\Omega-\alpha)^{3/2}\left(4e^w-3E_{\frac{5}{2}}(-w)\right)\right)}{\sqrt{\Omega-\alpha}} + 4(\alpha - \Omega)e^{2w} \right)}{2\pi\beta\mathrm{erfc}^2\left(\sqrt{-w}\right)} \tag{5}$$

Our next task is to determine an appropriate parametrization of the modified Levy distribution. We accomplish this task by fitting a model to empirical measurements. The "training" data are sorted into groups W_{p,t_N} which are indexed by the number of processes sharing a core, p, and the nominal execution time, t_N (see Equation (6a)). Within each group, the execution time noise is computed for each observation as in Equation (6b).

$$W_{p,t_N} = \bigcup_i obs_i \in p, t_N \tag{6a}$$

$$\Delta = t_A - t_N \tag{6b}$$

Separately for each group W, Maximum Likelihood (ML) techniques are used to find the best parameters for a number of distributions, including the modified Levy distribution that we are proposing. The quality of the distributions' fit to the empirical data is judged via an Anderson-Darling [2] goodness of fit test as shown in Table 3 (chosen because of the weight given to the tails of the distributions compared to other tests such as the Kolmogorov-Smirnov test [4]).

2.4   Parameterization

While the ML techniques used above can yield a parametrization for the modified Levy distribution that is well matched to the data, in general ML techniques are quite computationally expensive and also require substantial support to be effective. An alternative is to redefine the modified Levy distribution parameters, α, β, and Ω, in terms of a subset of the parameters in Table 4.

The selection of parameters from Table 4 is reduced based on the intuition that the nominal execution time and number of processes sharing a core will have the largest impact on the true execution time. Given the design of the minimal compute kernel, it is expected (and confirmed) that there are zero voluntary context swaps allowing the variable to be discarded. A Pearson correlation coefficient between the target variables and the training data (Table 4) quantifies the intuition about the remaining parameters.

Table 4. Correlation Between Target Predictors

	nv	t_N	p
Δ	.508	.771	-.0056

Table 4 summarizes the correlations within the training set between the execution time noise, Δ, and the other parameters. For the entire training set there is a weak correlation between the number of processes sharing a core and the execution time noise. There is a strong correlation between the nominal execution time and the noise. Not shown is the co-variance between the non-voluntary context swaps and the number of processes per core which implies a lack of independence. The models considered therefore consist only of the two independent parameters p and t_N.

Using simple linear regression to find coefficients for p and t_N that best fit the parameters for α, β and Ω found by ML, the relationships in Equation (7) are found with the following assumptions: $p \in \mathbb{Z} \wedge p > 1$ and $t_N \in \mathbb{R} \wedge t_N > 0$. To parametrize the modified Levy distribution as defined in Equation 3, several other constraints must be added, namely: Equation (7a) is expected to have a negative range for the entire domain, Equation (7b) is positive for the entire domain and Equation (7c) is greater than α for the entire domain. A limitation of these equations is the range of data used to create them. It is expected that α will not continue to decrease as $t_N \to \infty$ and the β, Ω parameters probably have limitations as well; however these equations are supported through the range of data specified in Section 2.2.

$$\alpha = 4.75 \times 10^{-9}p - 0.220t_N \tag{7a}$$

$$\beta = 4.19 \times 10^{-10}p + 0.007t_N \tag{7b}$$

$$\Omega = 3.19 \times 10^{-6}p + 0.742t_N \tag{7c}$$

Using Equations (7), which predict α, β and Ω based on p and t_N, the PDF and mean of the modified Levy distribution can now be described in terms of p and

t_N as shown in Equations (8) and (9), respectively. The variance of Equation (8) is a straightforward algebraic manipulation of Equation (5).

$$f_{mL}(x; p, t_N) = \frac{8.2 \times 10^{-6}(\sqrt{1p + 1.6 \times 10^7 t_N})e^{\frac{0.044p + 6.98 \times 10^5 t_N}{p - 4.6 \times 10^7 t_N - 2.1 \times 10^8 x}}}{(-4.8 \times 10^{-9}p + 0.22t_N + x)^{3/2}\text{erfc}\left(\sqrt{0.003 - \frac{0.003p}{p + 3.02 \times 10^5 t_N}}\right)} \tag{8}$$

$$\mu_{mL}[p, t_N] = \frac{(1.2 \times 10^{-10}p + 0.002t_N)\Gamma\left(-\frac{1}{2}, 0.003 - \frac{0.003p}{p + 3.02 \times 10^5 t_N}\right)}{\text{erfc}\left(\sqrt{0.003, -\frac{0.003p}{p + 3.02 \times 10^5 t_N}}\right)} \tag{9}$$
$$+ 4.8 \times 10^{-9}p - 0.22t_N$$

3 Results

How well does the modified Levy distribution approximate the actual BCETV observed while executing a nominally deterministic compute bound kernel? We will focus our evaluation on the PDF expressed in terms of processor sharing, p, and nominal execution time, t_N, presented above as Equation (8).

The Anderson-Darling (AD) goodness of fit test of Table 3 is, frankly, not very promising. Yet, we already know from Table 3 that the modified Levy is the best out of the listed distributions used to model the training data. It is not at all surprising that our overall p-value when using AD is not very high ranging from 0 to 0.73. What is welcome news is that AD is not the only metric available, as it is relatively ineffective at identifying portions of the parameter space that have a good vs. a poor fit.

A second measure of how well the modified Levy distribution fits empirical data is how well the moments match. When comparing the mean of the empirical data sets to that predicted by Equation (9), the differences are effectively below our ability to differentiate based on the techniques described in Section 2.2 (i.e., the difference is $\ll 10^{-12}s$). Comparing the variance for the modified Levy vs. the empirical measurements results in an r-squared value of 0.69, which indicates a reasonable degree of correlation between model and data, but the alignment between the two is clearly not perfect.

While the above quantitative assessments of the modified Levy distribution's match with empirical measurements make it clear that the model is far from perfect, we must keep in mind the fact that modelers can often exploit individual models that are far from perfect, and given prior use of models that are much more divergent from reality than our proposed modified Levy distribution there is the real potential for benefit from the ability to use a distribution that more closely matches empirical measurement than previous models and also has relatively simple expressions of its first two moments.

We continue the assessment of how well the modified Levy distribution characterizes the noise in execution times by presenting QQ-plots for three distributions relative to the empirical data (see Figure 6). The first column of plots is the modified Levy distribution of Equation (8), the second column is a Gaussian distribution, and the third column is a Cauchy distribution. The latter two distributions are parametrized by fitting to the data using ML techniques. For each distribution, 4 distinct QQ-plots are shown, separating the processor sharing variable, p, into quartiles. The first (top) row represents the range $1 \leq p \leq 5$, the second row represents the range $6 \leq p \leq 10$, the third row represents the range $11 \leq p \leq 15$, and the fourth (bottom) row represents the range $16 \leq p \leq 20$.

First consider the results in Figure 6(g) and (j), which include the modified Levy distribution and significant processor sharing. Here, we see quite nice alignment between the model and the empirical data, the best evidence yet that the modified Levy is a good execution time noise model. Next consider the results in Figure 6(a) and (d), which include the modified Levy distribution and little processor sharing. In this case, there is reasonably good alignment at the low end of the range, but the empirical data has slightly less variation than the model at the high end of the range. Finally, note that the alignment between model and empirical data is noticeably worse for both the Gaussian and the Cauchy distributions across the entire range of p.

From the above we conclude that the modified Levy distribution is a relatively good proxy for BCETV. The distribution of BCETV can in turn be used in many ways. To demonstrate the utility of BCETV, we explore the mean queue occupancy (MQO) of a single queue system when noise is added as in Figure 2. The single queue system operates as two threads with one way communication that is designed to have an exponentially distributed inter-arrival and service time distribution (i.e., workload is dependent upon an exponential random number source). A simple model for MQO is the $M/M/1$ queueing model, it expects the inter-arrival times to be exponentially distributed. We posit that the farther from this distribution the actual system is, the greater the model's predictions will differ from empirical reality. The Kullback-Leibler (KL) divergence [8] is a measure of the divergence between two distributions (zero being a perfect match). We are interested in how far the distributional lower bound as predicted by the convolution of the exponential distribution and Equation 8 differs from that expected by the $M/M/1$ MQO model. With a divergence of zero we should expect to find a very close match between modeled and experimental MQO. At higher divergences (the exact amount is an open question) we don't expect the $M/M/1$ model to be very accurate. Figure 7 is a summary of median KL divergences (y-axis) separated by percent model accuracy (calculated as $\frac{|\text{modeled MQO} - \text{measured MQO}|}{\text{measured MQO}} \times 100$, x-axis) for 6000+ separate executions of the single queue system described above on the platforms shown in Table 2. It shows that lower KL divergence (green bar) between the expected exponential and that convolved with the BCETV distribution, is associated with more accurate MQO predictions. This implies that BCETV can be used as a predictor for model choice (at least with a Markovian arrival process).

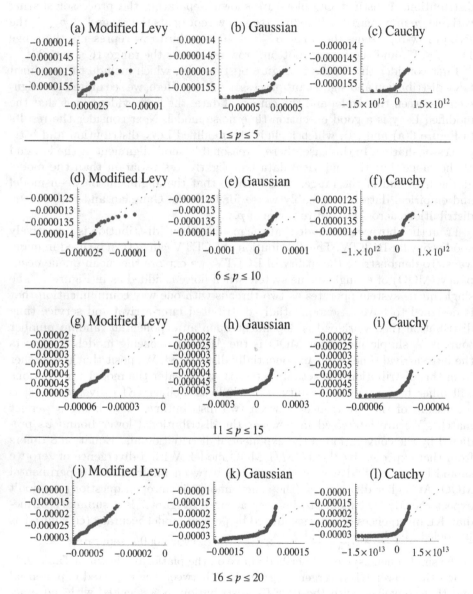

Fig. 6. QQ-plots comparing empirical data (vertical axis) to the analytic distributions (horizontal axis). The dashed line shows the ideal response.

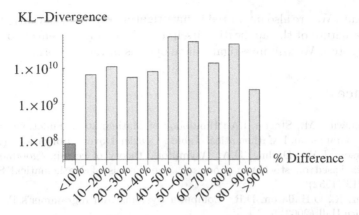

Fig. 7. The y-axis shows the median KL divergence between the $M/M/1$'s expected exponential inter-arrival distribution and the lower bound predicted by convolving the exponential distribution with Equation 8. The x-axis is the percent difference between the mean queue occupancy predicted by an $M/M/1$ model and the actual measurements from a single queue system designed to have a perfectly exponential workload. The lowest KL divergence (green bar) is associated with more accurate predictions.

4 Conclusions and Future Work

We've demonstrated a noise model that appears to work far better than a simple Gaussian assumption, in fact far better than multiple other distributions. It has also been shown to work for at least two differing platform types (see Table 2) using the same fair scheduling algorithm. First, we've shown expressions for the PDF and first two moments of a Levy distribution that has been modified to have bounded moments. Through empirical data collection, a model is derived that can be used to parametrize the modified Levy distribution relatively well without resorting to computationally expensive parameter fitting.

In Figure 6 we showed how well the quantiles of the modified Levy distribution match to the quantiles of the empirical data. We've also noted that the fit between the model and empirical data gets better as more processes are added per core. This is in keeping with our original assumption that a single process on a single core should exhibit its native distribution, in our case purely deterministic, or close to the nominal mean, t_N. The models demonstrated here are only validated over the range of empirical data that we've collected. For future work, we would like to extend the parameter estimators for $p > 20$ and higher nominal execution times t_N.

One concern with our approach is also one of its strengths, that it is based on wide empirical sampling. Is this noise model really applicable to multiple hardware types, or were our choices simply judicious? Could other parameters in addition to nominal execution time, t_N, and the number of processes per core, p, provide a better estimate on other platforms (e.g., alternative instruction sets). One potential application of this noise model is as a minimal expected noise for all workloads. We are also interested in investigating variability in execution time due to the nature of the application itself, e.g., including the effects of caching, branching, etc.). We will investigate these options in future work.

References

[1] Abramowitz, M., Stegun, I.A.: Handbook of Mathematical Functions: with Formulas, Graphs, and Mathematical Tables. Courier Dover Publications (2012)

[2] Anderson, T.W., Darling, D.A.: Asymptotic theory of certain "goodness of fit" criteria based on stochastic processes. The Annals of Mathematical Statistics, 193–212 (1952)

[3] Bryant, R., O'Hallaron, D.R.: Computer Systems: A Programmer's Perspective. Prentice Hall (2003)

[4] Chakravarty, I., Roy, J., Laha, R.: Handbook of Methods of Applied Statistics. McGraw-Hill (1967)

[5] Edgar, S., Burns, A.: Statistical analysis of WCET for scheduling. In: Proc. of 22nd IEEE Real-Time Systems Symposium, pp. 215–224 (2001)

[6] Engblom, J., Ermedahl, A.: Pipeline timing analysis using a trace-driven simulator. In: Proc. of 6th Int'l Conf. on Real-Time Computing Systems and Applications, pp. 88–95 (1999)

[7] Jain, R.: The Art of Computer Systems Performance Analysis. John Wiley & Sons (1991)

[8] Kullback, S., Leibler, R.A.: On information and sufficiency. The Annals of Mathematical Statistics, 79–86 (1951)

[9] Li, T., Baumberger, D., Hahn, S.: Efficient and scalable multiprocessor fair scheduling using distributed weighted round-robin. ACM SIGPLAN Notices 44(4), 65 (2009)

[10] Mazouz, A., Touati, S.A.A., Barthou, D.: Study of variations of native program execution times on multi-core architectures. In: Proc. of Int'l Conf. on Complex, Intelligent and Software Intensive Systems, pp. 919–924 (2010)

[11] Nolan, J.: Stable Distributions: Models for Heavy-tailed Data. Birkhäuser (2003)

On the Predictive Properties of Performance Models Derived through Input-Output Relationships

Mahmoud Awad and Daniel A. Menascé

Computer Science Department, George Mason University
Fairfax, VA 22030, USA
{mawad1,menasce}@gmu.edu

Abstract. Building an analytical performance model is a challenge when little is known about the functionality and behavior of the system being modeled and/or when obtaining model parameters through measurements is difficult. This paper addresses this problem by presenting an approach that derives analytic model parameters by observing the input-output relationships of a real system. More specifically, input (i.e., arrival rates for each job class) and output (i.e., average response time for each job class) measurements are used to estimate the per class service demands and number of servers for a Queuing Network model of the system. This model, called the computed model (CM), provides the same output values for the same input values used to derive the CM. The important question is whether the CM has predictive power, i.e., can the CM predict the output values that would be observed in the real system for different values of the input? The CM's parameters are obtained by solving a non-linear optimization problem. The paper shows through experiments that the CM is relatively robust and has predictive power over a range of input values.

Keywords: Queuing network models, parameter estimation, non-linear optimization.

1 Introduction

Analytical performance models, such as Queuing Network (QN) models, are essential for performance prediction as well as understanding the qualitative characteristics of a computer system. Developing these models requires intimate knowledge of the computer system and the availability of a number of performance measurements to estimate the model's input parameters. However, there are cases when performance prediction is needed for computer systems whose functionality and behavior are not fully understood.

For example, Internet data centers with virtualized environments, such as cloud computing providers, are capable of hosting multiple heterogeneous application systems of different sizes and complexities. Therefore, it is practically impossible for such data centers to adequately understand the functionality offered by all hosted application systems in order to develop precise performance models.

A. Horváth and K. Wolter (Eds.): EPEW 2014, LNCS 8721, pp. 89–103, 2014.
© Springer International Publishing Switzerland 2014

The issue we address in this paper is the ability to derive QN models for computer systems where little is known about the internal architecture or behavior of the system. We view the system as a "grey" box where input parameters (e.g., arrival rates) and output parameters (e.g., average response time) are known, in addition to some internal structural information about the system (e.g., number and function of each layer in a multi-tiered system, but not necessarily the number of servers of each type). We then derive a QN model, which we call the Computed Model (CM), that closely approximates the computer system. Such model will be of benefit only if it proves capable of predicting the behavior of the real system as the workload intensity changes over time.

Our approach for deriving QN performance models uses a non-linear optimization technique to determine the parameters of the CM. We conducted a number of experiments to evaluate the ability of the CM to predict the behavior of the real system as workload intensity changes. In order to test the viability of our approach, we initially, conducted a number of controlled experiments with an analytical QN model acting as a proxy for a real system. Our results showed that our approach is capable of producing computed QN models that have a robust predictive power. We then conducted two experiments using Apache OFBiz (Open For Business) ERP system; one in which the number of servers per tier is known a priori (Static-N) and one in which the number of servers per tier is inferred by the optimization technique (Variable-N). Both OFBiz experiments show results consistent with the controlled experiments.

A few prior efforts are related to our research. It is important to note that this paper is not about parameterizing QN models of real systems. There is a vast body of literature on that. In that context, the system components are totally visible to the modeler and therefore can be instrumented. What is of interest to the work in this paper is a situation where the modeler either has no access, is unwilling, or does not have the resources to conduct measurements on the internal components of a computer system. In that case, the work in [2,3] used a customized non-linear optimization technique to approximate unknown model parameters when the queuing model is already known. However, their technique is applied to a limited set of single queue models. Also, [11] proposed a black-box approach to calculating unknown input parameters in open multi-class queuing networks given that some service demands are known.

Our optimization technique is simple and efficient and can be applied to open multi-class QN models, where the workload intensity and the response time of the modeled system are known, in addition to minimal internal structure information.

Autonomic computing environments can benefit the most from our approach of deriving performance model parameters dynamically [6]. These environments rely heavily on optimal resource allocation through performance prediction, where the internal architecture of the hosted application systems is not fully understood. The technique presented here does not apply when one is interested in predicting metrics internal to a system (e.g., CPU and disk queue lengths)

The rest of the paper is organized as follows. Section 2 includes the problem definition and notation used throughout the rest of the paper. Section 3 discusses the methodology and algorithm used. The next section shows experimental results. Section 5 discusses some related work. Finally, section 6 presents discussions and concluding remarks.

2   Problem Definition

In model-based performance engineering, real computer systems are abstracted using analytical models, which can be used by performance engineers to answer "what if" questions related to predicting system performance. Queuing network (QN) models have been used quite often for that purpose. Such models have two types of parameters: workload intensity (e.g., arrival rates) and resource service demands (i.e., the average total time spent by a transaction using a resource). Service demands do not include the time waiting to use a resource. See e.g. [12] for details on QN models.

As discussed before, in virtualized environments, the internal architecture of hosted application systems may not be readily available. However, it is a common practice for such environments to provide monitoring tools that can easily record and analyze input and output parameters, such as the transaction arrival and departure rates as well as the time taken to process each transaction. Such monitoring is readily available in operating systems as well as virtualization software, and should be adequate, in our opinion, to develop an overall model that approximates the behavior of the application system.

Figure 1 demonstrates the problem we address in this paper. To derive an approximate analytical model of the system, we treat it as a "gray box," where the input and output parameters are known, as well as minimal information about the internal structure of the system (e.g., number of tiers in a multi-tiered system). The QN Parameter Estimator takes average arrival rates λ (input) and average response times T^{AS} (output) and establishes a relationship between them in order to estimate the parameters of the Computed QN Model (CM). The relationship between the input and output parameters is formulated as a non-linear optimization problem that can be solved using a non-linear solver.

After the Computed QN Model is parameterized, the behavior of the actual system and the corresponding QN model need to be compared frequently to ensure that the QN model accurately represents the actual system, which is important if the model is to be used for performance prediction. This comparison process is demonstrated in Figure 2, where the observed response time, T^{AS}, of the actual system and the computed response time, T^{CM}, of the Computed QN Model are compared for the same arrival rate. The question we pose in our experiments is how accurate is the response time of the CM when compared to the observed response time of the actual system.

This paper considers the problem of deriving and parameterizing a QN model given that the arrival rate (input) and the response time (output) of the actual system can be measured at regular intervals. We focus primarily on multi-tier

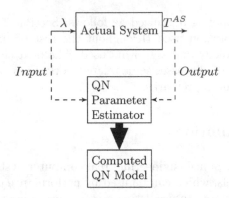

Fig. 1. Parameterizing the Computed QN Model

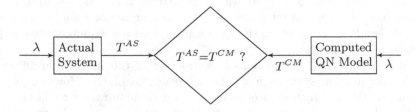

Fig. 2. Maintaining the Accuracy of the Computed QN Model

application systems, in which transactions are processed by one of several servers at each tier and passed on to the next tier. This architecture is typical of online transaction processing systems such as e-commerce application systems. However, our approach can be applied to a host of architectures and application systems.

Figure 3 shows a sample computed QN Model that would be used to represent an actual system with a 3-tier architecture. We will use this architecture to demonstrate our methodology and experimental results. It would be obvious to one skilled in QNs that our approach can be generalized to any number of tiers. In this architecture, transactions are submitted to the web server tier, which may consist of a number of servers that are load balanced. Transactions are passed from the web server tier to the application server tier, which may also consist of a number of load balanced servers. If the transaction needs to access the database, it will be passed on to the database server tier, which will process the transaction and return the results back to the application server tier and then to the web server tier.

We assume throughout the paper that the number of tiers in the actual system is known (we call this internal structural information). But, we assume we do not know the number of servers in each tier, their internal components, nor the service demands at these components for each transaction class. In order to parameterize the computed QN Model, we need to find the service demands for each tier for each class. Consider the following notation:

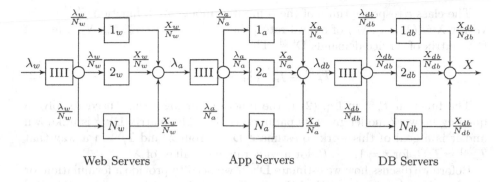

Fig. 3. Computed QN Model Topology in a 3-tier Architecture

- R: number of transaction classes in the queuing network
- λ_r: arrival rate of class r transactions $(r = 1, \cdots, R)$
- N_w, N_a, N_{db}: number of servers in the web, application, and database tiers, respectively
- $D_{w,r}^{\mathrm{AS}}, D_{a,r}^{\mathrm{AS}}, D_{db,r}^{\mathrm{AS}}$: average class r $(r = 1, \cdots, R)$ service demands at each of the three tiers for the actual system.
- $D_{w,r}^{\mathrm{CM}}, D_{a,r}^{\mathrm{CM}}, D_{db,r}^{\mathrm{CM}}$: average class r $(r = 1, \cdots, R)$ service demands at each of the three tiers for the computed model
- $\mathbf{D}^{\mathrm{AS}}$: matrix of service demands for the actual system. The rows correspond to the web server, application server, and database server tier, respectively, and the columns correspond to the classes.
- $\mathbf{D}^{\mathrm{CM}}$: matrix of service demands for the computed system. The rows correspond to the web server, application server, and database server tier, respectively, and the columns correspond to the classes.
- $T_{w,r}, T_{a,r}, T_{db,r}$: average class r $(r = 1, \cdots, R)$ response times at each of the three tiers
- T_r^{AS}: average class r $(r = 1, \cdots, R)$ response time for the actual system
- T_r^{CM}: average class r $(r = 1, \cdots, R)$ response time for the computed model

3   Methodology

The goal of our methodology is to derive service demands at all tiers and all classes, for the Computed Model (CM) using only the inputs (average arrival rates) and outputs (average response times) of the Actual System (AS).

The average response time of the Actual System for class r transactions is a function f_r^{AS} of the vector $\boldsymbol{\lambda} = (\lambda_1, \cdots, \lambda_R)$ of average arrival rates, the vector $\boldsymbol{N} = (N_w, N_a, N_{db})$, and of the matrix of service demands $\mathbf{D}^{\mathrm{AS}}$.

Hence,

$$T_r^{\mathrm{AS}} = f_r^{\mathrm{AS}}(\boldsymbol{\lambda}, \boldsymbol{N}, \mathbf{D}^{\mathrm{AS}}). \tag{1}$$

The values of $T_r^{\mathrm{AS}}, r = 1, \cdots, R$, can be obtained using standard measuring tools.

The class r response times of the computed model are a function f_r^{CM} of the vector $\boldsymbol{\lambda} = (\lambda_1, \cdots, \lambda_R)$ of arrival rates, the vector $\boldsymbol{N} = (N_w, N_a, N_{db})$, and of the matrix of service demands $\mathbf{D}^{CM}$. Thus,

$$T_r^{CM} = f_r^{CM}(\boldsymbol{\lambda}, \boldsymbol{N}, \mathbf{D}^{CM}). \tag{2}$$

The function f_r^{CM} in Eq. (2) is the function (or algorithm) used to solve a queuing network model given its parameters [12]. The matrix $\mathbf{D}^{CM}$ is unknown and it is a goal of this work to estimate $\mathbf{D}^{CM}$ from $\boldsymbol{\lambda}$ and T_r^{AS} in a way that $T_r^{CM} \approx T_r^{AS}$, for $r = 1, \cdots, R$, for a wide range of values of $\boldsymbol{\lambda}$.

Before we discuss how we estimate $\mathbf{D}^{CM}$, we need to provide a formulation for the function f_r^{CM} for the QN of Fig. 3. We use Seidmann's approximation [14] to model the multiple-server queues in this QN.

This approximation replaces a multiple-server queue by a sequence of a delay device and a load-independent queuing device with properly adjusted service demands. Then, the response time of an N-server single queue with service demand equal to D in each server is approximated as

$$T = D\frac{N-1}{N} + \frac{D/N}{1 - \lambda \times D/N}. \tag{3}$$

Therefore, the function f_r^{CM} with input parameters $\boldsymbol{\lambda}, \boldsymbol{N}$, and $\mathbf{D}^{CM}$ is

$$T_{w,r} = D_{w,r}^{CM}\frac{N_w - 1}{N_w} + \frac{D_{w,r}^{CM}/N_w}{1 - \sum_{r=1}^{R} \lambda_r D_{w,r}^{CM}/N_w} \tag{4}$$

$$T_{a,r} = D_{a,r}^{CM}\frac{N_a - 1}{N_a} + \frac{D_{a,r}^{CM}/N_a}{1 - \sum_{r=1}^{R} \lambda_r D_{a,r}^{CM}/N_a} \tag{5}$$

$$T_{db,r} = D_{db,r}^{CM}\frac{N_{db} - 1}{N_{db}} + \frac{D_{db,r}^{CM}/N_{db}}{1 - \sum_{r=1}^{R} \lambda_r D_{db,r}^{CM}/N_{db}} \tag{6}$$

$$T_r^{CM} = T_{w,r} + T_{a,r} + T_{db,r} \tag{7}$$

The problem of obtaining $\mathbf{D}^{CM}$ can be cast as the following non-linear optimization problem.

Find the service demands in $\mathbf{D}^{CM}$ (i.e., the values of the variables $D_{w,r}^{CM}, D_{a,r}^{CM}, D_{db,r}^{CM} \forall r$) that minimize MAXDIFF, the maximum value of the absolute differences between the response times of the actual system and that of the computed model.

$$\text{Minimize MAXDIFF} = \max_{r=1}^{R} | T_r^{AS} - T_r^{CM} | \tag{8}$$

subject to the following constraints:

1. $D_{w,r}^{CM}, D_{a,r}^{CM}, D_{db,r}^{CM} \geq 0 \quad r = 1, \cdots, R$
2. $D_{w,r}^{CM} + D_{a,r}^{CM} + D_{db,r}^{CM} \leq T_r^{CM} \quad r = 1, \cdots, R$

3. $\sum_{r=1}^{R} \lambda_r \frac{D_{w,r}^{\mathrm{CM}}}{N_w} < 1,\ \ \sum_{r=1}^{R} \lambda_r \frac{D_{a,r}^{\mathrm{CM}}}{N_a} < 1,\ \ \sum_{r=1}^{R} \lambda_r \frac{D_{db,r}^{\mathrm{CM}}}{N_{db}} < 1\ \ \ r = 1,\cdots,R$

4. $T_r^{\mathrm{CM}} = T_{w,r} + T_{a,r} + T_{db,r}$ where

$$T_{w,r} = D_{w,r}^{\mathrm{CM}} \frac{N_w-1}{N_w} + \frac{D_{w,r}^{\mathrm{CM}}/N_w}{1-\sum_{r=1}^{R}\lambda_r D_{w,r}^{\mathrm{CM}}/N_w},\ T_{a,r} = D_{a,r}^{\mathrm{CM}} \frac{N_a-1}{N_a} + \frac{D_{a,r}^{\mathrm{CM}}/N_a}{1-\sum_{r=1}^{R}\lambda_r D_{a,r}^{\mathrm{CM}}/N_a},$$

and

$$T_{db,r} = D_{db,r}^{\mathrm{CM}} \frac{N_{db}-1}{N_{db}} + \frac{D_{db,r}^{\mathrm{CM}}/N_{db}}{1-\sum_{r=1}^{R}\lambda_r D_{db,r}^{\mathrm{CM}}/N_{db}}$$

The first constraint says that all service demands must be non-negative, the second constraint says that the response time of each class must be at least equal to the sum of all service demands at all tiers for transactions of that class (this is the zero congestion case). The third constraint indicates that the utilization of the web server tier, application tier, and database tier have to be less than 100%. Finally, the fourth constraint provides the function f_r^{CM} used to compute T_r^{CM} as a function of $\boldsymbol{\lambda}, \boldsymbol{N}$, and the service demands in $\mathbf{D}^{\mathrm{CM}}$.

The above discussion assumes that the number of servers per tier is known a priori. However, one can extend this formulation, as done in the experiments, by including N_{ws}, N_a, and N_{db} as decision variables with the following constraints: $N_{ws} \in \mathbb{N}, N_a \in \mathbb{N}$, and $N_{db} \in \mathbb{N}$ (where it is understood that $0 \notin \mathbb{N}$).

This non-linear optimization problem can be solved using available solvers, including Microsoft's Excel Solver Add in that uses the Generalized Reduced Gradient (GRG2) method or NEOS solvers (www.neos-server.org/neos/solvers/). We used the Excel Solver in the results reported in this paper.

The solution of this optimization problem provides the necessary service demands to solve the QN model given the arrival rates measured in the actual system. As discussed at the outset of the paper, the question of interest is whether the computed model CM has predictive power over a range of arrival rate values. Given a certain threshold ϵ for the maximum percent absolute relative difference (MPARD) between the actual response time and that predicted by the computed model, the process of recomputing the service demands $\mathbf{D}^{\mathrm{CM}}$ should be repeated when the MPARD exceeds the threshold. Therefore, the computed model may have to be re-calibrated when

$$\mathrm{MPARD} = \max_{r=1}^{R}\{|\ (T_r^{\mathrm{AS}} - T_r^{\mathrm{CM}})/T_r^{\mathrm{AS}} \times 100\ |\} > \epsilon. \tag{9}$$

4 Experimental Results

The first subsection of this section discusses experiments, called "controlled experiments," in which we use an open QN model as a proxy for the actual system. The following subsection reports on experiments conducted with a real system.

4.1 Controlled Experiments

The controlled experiments use an open QN model, referred to as the Actual Model (AM), as a proxy for an actual system. The parameters of the AM, such as the service demands at all tiers, are known. Note that these known parameters of the AM are used only for the purpose of solving the AM and obtaining its response time as a proxy for measuring the response time in the actual system. The service demands of the AM are not used in any way to derive parameters for the CM.

The actual system (represented by the Actual Model) is a 3-tier web-based application system with two classes of transactions and includes a load balancer at each tier (See Fig. 3). In our experiments, we test the ability of the CM to predict the response time of the AM (i.e., the proxy for the AS) with varying average arrival rates and number of load-balanced servers per tier. This is a typical configuration in elastic cloud computing environments in which resources are allocated and de-allocated depending on the workload.

Figures 4-5 show results for 3 servers and 5 servers per tier, respectively. The top graph in each figure shows the value of MPARD (see Eq. (9)) versus the scaling factor used to vary average arrival rates. The scaling factor is a multiplier used to scale up by the same factor the arrival rate of both classes. The initial arrival rates are 0.3 tps and 0.4 tps for classes 1 and 2, respectively. The scaling factor varies over a very wide range: up to 180 for the 2-server case, 210 for the 3-server case, and 400 for the 5-server case. For example, a scaling factor of 20 indicates that the arrival rates for classes 1 and 2 are 6 ($= 20 \times 0.3$) tps and 8 ($=20 \times 0.4$) tps, respectively. The threshold ϵ in the experiments is set to 3% (a low value for the threshold). As shown in the MPARD graphs, re-calibration was able to bring the error rate back to zero after it first surpassed the threshold ϵ.

As Figs. 4-5 demonstrate, the number of times that MPARD exceeds the threshold is pretty low despite the wide variation of the scaling factor. For example, the threshold was exceeded twice for the 3-server case and six times for the 5-server case.

The three bottom graphs of each figure show the average response times with increasing scaling factors, which illustrates the ability of the CM to predict the response times of the AM. The figures show two curves for each transaction class; one for AM and another for CM. The curves show that the CM is capable of tracking very closely (i.e., within the 3% threshold) the response times of the AM. The following ranges of the scaling factor exhibited no need for recalibration: (a) 3-server case: 1-149, 160-200; and (b) 5-server case: 1-50, 60-90, 100-130, 140-290, 300-360, 370-390.

Table 1 shows the service demands at each of the three tiers for classes 1 and 2 for the AM and CM models, 5 servers per tier, and a scaling factor of 180. The table also shows the response times for each class for AM and CM. The results are very close even though the computed and actual service demands are significantly different. The value of MAXDIFF is 1.5×10^{-4}.

We also observed that the timing and frequency of model calibration is hard to predict as the number of servers per tier increases because the individual service demands of the actual system and computed model may be different, as shown in Table 1. Therefore, when the curves for AM and CM response times take longer time to diverge, that is an indication that the corresponding service demands for the different servers in AM and CM are accurate. In Figure 4,

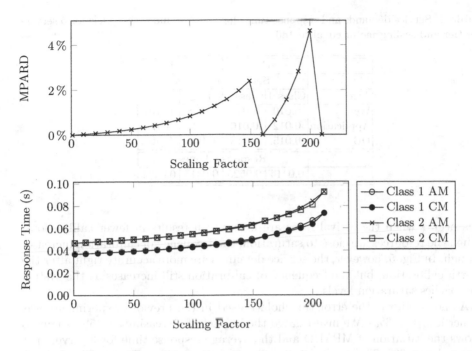

Fig. 4. 3 Servers per Tier. Top: MPARD. Bottom: Response Times for AM and CM for Classes 1 and 2. $\epsilon = 3\%$.

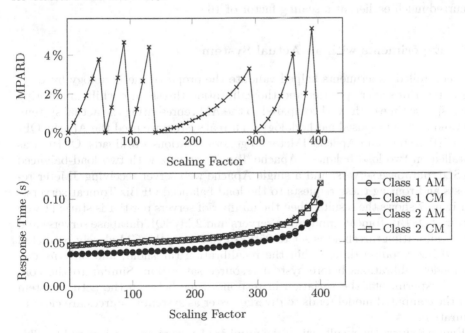

Fig. 5. 5 Servers per Tier. Top: MPARD. Bottom: Response Times for AM and CM for Classes 1 and 2. $\epsilon = 3\%$.

Table 1. Service demands and response times for the AM and CM models for 5 servers per tier and scaling factor equal to 180

	AM		CM	
	Service Demands			
Tier	Class 1	Class 2	Class 1	Class 2
Web	0.010	0.013	0.008	0.016
Application	0.012	0.016	0.021	0.015
DB	0.015	0.018	0.008	0.016
	Response Times			
	0.0414	0.0525	0.0415	0.0524

this occurs when the arrival rates are low, which results in fewer calibrations. When the resources are close to saturation, the frequency of calibration increases though. In Fig. 5, however, the service demands are more accurate right after the fourth calibration, but the frequency of calibration still increases as the system approaches saturation levels.

As noted above, the error threshold ϵ used in the previous experiments was rather low (i.e., 3%). We investigated the impact of increasing ϵ to 5%. Figure 6 shows the variation of MPARD and the average response time for 3 servers per tier and $\epsilon = 5\%$. The corresponding figure for $\epsilon = 3\%$ is Fig. 4. The first re-calibration for $\epsilon = 5\%$ occurred for a scaling factor of 200 while for $\epsilon = 3\%$ it occurred much earlier, at a scaling factor of 160.

4.2 Experiments with an Actual System

The controlled experiments helped validate the proposed methodology by varying number of servers per tier and the calibration threshold ϵ while monitoring the response time of the CM compared to the AM representing the actual system. To validate the proposed methodology on a real system we used the Apache OF-Biz ERP system and Apache JMeter to generate various workloads. OFBiz was installed on two load-balanced Apache Tomcat servers with two load-balanced MySQL database servers, and a single Apache web server receiving JMeter requests and routing these requests to the load-balanced OFBiz Tomcat servers.

Figure 7 shows the results when the number of servers per tier is static (1 web server, 2 OFBiz Tomcat application servers and 2 MySQL database servers) and the recalibration threshold is set to 25%. In this case, the CM closely predicted the OFBiz response time within the recalibration threshold, and only needed six model calibrations before system resource saturation. Similar to the controlled experiments, the variation in response time between the actual system and the computed model tends to diverge faster as system resources are close to saturation.

Figure 8 shows the results when the number of servers per tier is variable. This test case investigates the ability of our methodology to infer more information about the system architecture components by predicting the optimal number of

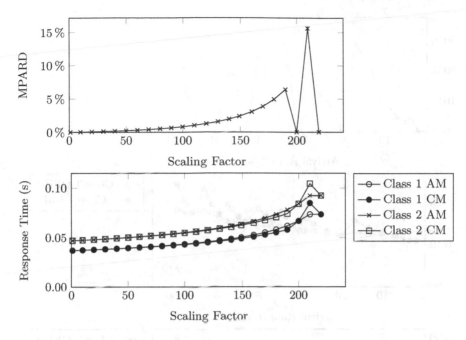

Fig. 6. 3 Servers per Tier. Top: MPARD. Bottom: Response Times for AM and CM for Classes 1 and 2. $\epsilon = 5\%$.

servers per tier that should be used in the CM to accurately represent the actual system. In this experiment, the optimizer was used to perform six model calibrations for the CM, and predicted the following number of web servers, application servers and database servers for each of the six calibrations, respectively: (1,1,2), (1,2,3), (1,2,3), (2,2,2), (2,2,2), (1,2,2), (1,1,2).

5 Related Work

Much of the related work in the fields of performance engineering and capacity planning is focused on dynamic resource allocation when a performance model is fully or partially known a priori. In this paper, we treat the system and its components as black boxes and we try to establish a relationship between system input and output in order to estimate and parameterize an analytical model that closely approximates the behavior of the system.

Some of the prior work that tackled the parameterization of analytical models includes [2,3,4], where the problem of estimating known model parameters is treated as an optimization problem that is solved using derivative-free optimization. The objective function to be optimized is based on the distance between the observed measurements and the corresponding points derived from the model. The authors point out that the main problem is determining how to couple these two sets of points in order to arrive at an objective function to be minimized. The proposed approach is applied to a small set of single queue models.

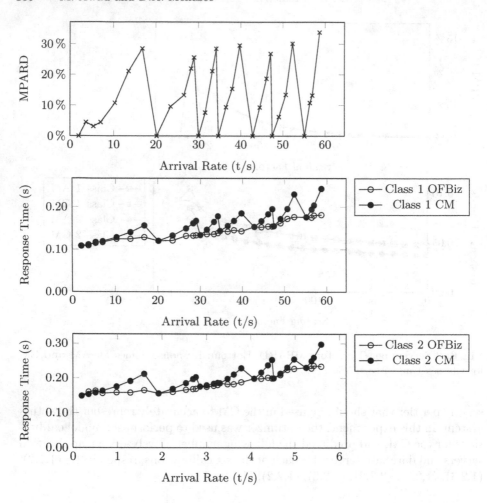

Fig. 7. Static N - Top: MPARD vs. Arrival Rate. Middle: Class 1 Transactions. Bottom: Class 2 Transactions. $\epsilon = 25\%$.

Menascé tackled the issue of model parameterization when some input parameters are already known [11]. The author proposed a closed-form solution to the case when a single service demand value is unknown, and a constrained non-linear optimization solution when a feasible set of service demands are unknown. However, that work did not propose a solution when none of the service demands are known a priori.

In [9], the authors presented a survey of performance modeling approaches focusing mainly on business information systems. The authors described the general activities involved in workload characterization, especially estimating service demands, and the various methods and approaches used to estimate it. Some of these methods include general optimization techniques, linear regression, and Kalman filters.

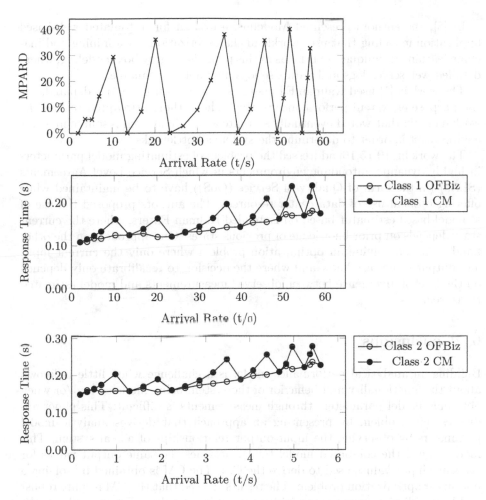

Fig. 8. Variable N - Top: MPARD vs. Arrival Rate. Middle: Class 1 Transactions. Bottom: Class 2 Transactions. $\epsilon = 25\%$.

In [7], the authors presented a method for extracting architecture level performance models in distributed component-based systems using tracing information and instrumentation to infer system components, their connections and the (probabilistic) dependency of their parameters. In contrast, our approach does not require such knowledge of system components or their relationships.

In [13], the authors presented an iterative methodology for building performance models in virtualized environments with a focus on the I/O function of storage systems. The method implemented in that paper focused on the storage component of a particular IBM mainframe system. Our approach addresses a higher level of system abstraction where the internal structure of individual servers, such as the storage system, is unknown.

In [8], the authors presented Modellus; a system for automated web-based application modeling that uses workload characterization, data mining and machine learning techniques. Our focus in this paper is on gray box modeling, where detailed web server logs and database logs may not be available.

The work in [1] used Kalman Filters to estimate resource service demands for the purpose of system performance testing. The authors attempted to find the workload mix that would eventually saturate a certain system resource in a test environment in order to determine the system's bottlenecks.

The work in [10,15,16] addressed the problem of estimating model parameters in highly dynamic autonomic environments in which Service Level Agreements (SLAs) (in the form of Quality of Service (QoS)) have to be maintained while offering optimal use of data center resources. The authors proposed the use of a model-based estimator based on Extended Kalman Filters, where the current state depends on prior knowledge of previous states. Our approach, on the other hand, relies on solving an optimization problem where only the current input and output values are know and where the decision to recalibrate only depends on the level of divergence between observed measurements and model estimated measurements.

6   Conclusions

Building an analytical performance model is a challenge when little is known about the functionality and behavior of the system being modeled and/or when obtaining model parameters through measurements is difficult. This paper addresses this problem by presenting an approach that derives analytic model parameters by observing the input-output relationships of a real system. This model, called the computed model (CM), provides the same output values for the same input values used to derive the CM. The CM is obtained by solving a non-linear optimization problem. The results showed that the CM is quite robust and has predictive power over a wide range of input values. For example, as the arrival rate of transactions for both classes was scaled by a factor varying from 1 to 400, the response times predicted by the CM only exceed the 3% error threshold six times. When the error threshold was increased to 5% (still a very low value), only two re-calibrations were needed. The ability of the CM to model the number of servers per tier of a multi-tier system is of a particular interest since it proves the ability of the CM to model system components previously unknown to it by knowing only the input and output parameters of a real information system, such as Apache OFBiz.

References

1. Barna, C., Litoiu, M., Ghanbari, H.: Autonomic load-testing framework. In: Proc. 8th ACM Intl. Conf. Autonomic Computing, pp. 91–100 (2011)
2. Begin, T., Baynat, B., Sourd, F., Brandwajn, A.: A DFO technique to calibrate queuing models. Computers & Operations Research 37(2), 273–281 (2010)

3. Begin, T., Brandwajn, A., Baynat, B., Wolfinger, B.E., Fdida, S.: Towards an automatic modeling tool for observed system behavior. In: Wolter, K. (ed.) EPEW 2007. LNCS, vol. 4748, pp. 200–212. Springer, Heidelberg (2007)
4. Begin, T., Brandwajn, A., Baynat, B., Wolfinger, B.E., Fdida, S.: High-level approach to modeling of observed system behavior. ACM SIGMETRICS Performance Evaluation Review 35(3), 34–36 (2007)
5. Bennani, M.N., Menascé, D.A.: Assessing the robustness of self-managing computer systems under highly variable workloads. In: Intl. Conf. Autonomic Computing, pp. 62–69 (2004)
6. Bennani, M.N., Menascé, D.A.: Resource Allocation for Autonomic Data Centers Using Analytic Performance Models. In: 2005 IEEE Intl. Conf. Autonomic Computing, Seattle, WA, June 13-16 (2005)
7. Brosig, F., Huber, N., Kounev, S.: Automated extraction of architecture-level performance models of distributed component-based systems. In: 26th IEEE/ACM Intl. Conf. Automated Software Engineering (ASE), pp. 183–192 (2011)
8. Desnoyers, P., Wood, T., Shenoy, P., Singh, R., Patil, S., Vin, H.: Modellus: Automated modeling of complex internet data center applications. ACM Tr. on the Web (TWEB) 6(2) (2012)
9. Kounev, S., Huber, N., Spinner, S., Brosig, F.: Model-based techniques for performance engineering of business information systems. In: Shishkov, B. (ed.) BMSD 2011. LNBIP, vol. 109, pp. 19–37. Springer, Heidelberg (2012)
10. Litoiu, M., Woodside, M., Zheng, T.: Hierarchical model-based autonomic control of software systems. ACM SIGSOFT Software Engineering Notes 30(4), 1–7 (2005)
11. Menascé, D.: Computing missing service demand parameters for performance models. In: Proc. 34th Intl. Computer Measurement Group Conf., pp. 7–12 (2008)
12. Menascé, D., Dowdy, L., Almeida, V.: Performance by Design: Computer Capacity Planning By Example. Prentice Hall (2004)
13. Noorshams, Q., Rostami, K., Kounev, S., Tuma, P., Reussner, R.: I/O Performance Modeling of Virtualized Storage Systems. In: IEEE 21st Intl. Symp. Modeling, Analysis & Simulation of Computer and Telecommunication Systems (MASCOTS), pp. 121–130 (2013)
14. Seidmann, A., Schweitzer, P., Shalev-Oren, S.: Computerized Closed Queueing Network Models of Flexible Manufacturing, Large Scale System. J. North Holland 12, 91–107 (1987)
15. Woodside, M., Zheng, T., Litoiu, M.: The use of optimal filters to track parameters of performance models. In: Second Intnl. Conf. Quantitative Evaluation of Systems, pp. 74–83 (2005)
16. Zheng, T., Yang, J., Woodside, M., Litoiu, M., Iszlai, G.: Tracking time-varying parameters in software systems with extended Kalman filters. In: Proc. 2005 Conf. of the Centre for Advanced Studies on Collaborative Research, pp. 334–345 (2005)

Deriving Work Plans for Solving Performance and Scalability Problems

Christoph Heger and Robert Heinrich

Karlsruhe Institute of Technology, 76131 Karlsruhe, Germany
{christoph.heger,robert.heinrich}@kit.edu

Abstract. The performance of an enterprise application (e.g. response time, throughput, or resource utilization) is an important quality attribute that can have a significant impact on a company's success. When a performance problem such as a performance bottleneck has been detected, the root cause identified and a solution proposed, developers have to identify the elements of the application often manually that will undergo changes and determine how these elements must be changed in order to implement the solution. Many existing approaches are able to identify the elements that have to be modified but only few are able to determine the necessary types of changes on these elements. Neither of the approaches supports developers with a work plan sketching the implementation steps. In this paper, we propose an approach to point developers the way torwards an implementation of a performance or scalability solution with an ordered set of work activities. Rules are used to derive a work plan sketching the implementation of a solution for the particular application based on an initial set of work activities. The rule-based approach identifies impacted elements and determines how they should be changed. We demonstrate the proposed approach with a solution of a performance bottleneck as an example.

Keywords: Software Performance Engineering, Solution Implementation Support, Rules, Impact Propagation.

1 Introduction

The performance (i.e. resource usage, timing behaviour and throughput) of an information system directly influences the total cost of ownership (TCO) as well as the user satisfaction which are highly business critical metrics. Performance problems can be caused by various modifications such as error corrections, improvements, extensions or variations in user quantity and behaviour as well as changing requirements. After a performance problem has been detected and the root cause has been isolated, a performance expert often proposes a solution in form of abstract work activities like splitting an interface that have to be implemented in order to solve the problem. A solution can affect various parts of the application such as the architecture, implementation and/or configuration. Therefore, all elements (including possible side-effects) have to be identified that

A. Horváth and K. Wolter (Eds.): EPEW 2014, LNCS 8721, pp. 104–118, 2014.

will undergo changes as well as the necessary types of changes for these elements. This time-consuming task is often done manually by developers or performance experts in advance to get an understanding of the implementation effort of a solution. Such cost factors can then be considered to select the most appropriate solution when a variety of solutions exists.

Currently, developers are not supported in this regard at the implementation level. Existing approaches for solving performance problems, for example [10,5,28,31], are model-based. Only Jing Xu considers in [31] to use the effort estimation of the designer for the necessary design model changes in selecting a solution among alternatives and suggests what should be changed, and in what way, but not how to do it. The approaches neither consider an existing code base nor support of the developer for implementing a solution at the implementation level. Existing impact propagation approaches on the other hand can identify and classify following changes and the impacted elements (side-effects) based on an initial set of changes [19] but neither of the approaches determines a work plan for developers describing the essential work activities of a change's implementation.

In light of these observations, we propose an approach to support developers in implementing a performance or scalability solution at the implementation level by deriving a work plan for the particular solution and application based on an initial set of work activities as a part of Vergil. Vergil is an approach to guide developers from a performance or scalability issue to solutions, by providing hypotheses about what to change, evaluating the changes in the context of the particular application and ranking the solutions to support developers in making a decision. The two major challenges in building a work plan are to identify the impacted elements and to determine how they are actually impacted; respectively how they should be changed. Vergil considers changes on the architecture level, implementation level and configuration level of the application. Architecture level changes are evaluated with architecture level performance models, like the Palladio Component Model [7] or Marte [1] for UML, in contrast to implementation level changes that are evaluated with a system under test and measurement-based experiments. Due to the different levels of abstraction, work activities have to be traced down from the architecture level to the implementation level while developers are often most familiar with the implementation artifacts of an application.

The proposed approach for building work plans uses rules to propagate the impact within and between the architecture performance model and the source code model of an application. The impact propagation uses a correspondence model describing the equality of elements in the architecture performance model and the source code model. Work plans describe which elements of the application are impacted and how they should be changed. This offers three benefits for developers: (1) they are aware of how to change the application, (2) they are able to estimate the implementation effort based on work activities, and (3) they can discuss alternative solutions based on the type and number of impacted elements, and the types of changes required to implement a certain solution.

We demonstrate the approach using the solution of the "GOD" class antipattern [25] as an example in the context of the MediaStore [7] application. We use the Palladio Component Model as the architecture performance model and Java as programming language. Both are established and relevant technologies in industrial practice. Overall, in this paper, we make the following contributions:

1) We propose an approach to derive work activities sketching the implementation of a solution based on impact propagation between model elements.

2) We demonstrate the applicability of the approach with an example.

The remainder of the paper is structured as follows: In Section 2, we present the foundations of our approach. In Section 3, we give an overview of Vergil to position the content of this paper in the overall approach. In Section 4, we introduce the MediaStore example. We present the approach for building Vergil's work plans in Section 5. In Section 6, we discuss related work and conclude the paper in Section 7.

2 Foundations

The concept of work plans is inspired by the idea of the Karlsruhe Architectural Maintainability Prediction (KAMP) [26,27] approach to estimate the change effort based on a work plan. The goal of KAMP is to compare architectural design alternatives, which are represented by instances of the Palladio Component Model (PCM) [7], by estimating the effort of a change request in the context of each alternative. PCM is a software architecture simulation approach to analyze software at the model level for performance bottlenecks and scalability issues. It enables software architects to test and compare various design alternatives without the need to fully implement the application or buying expensive execution environments. PCM has already been used to detect and solve performance problems [28].

KAMP combines a top-down phase to determine the work activities and to create the work plan, with a bottom-up phase in which developers assign an effort estimation to each work activity in the plan. A work plan is a hierarchical structured collection of work activities and is stepwise refined into small tasks by identifying resulting changes and describing high-level changes on a lower level of abstraction. KAMP relies on the following assumptions: (a) change efforts must take into account all artifacts of system development and operation. Focusing only on code is not sufficient, (b) there are specified change scenarios, and (c) it is easier to estimate costs of small specific tasks than of coarse-grained tasks. We also rely on this assumptions.

The change estimation process starts with identifying model elements directly affected by a change request, such as an interface or component. To identify resulting changes, the direction of the change propagation is reversed to the direction of the reference-relation of architectural elements (architectural elements which refer to (or use) other elements are related by a reference-relation). In the case of include-/contains-relations (architectural elements which are contained in each other), a change of the inner element is propagated to the outer element and any change of the outer element is propagated to inner elements resulting in a refined set of work activities. Additionally, KAMP considers also other work activities like running unit or integration tests or deploying components [26].

The major difference between KAMP and our proposed approach is that KAMP operates on models only while our approach also considers the code basis. This allows for traceability and change impact propagation covering both, code and models. KAMP results in an unsorted set of tasks that have to be executed to realize a certain change request. As KAMP targets at effort estimation, there is no need to arrange the tasks in an order. In contrast, the proposed approach aims at creating a work plan that can be followed by developers through an ordered list of work activities.

3   Vergil Overview

We are currently developing the Vergil approach. The main goal of Vergil is the provisioning of solutions (e.g. to split an interface or to move functionality to a certain component) to developers for solving performance and scalability problems. Vergil combines the advantages of model-based performance improvement approaches like [10,31,5] and extends them with the introduction of measurement-based performance problem solution at the implementation level by means of monitoring-driven testing techniques and work plans sketching the implementation of the solution at the implementation level. The knowledge of how to change a system is formalized in rules (henceforth called change hypotheses) that are explored, evaluated and rated. The process consists of four major activities as shown in Figure 1 as BPMN diagram [14]. In the context of this paper, we are focussing only on the concepts of the work plans and the activities *Propagate Work Activities* and *Estimate Effort of Work Plans*.

The process starts with the *Extract Models* activity [22] that takes the source code of the application as input. The source code is parsed into the Source Code Model (SCM). An architecture performance model (APM) is extracted from the source code or when such a model already exists an existing one is imported. The APM provides an architectural view of the application and is used to evaluate architectural change hypotheses in the remainder of the process. During extraction, a correspondence model (COM) is build that links model elements in the APM and SCM that correspond to each other like interfaces. The extracted models are forwarded to the *Explore Change Hypotheses* subprocess.

The *Explore Change Hypotheses* subprocess takes the SCM, COM, APM and the Performance Problem Model (PPM) as input. The performance problem is formalized as symptom trace through the application and results in the PPM. A symptom like high response time or high CPU utilization references an SCM element where it can be observed such as a method and a description of the workload and usage profile as formalized in [29]. The subprocess starts with the *Test Change Hypotheses* activity that takes the change hypotheses, the performance requirements, the test environment and the models as input. In this activity, the applicability of change hypotheses is tested and the effect of change

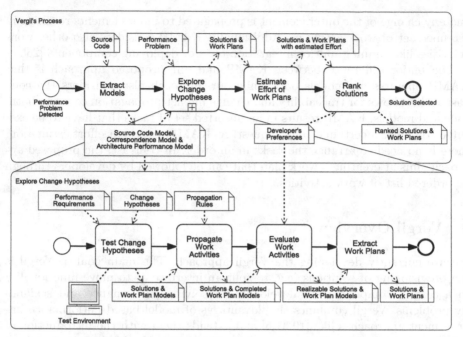

Fig. 1. Vergil Overview

hypotheses is evaluated to build solutions [31]. Change hypotheses provide the knowledge about what can be changed to solve a performance problem. It consists of a precondition that must be fulfilled in order to be applicable and a set of transformation rules that apply the changes on the defined level of abstraction (e.g. APM, SCM, etc.). Each change hypothesis also has a postcondition testing if the desired effect has taken place as well as the work plan model template. A condition can consist of any number of structural (on the SCM, PPM and APM model) and behavioral (on measurement or prediction results) conditions testing static and dynamic requirements of the hypothesis. The conditions are rules expressed in first-order logic. First-order logic has already been used before in literature to formalize performance antipatterns [11]. The exploration algorithm selects sets of change hypotheses with fulfilled precondition and evaluates their effect through instantiating the changes in the context of the particular application and on the hypothesis' level of abstraction (e.g. APM, SCM, etc.) and evaluates the performance. To give an example, two approaches for automated model refactoring for solving performance problems are presented in [4] and [31]. The performance evaluation is done by calibrating and simulating the APM instance or executing measurements with the system under test. The postcondition is tested with the returned results and if fulfilled, the impacted elements of the application are identified as well as how they are actually impacted building the Work Plan Models (WPMs) and its initial work activities based on the templates. The performance evaluation and the postcondition also ensure, that the changes do not lead to a performance degradation [3]. If the evaluation results of a single change hypothesis do not satisfy the performance requirements (for example response time, throughput, or CPU utilization), the current algorithm uses backtracking (as suggested by Arcelli and Cortellessa [3]) to find composite solutions (combinations of two or more change hypotheses) that fulfill the performance requirements. The solutions and the corresponding WPMs fulfilling the requirements are forwarded to the *Propagate Work Activities* activity. Propagation rules and the WPM of each solution are used to identify all impacted elements of a solution and to determine the necessary type of work activity for each element to complete the WPM.

The solutions and completed WPMs are forwarded to the *Evaluate Work Activities* activity. In this activity, the work activities and their referenced elements are validated that they can be changed [3] based on the information in the developer's preferences. In the developer's preferences, developers express what they prefer to change (e.g. configuration of a component, implementation etc.) and what they are unwilling to change (or cannot change), for example the architecture or particular parts of the application. If elements are impacted that the developer is unwilling to change or cannot change such as legacy systems or the architecture, the solution is discarded. Arcelli and Cortellessa raised the concern to take constraints such as costs and legacy constraints (e.g. the database cannot be changed) into account when proposing solutions [3]. The remaining solutions and WPMs are forwarded to the *Extract Work Plans* activity in which the WPMs are translated from their graph-based structure into an ordered list of work activities for the developers. The solutions and work plans are then forwarded to the *Estimate Effort of Work Plans* activity.

In the *Estimate Effort of Work Plans* activity, the developers are asked for an estimation of the effort they will need to complete the work plan. The effort estimation is a manual task done by the developers themself. The effort can vary between individual developers depending on their knowledge, experience and practice. Hence, it is possibly unreliable when done automatically. The effort is provided as unitless quantities, leaving the decision of the concrete unit of measurement by the developers, and is provided for all atomic work activities. The unit of measurement has to be the same for all work activities. The effort can be estimated, for example, in person (-hours, -days, or -months) [31], or story points (in the context of agile development such as SCRUM [24]) as well as function points which is also an established means for effort estimation [2]. In the case of experienced development teams, that have historic data from previous development projects for cost model calibration, the usage of a cost model such as COCOMO [8] is also possible. The process uses the given unitless quantities of each atomic work activity such as adding a method to an interface as input to compute the total effort estimation for a work plan. The estimated effort addresses the concern of taking the costs of solution alternatives into account [3,31]. The solutions and the work plans with estimated effort are forwarded to the *Rank Solutions* activity.

In the *Rank Solutions* activity, a multi-criteria decision analysis to rank the solutions is done. This activity addresses the issue of taking costs and constraints into account in deciding on an appropriate solution when a variety of choices exist [3,31]. The solution rating is done with a combination of the Simple Multi-Attribute Rating Technique (SMART) [12] and the Analytic Hierarchy Process (AHP) [23] taking the performance impact, cost factors, constraints and the developer's preferences into account. Developers are then able to discuss the proposed solutions based on the work plans, the impacted elements—and how they are actually impacted, the costs, and the estimated performance improvement. The selected solution and its work plan are the final result of the process.

4   MediaStore Example

One of the 14 notion- and domain-independent software performance antipattern defined by Smith and Williams is the "GOD" class [25]. The antipattern describes the problem of poorly distributed application intelligence when one class is performing all the work or contains all the application's data. A proposed solution [25] is to employ the locality principle and to move the functionality close to where it is needed. We have already investigated the "GOD" class antipattern and shown how it is automatic detectable with systematic experiments based on measurements in our previous work [30].

In this section, we use the "GOD" class as motivating example with the MediaStore [7] application. The MediaStore allows its users to upload and store audio files as well as to download audio files encoded in a less or equal audio bit rate compared to the uploaded one. The MediaStore is implemented with Java Enterprise Edition. The "GOD" class in the example is the `MediaStoreBean` that is accessed from the `WebGUIBean` and provides all the functionality as shown in Figure 2 (the Java source code elements are shown in UML notation for the sake of illustration). We omitted other components of the MediaStore application which are irrelevant for the example for the sake of simplicity. A detailed overview on the application is given in [7]. Figure 2 is devided into the *Current State* and *Current Deployment* showing the state of the application with the problem and the *Target State* and *Target Deployment* showing the application with the solution. The `WebGUIBean` and the `MediaStoreBean` are deployed on different servers. The `WebGUIBean` has to communicate with the `MediaStoreBean` to register or login users causing high response times for both operations. The change hypothesis of Vergil is to split the interface `IMediaStoreBean` and to move the functionality closer to the `WebGUIBean`. The hypothesis evaluates the changes on a higher level of abstraction for simplicity and uses a PCM of the MediaStore application. The automated refactoring of architectural models for solving performance problems has been shown, for example, in [4]. The change hypothesis provides the work activities to split the `IMediaStoreBean` interface, to add a new interface (in this example the `IUserManagementBean` interface), to move the methods `register` and `logIn` to the new interface, and to update the deployment descriptor of the application. The propagation of the provided work activities in the source code adds the work activities to split the `MediaStoreBean` class, to add a new class (in this example the `UserManagementBean`), to move

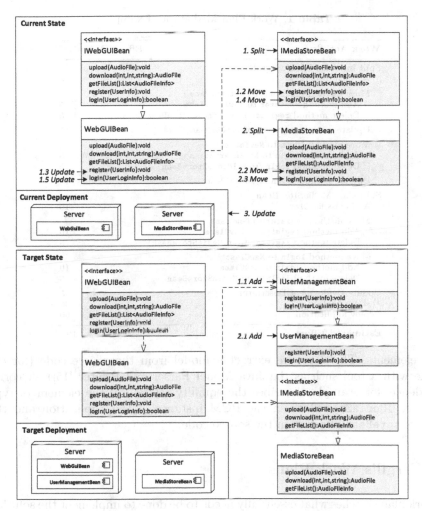

Fig. 2. MediaStore Example Overview

the methods implementing interface methods to the new class, and to update the methods of the WebGUIBean to call the corresponding methods of the new interface. The necessary work activities and their proposed order are shown in Figure 2. The determined work plan for the refactoring is shown in Table 1.

In this work plan, there are composite activities (*split*, *move*), that can consist of other composite activities or atomic activities (*add*, *update*, *delete*). All composite activities are broken down until they are expressed by atomic activities. The follow-up activities (*test* and *deploy*) result from the atomic and composite work activities.

In the scenario above, we initially have an architectural change proposed and embodied in the change hypothesis. Vergil uses the PCM as starting point for building the work plan of architectural changes. We propagate the impact to the implementation using the correspondence between model elements and source

Table 1. Work Plan MediaStore Example

Work Activity	Effort [Minutes]
Split interface IMediaStoreBean	
Add `NewInterface`	10
Move method `register` to `NewInterface`	
Add method `register` to `NewInterface`	5
Delete method `register` from `IMediaStoreBean`	5
Update method `register` of `WebGUIBean`	15
Move method `logIn` to `NewInterface`	
Add method `logIn` to `NewInterface`	5
Delete method `logIn` from `IMediaStoreBean`	5
Update method `logIn` of `WebGUIBean`	15
Split class MediaStoreBean	
Add class `NewClass`	10
Move method `register` to `NewClass`	
Add method `register` to `NewClass`	10
Delete method `register` from `MediaStoreBean`	5
Move method `logIn` to `NewClass`	
Add method `logIn` to `NewClass`	10
Delete method `logIn` from `MediaStoreBean`	5
Update Deployment Descriptor	5
Test application	30
Deploy application	60
Estimated Effort	**195**

code elements. Therefore, we extract a model from the source code (for example with a tool such as the Java Model Parser and Printer [15]). A correspondence, for example, describes the equality relation of the element of type interface IMediaStoreBean in the PCM instance of the application and the IMediaStoreBean interface in the source code.

5   Vergil's Work Plans

A work plan sketches what essentially needs to be done to implement the solution by modelling abstract work activities without prescribing (and limiting the developer) on how the solution is concretely implemented in the application. A work plan is an ordered set of work activities. A work acitivity can be atomic such as add, delete, or update an element like a class, interface, or method, or composite such as split, move, merge, swap, or replace elements. The concept of work activities is inspired by the taxonomy of change types which has been introduced by Lehnert et al. in [18]. This taxonomy is similar to the work activity concept used in KAMP. Both approaches consider a graph-based representation of the software artifacts and use atomic operations and composite operations to categories modifications. The formalization of the work plan is shown in Figure 3. The work plan also lists follow-up activities such as redeployment or testing activities.

A composite activity can be composed of other composite and atomic activities. Refinement rules are used to break composite activities in the work plan down into atomic activities. For example, the composite work activity "Move" to move a method from one interface to a new one consists of the atomic work

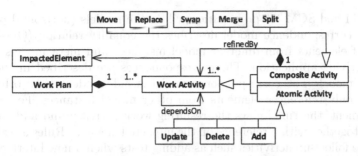

Fig. 3. Work Plan meta-model

activities "Add" to add the method to the new interface and "Delete" to delete the method from the old interface. The evolution of the work plan model through the application of such a refinement rule for the latter work activitiy is shown as an example in Figure 4. The underlying graph transformation rule matches elements of type Method in the SCM that are referenced from a *Move* work activitiy but are not already referenced by a *Delete* work activitiy. For all matches, the rule adds the *Delete* work activity as refinement expressed through the added *refinedBy* relation to the work plan.

The *refinedBy* and *dependsOn* relations between work activities in the work plan model instance are used to extract the ordered list of work activities for the developers. An activity like "Split" that has a *refinedBy* relation to an "Move" activity is added as child of that activitiy in the work plan. An activitiy like "Split" that has a *dependsOn* relationship to another "Split" activity is added after that activity in the work plan.

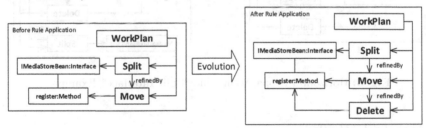

Fig. 4. Refinement Rule "Delete Method" Example

The initial work activities are provided by the change hypothesis as work plan template and the impacted elements are determined by the instantiation of the changes.

The impact on an element can cause an impact on other elements that are in a relation. To identify all elements and how they are impacted, Vergil uses impact propagation rules to identify additional impacted elements that are in a relation to already impacted elements and the work aktivities in the work plan referencing the impacted elements. The propagation uses the work plan model as starting point and correspondences of elements in different model instances for propagation between APM and SCM (vertical propagation) and the relation within APM and SCM for propagation wihtin the models (horizontal propagation). The correspondence model describes the equality relation (One-To-One relation) of elements from different model instances and meta models but the same underlying application. The correspondences are described in the correspondence model. The correspondence model is independent from other meta models. It only references elements from other model instances. For each impacted element, the rule knows the resulting work activitiy and adds it to the work plan together with a reference on the impacted element. Rules are also used to conclude follow-up activities such as adding tests when a new interface is created or the redeployment of components when elements of the implementation of that component are impacted.

For example, when an interface in SCM is referenced from a *Split* work activity, then the rule knows that a class implementing that interface has to be splitted too. Figure 5 shows the evolution of the work plan model through applying the rule for the split interface work activity as an example. The rule matches all classes that are not referenced from a *Split* work activity and that implement the interface which is referenced from a *Split* work activity. For all matches, the rule adds the *Split* work activity to the work plan model and a reference to the impacted class.

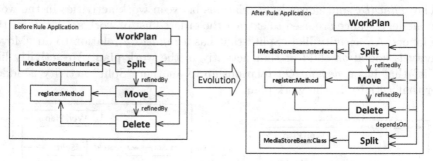

Fig. 5. Horizontal Propagation Rule "Split Interface" Example

6   Related Work

In this section, we review impact propagation approaches for their support of determining work plans to implement solutions and we discuss common effort estimation approaches.

Impact Propagation. Most approaches that have been proposed to assess the propagation of change impacts are limited to source code as shown by a recent study [17] and are often able to identify impacted elements only. Some of the proposed rule-based approaches, such as the approaches of Keller et al. [16] or Briand et al. [9], are able to classify how the impacted element is actually impacted and has to be changed while others are not able to detect impacts in heterogeneous software artefacts like source code and models. The research conducted by Lindvall and Sandahl [20] outlines the applicability of traceability relations for impact analysis. The use of correspondence relations between different views and viewpoints has been shown by Eramo et al. in [13]. A more detailed review of existing approaches and their limitations is provided in [19]. An introduction to the topic of change impact analysis discussing techniques and problems is given in [6].

The impact analysis approach of Lehnert et al. [19] combines impact analysis, multi-perspective modelling, and horizontal traceability to determine further change propagation. The impact propagation technique is based on the type of dependency which exists between EMF-based models and the type of change which is applied on one of the model elements. The underlying hypothesis is that the change type, dependency type and the types of involved artifacts determine if and how a change ripples to related artifacts. Therefore, a set of impact propagation rules is used to identify all impacted elements in a recursive manner. The rules are derived, as example, from relations defined in the meta-models of artifacts such as inheritance-relations between classes or implementation-relations between classes and interfaces as well as from Correspondences according to design methodologies such as the equivalence between UML and Java classes or between UML and Java packages [19]. The type of changes is expressed through a taxonomy of change types comprising of atomic operations (*add, delete, update*) and composite operations (*move, replace, split, merge, swap*), where the latter can consist of sequences of atomic and composite operations. Each rule receives the changed element, the type of change, and a list of all related elements as input. From this input, a list of all impacted element(s) along with the resulting change type(s) is computed. This output is then again fed into the impact analysis process. The propagation takes the dependency relation into account to limit the size of impact sets. The rules determine how impacted elements are actually impacted [19].

The main objective of impact propagation approaches is to identify the impacted elements. Few of them also deal with the determination of how an element must be changed. Neither of the approaches has the objective to build a model that abstractly models the implementation of the changes with the necessary work activities. Our proposed approach extends the concepts of existing impact propagation approaches, i.e. [19], by building such a model describing the refactoring of the application that can be transfered into a work plan for developers.

Effort Estimation. Three categories of related work on change effort estimations can be distinguished – complexity metrics (esp. of source code), estimations based on the extent of changes in requirements, and architecture-based procedures. One of the most common complexity metric for software systems is the cyclomatic complexity [21]. Complexity metrics only take into account the structure of the system. They can be elicited automatically but their conversion into effort or costs is unclear and empirically not consistent proven. There are various approaches to initial effort estimation of software development projects based on requirements, cf. algorithmic effort estimation and effort-based project planning [8]. Changes to requirements are triggers for changes in the system but drawing inference from the extent of changes in requirements about efforts for implementing the changes is not possible without considering the system architecture. Some existing approaches target at scenario-based software architecture analysis but lack a formalized architecture description or are limited to software development but do not take into account management tasks. An overview of related work on architecture-based effort estimation is given in [26]. KAMP combines several strengths of existing approaches. It makes explicit use of formal software architecture models, provides guidance and automation via tool support, and considers development as well as management effort. KAMP evaluates maintainability for concrete change requests. It estimates change efforts using semi-automatic derivation of work plans, bottom-up effort estimation, and guidance in investigation of estimation supports (e.g., design and code properties, team organization, development environment, and other influence factors).

In our proposed approach, we use the work plan-based concept of KAMP as the foundation for effort estimation by developers to consider cost factors in making a decision among solution alternatives as well as to build a model of the refactoring. We apply the concept of work activities not only on the architecture performance model as it is currently provided by KAMP but also on implementation level artifacts. We also introduce relations in the work plan model that are used to extract an ordered list of work activities.

7 Conclusion

Sketching the implementation of a solution is an essential part of guiding developers to the solution of performance and scalability problems. Vergil's work plans aim to provide this support for the proposed solutions. The work plans as an ordered list of work activities guide developers without prescribing how the implementation is concretely realized. Developers are able to discuss and compare solution alternatives based on the impacted elements and the necessary type of change, they are aware of how they have to change the application and which parts of the application are affected and they can estimate the implementation effort for each solution before making a decision on which solution will be implemented. Vergil uses rules and a graph-based representation of the application to determine the necessary work activities and to build the work plans. We demonstrated the approach with an example while the validation of the approach is part of our current research. We plan to conduct an emperical case study to validate the approach with a group of developers at SAP.

Due to the graph-based approach, we are not limited to a particular architecture performance model like the Palladio Component Model or to a specific programming language like Java as long as a graph-based representation is available. Rules may have to be adjusted and new ones have to be created in order to take other models such as UML or Entity Relationship diagrams into account that describe the application from a different perspective.

We plan to implement the proposed approach in the context of Vergil's framework that is currently under development. Vergil's automated work plan builder will then be used in the validation of the overall approach.

Acknowledgements. We would like to thank Alexander Wert, Roozbeh Farahbod, Michael Langhammer and Max Kramer for the discussions and their feedback. We would also like to thank Jonas Kunz and Sven Kohlhaas for their support. This work is supported by the German Research Foundation (DFG), grant RE 1674/6-1, and the Priority Programme SPP1593: Design For Future – Managed Software Evolution.

References

1. UML Marte, http://www.omgmarte.org
2. Albrecht, A.J.: Measuring application development productivity. In: Proceedings of the Joint SHARE/GUIDE/IBM Application Development Symposium, vol. 10 (1979)
3. Arcelli, D., Cortellessa, V.: Software model refactoring based on performance analysis: Better working on software or performance side? In: FESCA (2013)
4. Arcelli, D., Cortellessa, V., Di Ruscio, D.: Applying model differences to automate performance-driven refactoring of software models. In: Balsamo, M.S., Knottenbelt, W.J., Marin, A. (eds.) EPEW 2013. LNCS, vol. 8168, pp. 312–324. Springer, Heidelberg (2013)
5. Arcelli, D., Cortellessa, V., Trubiani, C.: Antipattern-based model refactoring for software performance improvement. In: Proceedings of the 8th International ACM SIGSOFT Conference on Quality of Software Architectures. ACM (2012)
6. Arnold, R.S., Bohner, S.A.: Software Change Impact Analysis. IEEE Computer Society Press (1996)
7. Becker, S., Koziolek, H., Reussner, R.: The palladio component model for model-driven performance prediction. Journal of Systems and Software 82(1) (2009)
8. Boehm, B., Clark, B., Horowitz, E., Westland, C., Madachy, R., Selby, R.: Cost models for future software life cycle processes: Cocomo 2.0. Annals of Software Engineering 1(1) (1995)
9. Briand, L.C., Labiche, Y., O'sullivan, L.: Impact analysis and change management of uml models. In: Proceedings of the International Conference on Software Maintenance, ICSM 2003. IEEE (2003)
10. Cortellessa, V., Di Marco, A., Eramo, R., Pierantonio, A., Trubiani, C.: Approaching the model-driven generation of feedback to remove software performance flaws. In: 35th Euromicro Conference on Software Engineering and Advanced Applications, SEAA 2009. IEEE (2009)
11. Cortellessa, V., Di Marco, A., Trubiani, C.: An approach for modeling and detecting software performance antipatterns based on first-order logics. Software and Systems Modeling (2012)
12. Edwards, W.: How to use multiattribute utility measurement for social decision making. IEEE Transactions on Systems, Man and Cybernetics 7(5) (1977)
13. Eramo, R., Pierantonio, A., Romero, J.R., Vallecillo, A.: Change management in multi-viewpoint system using asp. In: 2008 12th Enterprise Distributed Object Computing Conference Workshops. IEEE (2008)
14. O.M. Group. Business process model and notation, bpmn (2011)
15. Heidenreich, F., Johannes, J., Seifert, M., Wende, C.: Jamopp: The java model parser and printer. Techn. Univ., Fakultät Informatik (2009)
16. Keller, A., Schippers, H., Demeyer, S.: Supporting inconsistency resolution through predictive change impact analysis. In: Proceedings of the 6th International Workshop on Model-Driven Engineering, Verification and Validation, p. 9. ACM (2009)

17. Lehnert, S.: A review of software change impact analysis. Ilmenau University of Technology, Tech. Rep. (2011)
18. Lehnert, S., Farooq, Q., Riebisch, M.: A taxonomy of change types and its application in software evolution. In: 2012 IEEE 19th International Conference and Workshops on Engineering of Computer Based Systems (ECBS) (April 2012)
19. Lehnert, S., Riebisch, M., et al.: Rule-based impact analysis for heterogeneous software artifacts. In: 2013 17th European Conference on Software Maintenance and Reengineering (CSMR). IEEE (2013)
20. Lindvall, M., Sandahl, K.: Traceability aspects of impact analysis in object-oriented systems. Journal of Software Maintenance: Research and Practice 10(1) (1998)
21. McCabe, T.J.: A complexity measure. In: Proceedings of the 2nd International Conference on Software Engineering, ICSE 1976, Los Alamitos, CA, USA. IEEE Computer Society Press (1976)
22. Parsons, T.: Automatic Detection of Performance Design and Deployment Antipatterns in Component Based Enterprise Systems. PhD thesis, University College Dublin (2007)
23. Saaty, T.: The analytic hierarchy and analytic network processes for the measurement of intangible criteria and for decision-making. In: Multiple Criteria Decision Analysis: State of the Art Surveys. Springer, New York (2005)
24. Schwaber, K., Beedle, M.: Agile Software Development with Scrum. Pearson (2002)
25. Smith, C.U., Williams, L.G.: Software performance antipatterns. In: Workshop on Software and Performance (2000)
26. Stammel, J., Reussner, R.: Kamp: Karlsruhe architectural maintainability prediction. In: Proceedings of the 1st Workshop des GI-Arbeitskreises Langlebige Softwaresysteme (L2S2): Design for Future-Langlebige Softwaresysteme (2009)
27. Stammel, J., Trifu, M.: Tool-supported estimation of software evolution effort in service-oriented systems. In: Joint Proceedings of the First International Workshop on Model-Driven Software Migration (MDSM 2011) and the Fifth International Workshop on Software Quality and Maintainability (SQM 2011), vol. 708 (2011)
28. Trubiani, C., Koziolek, A.: Detection and solution of software performance antipatterns in palladio architectural models. In: ICPE (2011)
29. van Hoorn, A., Rohr, M., Hasselbring, W.: Generating probabilistic and intensity-varying workload for web-based software systems. In: Kounev, S., Gorton, I., Sachs, K. (eds.) SIPEW 2008. LNCS, vol. 5119, pp. 124–143. Springer, Heidelberg (2008)
30. Wert, A., Oehler, M., Heger, C., Farahbod, R.: Automatic Detection of Performance Anti-patterns in Inter-component Communications. In: Proceedings of the 10th International Conference on Quality of Software Architecture, QoSA 2014 (2014)
31. Xu, J.: Rule-based automatic software performance diagnosis and improvement. Performance Evaluation 69(11) (2012)

Dealing with Zero Density Using
Piecewise Phase-Type Approximation*

Ľuboš Korenčiak[1], Jan Krčál[2], and Vojtěch Řehák[1]

[1] Faculty of Informatics, Masaryk University, Brno, Czech Republic
{korenciak,rehak}@fi.muni.cz
[2] Saarland University – Computer Science, Saarbrücken, Germany
krcal@cs.uni-saarland.de

Abstract. Every probability distribution can be approximated up to a given precision by a phase-type distribution, i.e. a distribution encoded by a continuous time Markov chain (CTMC). However, an excessive number of states in the corresponding CTMC is needed for some standard distributions, in particular most distributions with regions of zero density such as uniform or shifted distributions. Addressing this class of distributions, we suggest an alternative representation by CTMC extended with *discrete-time* transitions. Using discrete-time transitions we split the density function into multiple intervals. Within each interval, we then approximate the density with standard phase-type fitting. We provide an experimental evidence that our method requires only a moderate number of states to approximate such distributions with regions of zero density. Furthermore, the usage of CTMC with discrete-time transitions is supported by a number of techniques for their analysis. Thus, our results promise an efficient approach to the transient analysis of a class of non-Markovian models.

1 Introduction

In the area of performance evaluation and probabilistic verification, discrete-event systems (DES) are a prominent modelling formalism. It includes models such as continuous-time Markov chains, stochastic Petri nets, or generalized semi-Markov processes. A DES is a random process that is initialized in some state and then moves from state to state in continuous-time whenever an event occurs. Every time a state is entered, some of the events get *initiated*. An initiated event then occurs after a delay chosen randomly according to its distribution function. When no restrictions on the distribution functions are imposed, analysis of these models is complicated [8,21], one often resorts to simulation [17]. When all the distributions F_e are exponential, the DES is then called

* This work is supported by the EU 7th Framework Programme under grant agreements 295261 (MEALS) and 318490 (SENSATION), Czech Science Foundation grant No. P202/12/P612, the DFG Transregional Collaborative Research Centre SFB/TR 14 AVACS, and by the CAS/SAFEA International Partnership Program for Creative Research Teams. Access to computing and storage facilities owned by parties and projects contributing to the National Grid Infrastructure MetaCentrum, provided under the programme "Projects of Large Infrastructure for Research, Development, and Innovations" (LM2010005), is greatly appreciated.

A. Horváth and K. Wolter (Eds.): EPEW 2014, LNCS 8721, pp. 119–134, 2014.
© Springer International Publishing Switzerland 2014

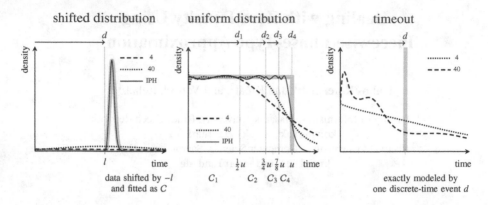

Fig. 1. Three usages of the discrete-time events for PH approximation. In the figures there are the densities (with thick grey lines), their standard PH approximations with 4 and 40 phases, and their IPH approximation with 30 phases. On the left, the discrete-time event d postpones the start of the CTMC C fitted to the area of positive density. On the right, the discrete-time event can be used directly, instead of its continuous approximation. In the middle, 3 discrete-time events split the support into 4 intervals with different approximations C_1, C_2, C_3, and C_4. Note that for a distribution with a steep change in density at its upper bound (such as the uniform distribution), PH fitting performs well on the first *half* of the support; logarithmic partitioning into intervals works better than equidistant.

a continuous-time Markov chain (CTMC) for which many efficient analysis methods exist [26,4] thanks to the memoryless property of the exponential distribution. Hence, an important method for analysing DES is to *approximate* it by a CTMC using *phase-type* (PH) approximation and to solve resulting CTMC analytically. Roughly speaking, each event e such that its distribution function is not exponential is replaced by a small CTMC C_e. This CTMC has a designated absorbing state such that the time it takes to reach the absorbing state is distributed as closely as possible to the given distribution function. A well known result [34] states that any continuous probability distribution can be fitted up to a given precision by the PH approximation. Nevertheless, the closer the approximation, the more states it requires in the CTMC. For some lower bounds on the number of required states see, e.g., [1,35,13,12].

In this paper we propose another approach for approximating probability distributions where phase-type requires extreme amount of states to be fitted precisely [35,12]. In particular, we deal with distributions often encountered in practice that we call *interval distributions* and that are supported on a proper subinterval of $[0, \infty)$. For example distributions of events that cannot occur before time $l > 0$ such as due to physical limits when *sending a packet*; or that cannot occur after time $u < \infty$ such as *waiting for a random amount of time* in a collision avoidance protocol; or that occur exactly after time $l = u$ such as *timeouts*. We address these interval distributions by an approach that we call *Interval phase-type (IPH)* approximation. The crucial point is that it allows to separate the discrete and the continuous nature of these distributions by enriching the output formalism. Along with the exponential distribution of the CTMC we allow discrete-time events (also called fixed-delay, deterministic, or timeout events) and denote it as d-CTMC.[1] As illustrated in Figure 1, the usage of discrete-time events for approximating a non-exponential distribution is threefold:

1. For an event e with occurrence time bounded from below by $l > 0$, an occurrence of a discrete-time event d splits the waiting into two parts – an initial part of length l where the event e cannot occur and the rest that can be more efficiently approximated by a CTMC C using standard PH methods.

2. For an event e with occurrence time bounded from above by $u < \infty$, a series of discrete-time events partition the support of its distribution into n subintervals. The system starts in the chain C_1 which is the standard PH approximation of the whole density. In parallel to movement in C_1 a discrete-time event d_1 is awaited with its occurrence set to the beginning of the second interval. If the absorbing state in C_1 is not reached before d_1 occurs, the system moves to C_2. The chain C_2 is fitted to the whole remaining density conditioned by the fact that the event does not occur before the beginning of the second interval. Similarly, another discrete-time event d_2 is awaited in C_2 with its occurrence set to the beginning of the third interval, etc. The last interval is not ended by any discrete-time event; occurrence of the event e thus corresponds to reaching any absorbing state in any of $C_1,\dots,C_n$.

3. An event with constant occurrence time ($\ell = u$) is directly a discrete-time event.

Example. As our running example, we consider the Alternating bit protocol. Via a lossy FIFO channel, a transmitter attempts to send a sequence of messages, each endowed with a one-bit sequence number – alternating between 0 and 1. The transmitter keeps resending each message until it is acknowledged by its sequence number (the receiver sends back the sequence number of each incoming message). As resending of messages is triggered by a timeout, setting an appropriate value for the timeout is essential in balancing the performance of the protocol and the network congestion. For a given timeout, one may ask, e.g., *what is the probability that* 10 *messages will be successfully sent in* 100*ms?* In the next section we show a simple DES model of this protocol. Subsequently, we show the CTMC model yielded by a PH approximation of individual events, and the d-CTMC model obtained by our proposed IPH approximation.

Our Contribution. We propose an alternative approach to PH approximation, resulting in a CTMC enriched with fixed-delay events. Our approach is tailored to interval probability distributions that are often found in reality and for which the standard continuous PH approximation requires a substantial amount of states. We performed an experimental evaluation of our approach. In the evaluation, we represent (1) the lower-bounded distributions by the distribution of the *transport time in network communication* and (2) the upper-bounded distributions by the *uniform* distribution. For both cases, we show that our approach requires only a moderate number of states to approximate these distributions up to a given error. Thus, for DES models with interval distributions our approach promises a viable method for transient analysis as also indicated by our experiments.

---

[1] Note that the formalism of d-CTMC is inspired by the previously studied similar formalisms of *deterministic and stochastic Petri nets* [32] and *delayed CTMC* [16].

Related Work. Already in the original paper of Neuts [34], the fixed-delay and shifted exponential distributions have been found difficult to fit with a phase-type approximation. This fact was explicitly quantified by Aldous and Shepp [1] showing that the Erlang distribution is the best PH fitting for the fixed-delay distributions. A notoriously difficult example of a shifted distribution is the data set measuring the length of eruptions of a geyser in the Yellowstone National Park [38] whose PH approximation has been discussed in, e.g., [3,13]. Also heavy tailed distributions often found in telecommunication systems are hard to fit; similarly to our method, separate fitting of the body and the tail of such distributions is used [14,23].

Apart from continuous PH fitting, there are several other methods applicable to analysis of DES with interval distributions. First, there are several symbolical solution methods [2,5,20,21] for direct analysis of DES with non-exponential events. Usually, *expolynomial* distributions are allowed; non-expolynomial distributions need to be fitted by expolynomials – a problem far less studied than standard PH fitting. Our approach can be understood as a specific fitting technique that uses a limited subclass of expolynomial distributions (resulting in models with a wider range of analysis techniques). Second, interval distributions can be efficiently fitted by discrete phase-type approximation [6]. Instead of a CTMC, this method yields a *discrete-time* Markov chain (DTMC) where each discrete step corresponds to elapsing some fixed δ time units. Note however that this method usually requires to discretize *all the events* of a DES into a DTMC. To analyse faithfully a DES with many parallel events one either needs to use a very small δ [40] or to allow occurrence of multiple events within each δ-time step [33,19], exponentially increasing the amount of states or transitions in the DTMC, respectively. Third, similarly to our approach, ideas for combining discrete PH approximation with continuous PH approximation have already appeared [27,18]. To the best of our knowledge, no previous work considers combining these two approaches on *one* distribution having both discrete and continuous "nature". Expressing the continuous part of such a distribution using continuous PH again decreases the coincidence of parallel discrete events discussed above. Note that with d-CTMC, one can freely combine continuous PH, discrete PH, and interval PH for approximation of different events of a DES.

Organization of the Paper. In Section 2, we define the necessary preliminaries. In Section 3, we describe the IPH approximation method and briefly review the analysis techniques for d-CTMC. The paper is concluded by an experimental evaluation in Section 4.

2   Preliminaries

We denote by $\mathbb{N}$, $\mathbb{Q}$, and $\mathbb{R}$ the sets of natural, rational, and real numbers, respectively. For a finite set X, $\mathcal{D}(X)$ denotes the set of all discrete probability distributions over X.

Modelling Formalisms.   There are several equivalent formalisations of DES. Here we define *generalized-semi Markov processes* that contain both CTMC and d-CTMC as subclasses. Let $\mathcal{E}$ be a finite set of *events* where each event is either a *discrete-time* event or a *continuous-time event*. To each discrete-time event e we assign its delay $delay(e) \in \mathbb{Q}$. To each continuous-time event e we assign a *probability density function*

$f_e : \mathbb{R} \to \mathbb{R}$ such that $\int_0^\infty f_e(x)\,dx = 1$. An event is called *exponential* if it is a continuous-time event with density function $f(x) = \lambda \cdot e^{-x\lambda}$ where $\lambda > 0$ is its rate.

Definition 1. *A generalized semi-Markov process (GSMP) is a tuple $(S, \mathcal{E}, \mathbf{E}, \mathrm{Succ}, \alpha_0)$ where*

- *S is a finite set of states,*
- *$\mathcal{E}$ is a finite set of events,*
- *$\mathbf{E} : S \to 2^{\mathcal{E}}$ assigns to each state s a set of events active in s,*
- *Succ : $S \times \mathcal{E} \to \mathcal{D}(S)$ is the successor function, i.e. it assigns a probability distribution specifying the successor state to each state and event that occurs there,*
- *$\alpha_0 \in \mathcal{D}(S)$ is the initial distribution.*

We say that a GSMP is a continuous-time Markov chain (CTMC) if every events of $\mathcal{E}$ is exponential. We say that a GSMP is a continuous-time Markov chain with discrete-time events (d-CTMC) if every event of $\mathcal{E}$ is either exponential or discrete-time.

The run of a GSMP starts in a state s chosen randomly according to α_0. At start, each event $e \in \mathbf{E}(s)$ is *initialized*, i.e. the amount of time $remain(e)$ remaining until it occurs is (1) set to $delay(e)$ if e is a discrete-time event, or (2) chosen randomly according to the density function f_e if e is a continuous-time event. Let the process be in a state s and let the event e have the minimal remaining time $t = remain(e)$ among all events active in s. The process waits in s for time t until the event e occurs, then the next state s' is chosen according to the distribution $\mathrm{Succ}(s, e)^2$. Upon this transition, the remaining time of each event of $\mathbf{E}(s) \setminus \mathbf{E}(s')$ which is not active any more is discarded, and each event of $\mathbf{E}(s') \setminus \mathbf{E}(s)$ is initialized as explained above. Furthermore if the just occurred event e belongs to $\mathbf{E}(s')$, it is also initialized. For a formal definition we refer to [8].

Example (continued) To illustrate the definition, Figure 2 shows on the left a simplified GSMP model of the Alternating bit protocol. The transmitter sending a message corresponds to the exponential event send. The whole remaining process of the message being transported to the receiver, the receiver sending an acknowledgement message and the acknowledgement message being transported back to the transmitter is modelled using one continuous-time event ack. In parallel with the event ack, there is a discrete-time event timeout and an exponential event err representing a packet loss.

To exemplify the semantics, assume the process is in the state sent with $remain(\text{timeout}) = 10$, $remain(\text{ack})$ is chosen randomly to 12.6 and $remain(\text{err})$ is chosen randomly to 7.2. Hence, after 7.2 time units the event err occurs and the process moves to the state lost with $remain(\text{timeout}) = 2.8$. After further 2.8 time units, the timeout elapses and the process moves to the state init where $remain(\text{send})$ is chosen randomly to 0.8. After this time, the process moves to send where $remain(\text{timeout})$ is again set to 10 and $remain(\text{ack})$ and $remain(\text{err})$ are again sampled according to their densities and so on. In the next section, we show the PH approximation of this model.

---

[2] For the sake of simplicity, when multiple events $X = \{e_1, \dots, e_n\}$ occur simultaneously, the successor is determined by the minimal element of X according to some fixed total order on $\mathcal{E}$. A more general definition [8] allows to specify different behaviour for simultaneous occurrence of any subset of events.

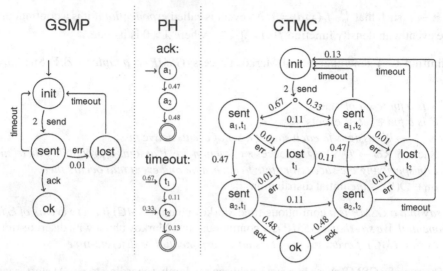

Fig. 2. On the left, there is a GSMP model of sending *a single* message using the the Alternating bit protocol. The set of events active in a state corresponds to the edges outgoing from that state. The event timeout is discrete-time with delay 10 ms, send is exponential with rate 2 meaning that it takes 0.5 ms on average to send a message, err is exponential with rate 0.01 corresponding to a packet being lost each 100 ms of network traffic on average, and ack is continuous-time with density displayed in Figure 4 on the left. In the middle, there are 2-phase PH approximations of events ack and timeout. On the right, there is a PH approximation of the GSMP model obtained roughly speaking as a product of the GSMP and the two PH components.

Continuous PH Approximation. Continuous PH can be viewed as a class of algorithms

- which take as input the number of phases $n \in \mathbb{N}$ and a probability density function f of a positive random variable, and
- output a CTMC C with states $\{0, 1, \ldots, n\}$ where 0 is an absorbing[3] state.

Any such CTMC C defines a positive random variable X expressing the time when the absorbing state 0 is reached in C. Let $\hat{f}$ denote the probability density function of X. A possible goal of a PH algorithm is to minimize the *absolute density difference* [7][4]

$$\int_0^\infty |f(x) - \hat{f}(x)| \, dx. \tag{Err}$$

Example (continued) When building a CTMC model of the Alternating bit protocol from the GSMP model, we need to approximate the non-exponential events ack and timeout. Their simple approximation and the whole CTMC model of the system is depicted in Figure 2 on the right. Observe that each state of the whole model needs to be enriched with the phase-number of every non-exponential event scheduled in this state. The events are then defined in a natural way on this product state space.

[3] We say that a state s is *absorbing* if there are no outgoing transitions, i.e. $\mathbf{E}(s) = \emptyset$.

[4] Note that there are PH methods that do not allow specifying the number of phases. For further metrics for evaluating quality of PH approximation, see, e.g., [7].

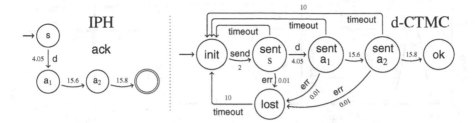

Fig. 3. On the left, there is an IPH approximation of the event ack using the algorithm IPH-shift[PhFit] with 3 phases. The discrete event d has delay 4.05. On the right, there is the whole d-CTMC with discrete-time events timeout and d. The model is obtained similarly as the CTMC in Figure 2.

In the next section we describe our extension of PH fitting with discrete-time events.

3 Interval Phase-Type Approximation

The Interval phase-type (IPH) approximation addresses the interval probability distributions which are supported on a proper subinterval of $[0, \infty)$. Similarly as above,

- it takes as input the number of phases $n \in \mathbb{N}$ and a probability density function f of a positive random variable, and
- outputs a d-CTMC $\mathcal{D}$ with states $\{0, 1, \ldots, n\}$ where 0 is absorbing.

The goal is again to minimize (Err) for $\hat{f}$ being the probability density function[5] of the random variable X expressing the time when the absorbing state 0 is reached in $\mathcal{D}$.

3.1 Constructing d-CTMC

As the first step in this alternative direction, we provide two basic techniques that significantly decrease the error for interval distributions (compared to standard PH algorithms that are by definition IPH algorithms as well). The first technique deals with interval distributions bounded from below.

Delay Bounded from Below. For an event that cannot occur before some $l > 0$ and for a given number of phases $n > 1$, our algorithm works as follows. Let $C = (S, \mathcal{E}, \mathbf{E}, \text{Succ}, \alpha_0)$ be a chain with $n - 1$ phases fitted by some other tool FIT to the density on the interval $[l, \infty)$. We output a d-CTMC $(S \uplus \{s_0\}, \mathcal{E} \uplus \{d\}, \mathbf{E}', \text{Succ}', \alpha_0')$ with n states that starts with probability $\alpha_0'(s_0) = 1$ in the newly added state s_0 in which only the newly added event d is scheduled, i.e. $\mathbf{E}'(s_0) = \{d\}$; the event d has delay $delay(d) = l$ and after it occurs, the chain moves according to the initial distribution of C, i.e. $\text{Succ}'(s_0, d) = \alpha_0$; $\mathbf{E}'$ and Succ' coincide with $\mathbf{E}$ and Succ elsewhere. A pseudo-code for this algorithm IPH-shift[FIT] is given in [28].

---

[5] For the error metrics (Err) we assume that the algorithm outputs a d-CTMC such that X has a density (which holds for our algorithms presented later).

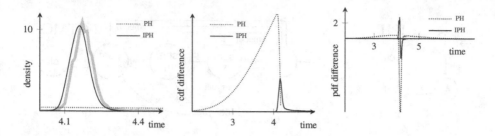

Fig. 4. The comparison of the approximations of the event ack using the algorithms PhFit and IPH-shift[PhFit] with 30 phases. On the left, there is the density of the original distribution as well as the both approximated densities. In the centre, there is for both approximations the difference of the original and the approximate cumulative distribution function. Notice that in point x the plot displays for each algorithm the error we obtain when measuring the probability that the event occurs within time x (rising as high as 0.5). On the right, there is for both approximations the difference of the original and the approximate density. The integral of this curve is the absolute density difference (Err) that we study.

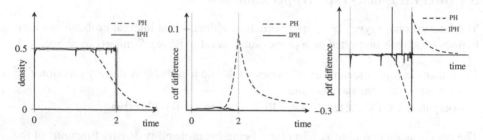

Fig. 5. The comparison of the approximations of the distribution uniform on $[0,2]$ using the algorithms PhFit and IPH-slice[PhFit]. It goes along the same lines as in Figure 4.

Example (continued). To obtain the d-CTMC approximation of the GSMP model of the Alternating bit protocol, we only need to approximate the event ack since timeout is a discrete-time event. To show an example of the technique, the approximation of the event ack using the algorithm IPH-shift[PhFit] as well as the whole resulting d-CTMC is depicted in Figure 3. Since IPH-shift is using the phase-type approximation only on the "simple" part of the density function, it gets much better results. For instance for 30 phases it yields approx. 4x smaller error compared to the best results of PH algorithms. In Figure 4 we provide a more detailed comparison.

Delay Bounded from Above. For an event that cannot occur after some $u < \infty$, our algorithm IPH-slice[FIT,p] slices the interval $[0,u]$ using discrete-time events into p subintervals $[0, \frac{1}{2}u], [\frac{1}{2}u, \frac{3}{4}u], [\frac{3}{4}u, \frac{7}{8}u], \ldots, [(1 - \frac{1}{2}^{p-2})u, (1 - \frac{1}{2}^{p-1})u], [(1 - \frac{1}{2}^{p-1})u, u]$. Their length decreases exponentially with the last two subintervals having the same length. Corresponding to these intervals, we build a sequence of components $C_1, \ldots, C_p$ that is traversed by a sequence of discrete-time events $d_1, \ldots, d_{p-1}$ as the time flows. The component of each subinterval $[a,b]$ has n/p phases and is fitted by FIT to the *conditional* density of the *remaining* delay given the event has not occurred on $[0,a)$.

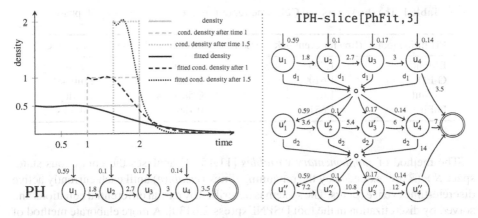

Fig. 6. On the left, the uniform distribution on [0,2] is sliced into three subintervals. With the solid line, there is the whole density and its PH approximation corresponding to the CTMC below. With the dashed and dotted line, there are the conditional densities given the event does not occur before 1 and 1.5, and their PH approximations. Their corresponding CTMC are the same as the CTMC below, only with rates $2x$ and $4x$ larger, respectively. This is clarified on the right, in the complete d-CTMC approximation with all 3 components (sharing the absorbing state).

Consider the example from Figure 6. The uniform distribution on [0,2] has density 0.5 in this interval and 0 elsewhere. When already 1.5 time units pass, the conditional density of the remaining delay equals 2 on [0,0.5] and 0 elsewhere.

This algorithm `IPH-slice[FIT,p]` is formally described in [28]. Example output of `IPH-slice[PhFit,3]` on the above mentioned uniform distribution is depicted in Figure 6. Similarly to the previous technique, it provides approximately 8x better results than the standard PH fitting as demonstrated in Figure 5. Note that we can easily combine the two techniques for distributions bounded both from below and above such as uniform on [5,6]. It suffices to apply `IPH-shift[IPH-slice[FIT,slices]]`.

Let us provide two remarks on this technique. First, notice that a standard fitting tool is applied on the conditional densities. However, a standard fitting tool tries to minimize the error also *beyond* the subinterval we are dealing with which may lead to suboptimal approximation *on* the subinterval. Modification of a PH algorithm addressing this issue might decrease the error of `IPH-slice` even more. Second, dividing the support of the distribution into subintervals of exponentially decreasing length is a heuristic that works well for distributions where the density does not vary much. For substantial discontinuities in the density, one should consider dividing the support in the points of discontinuity. Next, we briefly review the analysis methods for d-CTMC.

3.2 Analysing d-CTMC

The existing theory and algorithms applicable to analysis of d-CTMC are a crucial part of our alternative IPH approximation method. Extending the knowledge in this direction is out of scope of this paper, here we only summarize the state-of-the-art of transient and stationary analysis.

Table 1. The (Err) errors and CPU time for different PH tools fitting by 30 phases

PH fitting tool	(Err) for event ack	(Err) for uniform distribution	CPU time
EMpht	1.7957	1.8980	over one day
G-FIT	1.6100	0.1603	4 min 49 s
momfit	1.8980	0.5820	1 day
PhFit	1.6518	0.1868	4.33 s

The method of *supplementary variables* [11,15,31] analyses the continuous state-space $S \times (\mathbb{R}_{\geq 0})^{\mathcal{E}}$ extended by the remaining times $remain(e)$ until each currently active discrete-time event e occurs. The system is described by partial differential equations and solved by discretization in the tool DSPNExpress 2.0 [30]. A more elaborate method of *stochastic state classes* [37,2,22,21] implemented in the tool Oris [9] studies the continuous state-space model at moments when events occur (defining an embedded Markov chain). In each such moment, multidimensional densities over $remain(e)$ are symbolically derived. The embedded chain is finite iff the system is regenerative, approximation is applied otherwise.

If the d-CTMC has at most one discrete-time event active at a time (e.g. when only one event is approximated by IPH), one can apply the efficient method of *subordinated Markov chains* [32]. It builds the embedded Markov chain using transient analysis of CTMC, similarly to the analysis of CTMC observed by a one-clock timed automaton [10]. In the tool Sabre [16], this method is extended to parallel discrete-time events by approximating them using one discrete-time event Δ [18] that is active in all states and emulates other discrete-time events. An event e occurs with the $\lfloor delay(e)/delay(\Delta) \rfloor$-th occurrence of Δ after initialization of e. Note that this corresponds to discretizing time for the discrete-time events while leaving the exponential events intact.

As some of the methods are recent, no good comparison of these methods exists. Based on our preliminary experiments, we apply in Section 4 the tool Sabre.

4   Experimental Evaluation

In this section we evaluate the reduction of the state space and hence the reduction of the time needed for the analysis when using IPH compared to PH. Precisely, (1) we inspect the growth of the state space of both IPH and PH approximations when decreasing the tolerated error; (2) for a fixed tolerated error, we examine the growth of the state space of the PH approximation when increasing the shift of a shifted distribution; and (3) for a fixed model and a fixed PCTL property we compare the running time of the analysis of d-CTMC yielded by IPH and the running time of the analysis of CTMC yielded by PH when increasing the number of phases.

We consider the distributions from the previous sections, namely the shifted distribution of the event ack addressed by the IPH-shift algorithm and the distribution uniform on $[0, 2]$ addressed by the IPH-slice algorithm. The uniform distribution is specified simply by its formula whereas the density of the event ack is based on real data. Using the Unix ping command, we collected 10000 successful ICMP response

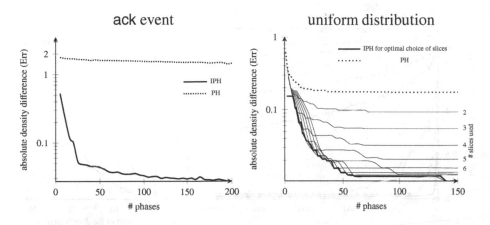

Fig. 7. The (logarithmically scaled) relationship between the size of the state space and the error obtained. For the uniform distribution, we show the results for different numbers of slices, each with the same number of phases. The plotted number of phases is the sum of the phases within all used slices. The error for the optimal number of slices is plotted in bold.

times of a web server (www.seznam.cz, the most visited web portal in the Czech Republic). The data set has mean 4.19 ms, standard deviation 0.314 ms, variance 0.0986, coefficient of variation 0.075 ms, and the shortest time is 4.06 ms (see Figure 4).

To get reliable results, we need to compare IPH with state-of-the-art tools for continuous PH fitting. For our experiments, we considered the tools *EMpht* [3], *G-FIT* [39], *momfit* [25], and *PhFit* [24] (an extended comparison including the tool *HyperStar* [36] is in [28]). We ran the tools to produce PH approximations of the two events with 30 phases (we chose such a small number of phases because for some tools it already took a substantial amount of time). Based on the results shown in Table 1, we have selected PhFit as the baseline tool. Most of the tools achieve similar precision, however PhFit significantly outperforms all others regarding the CPU time[6].

4.1 Growth of the State Space When Decreasing Error

In the first experiment, we focus on the size of the state space necessary to fit the distributions up to a decreasing error. The decreasing errors (Err) when increasing the number of phases, i.e. the state space, are shown in Figure 7. Both our IPH algorithms exhibit a fast decrease of the error (note that the scales are logarithmic). Observe that the continuous PH method does not perform particularly well on the event ack obtained as a real-world example since the absolute density difference of two densities can never exceed 2. For the uniform distribution, we show the results for different numbers of slices used in the IPH-slice algorithm. According to our experiments on the uniform distribution, a finer slicing with less phases in each slice is better than a coarser one with more phases in each slice, whenever each slice is fitted by at least 4 phases.

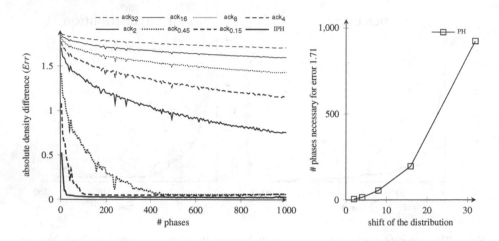

Fig. 8. The growth of the state space when increasing the shift. On the left, there is the dependence of the error on the size of the state space for distributions with different shifts. Note that the IPH fitting does not depend on the shift. On the right, there is the growth of the state space when increasing the shift and fixing the PH fitting error to 1.71.

4.2   Growth of the State Space When Increasing the Shift

In the second experiment, we analyse the growth of the state space when increasing the shift of a shifted distribution. In other words, *how much larger model we get when we try to fit with a fixed error an event with lower coefficient of variation?* We took the distribution of the ack event and shifted the data to obtain a sequence of events $ack_{0.15}, \cdots, ack_4, \cdots ack_{32}$ where ack_i has zero density on the interval $[0, i]$. Note that compared to ack we shifted the data in both directions as $ack \approx ack_4$. The results in Figure 8 confirm a quadratic relationship between the shift and the necessary number of phases for the PH approximation [16].

The quadratic relationship can be supported by the following explanation. Assume we want to approximate a discrete distribution with shift s by a PH distribution. Due to [1], the best PH distribution for this purpose is the Erlang distribution, the chain of k phases with exit rates k/s. Since (Err) does not work in this setting (density is not defined for discrete distributions), we use another common metric - matching moments. Here the goal is to exactly match the mean and minimize the difference of variance. Since the variance of the discrete distribution is zero, the error for k phases is the variance of the Erlang distribution, i.e. s^2/k. To get the same error for a discrete distribution with n-times increased shift $n \cdot s$, we need $n^2 \cdot k$ phases as $(n \cdot s)^2/(n^2 \cdot k) = s^2/k$.

4.3   Time Requirements and Error Convergence When Increasing State Space

So far, we studied how succinct the IPH approximations are compared to PH. One can naturally dispute the impact of IPH approximation by saying that the complexity of d-CTMC analysis is higher that the complexity of CTMC analysis. Here, we show an example where IPH in fact leads to a *lower* overall analysis time.

---

[6] The analysis has been performed on Red Hat Enterprise Linux 6.5 running on a server with 8 processors Intel Xeon X7560 2.26GHz (each with 8 cores) and shared 448 GiB DDR3 RAM.

Table 2. Probability of collision computed by unbounded reachability in CTMC derived using PH and in d-CTMC derived using IPH. The exact probability of collision is 0.7753.

PRISM on CTMC			Sabre on d-CTMC		
phases	result	CPU time	phases	result	CPU time
100	0.527	6.37 s	5	0.695	41s
200	0.541	14.92 s	10	0.730	1 min 24 s
500	0.562	47.74 s	20	0.745	3 min 37 s
1000	0.585	2 min 55 s	30	0.758	10 min 30 s
2000	0.629	9 min 20s			
3000	0.680	50 min 49s			
5030	0.705	3h 14 min			
10030	0.731	32 h 2 min			

We model two workstations competing for a shared channel. Each workstation wants to transmit its data for which it needs 1.2 seconds of an exclusive use of the channel. Each workstation starts the transmission at a random time. If one workstations starts its transmission when the other is transmitting, a collision occurs. Our goal is to compute the probability of collision. For the transmission initiations, we again used the ping command (for two different servers) and obtained two distributions with zero density in the first 4.1 seconds and the first 5.51 seconds, respectively.

We approximated the model using both PH and IPH and subsequently run analysis in the tools PRISM [29] and Sabre [16] that are according to our knowledge the best tools for analysing large CTMC and d-CTMC models, respectively [7] we used IPH approximation for both transmission initiating distributions and a discrete-time event for the 1.2 seconds of transmission. The probability of collision was computed by reachability analysis. In the CTMC model for PRISM, we used PH approximations for both transmission initiating distributions. Furthermore, as PRISM does not support nesting of time bounded until operator into until operator, we again needed to transform the problem into (unbounded) reachability analysis by incorporating the 1.2 seconds of transmission time in the model. We approximated the time by Erlang distribution with 1000 phases (using different number of phases causes at most 1% error in the result).

The results of our experiments are shown in Table 2. The exact probability of collision is 0.7753 as computed directly from the data sampled by ping using a LibreOffice spreadsheet. Due to some numerical errors in the version of Sabre that we used, we were not able to get a lower error than 2% even when using more than 30 phases. Note that the results we were able to obtain from PRISM have a more then twice as high error[8]. Moreover, the immense analysis times shown in Table 2 do not include the durations of PH approximations. The largest approximation we were able to obtain using PhFit was for 3000 phases as for 4000 phases it did not finish within 5 days. For 5030 and 10030 phases we thus constructed the approximations by concatenating an Erlang

---

[7] To eliminate the effects of the implementation, we also run the CTMC analysis in Sabre. However, it is much slower than PRISM. Full details are in [28].

[8] We did our best to make CTMC analysis as quick as possible, we used parameters -s -gs -maxiters 1000000 -cuddmaxmem 18000000 and set PRISM_JAVAMAXMEM to 200000m.

approximation of the shift with the 30 phases PH approximation of the remaining part (as it was obtained during IPH). Overall, the results indicate that for models that are sensitive to precise approximation of the distributions, the IPH approximation can lead to a significantly faster analysis compared to PH approximation.

5   Future Work

There are several directions for future work. First, a comparison of the existing algorithms [16,31,21] that can be applied to the transient analysis of d-CTMC would be highly welcome. Furthermore, for the best algorithm for d-CTMC, one can perform a more detailed comparison of its running times on the d-CTMC obtained by the IPH fitting with the analysis times of other available methods (such as the standard PH fitting). Second, we believe that further heuristics can increase the efficiency of IPH or its applicability to a wider class of distributions. Finally, our method justifies the importance of research on further analysis algorithms for d-CTMC.

Acknowledgement. We would like to thank Vojtěch Forejt, András Horváth, David Parker, and Enrico Vicario for inspiring discussions.

References

1. Aldous, D., Shepp, L.: The least variable phase type distribution is Erlang. Communications in Statistics. Stochastic Models 3(3), 467–473 (1987)
2. Alur, R., Bernadsky, M.: Bounded model checking for GSMP models of stochastic real-time systems. In: Hespanha, J.P., Tiwari, A. (eds.) HSCC 2006. LNCS, vol. 3927, pp. 19–33. Springer, Heidelberg (2006)
3. Asmussen, S., Nerman, O., Olsson, M.: Fitting phase-type distributions via the EM algorithm. Scandinavian Journal of Statistics 23(4), 419–441 (1996)
4. Baier, C., Haverkort, B., Hermanns, H., Katoen, J.-P.: Model-checking algorithms for continuous-time Markov chains. IEEE Trans. on Software Engineering 29(6), 524–541 (2003)
5. Bernadsky, M., Alur, R.: Symbolic analysis for GSMP models with one stateful clock. In: Bemporad, A., Bicchi, A., Buttazzo, G. (eds.) HSCC 2007. LNCS, vol. 4416, pp. 90–103. Springer, Heidelberg (2007)
6. Bobbio, A., Horváth, A., Scarpa, M., Telek, M.: Acyclic discrete phase type distributions: Properties and a parameter estimation algorithm. Performance Evaluation 54(1), 1–32 (2003)
7. Bobbio, A., Telek, M.: A Benchmark for PH Estimation Algorithms: Results for Acyclic-PH. Communications in Statistics. Stochastic Models 10(3), 661–677 (1994)
8. Brázdil, T., Krčál, J., Křetínský, J., Řehák, V.: Fixed-delay events in generalized semi-Markov processes revisited. In: Katoen, J.-P., König, B. (eds.) CONCUR 2011. LNCS, vol. 6901, pp. 140–155. Springer, Heidelberg (2011)
9. Bucci, G., Carnevali, L., Ridi, L., Vicario, E.: Oris: A tool for modeling, verification and evaluation of real-time systems. International Journal on Software Tools for Technology Transfer 12(5), 391–403 (2010)
10. Chen, T., Han, T., Katoen, J.-P., Mereacre, A.: Quantitative model checking of continuous-time Markov chains against timed automata specifications. In: LICS, pp. 309–318. IEEE (2009)
11. Cox, D.R.: The analysis of non-Markovian stochastic processes by the inclusion of supplementary variables. Math. Proceedings of the Cambridge Phil. Society 51, 433–441, 7 (1955)

12. Fackrell, M.: Fitting with matrix-exponential distributions. Stoch. Models 21(2-3), 377–400 (2005)
13. Faddy, M.J.: On inferring the number of phases in a Coxian phase-type distribution. Communications in Statistics. Stochastic Models 14(1-2), 407–417 (1998)
14. Feldmann, A., Whitt, W.: Fitting Mixtures of Exponentials to Long-Tail Distributions to Analyze Network. Perform. Eval. 31(3-4), 245–279 (1998)
15. German, R., Lindemann, C.: Analysis of stochastic Petri nets by the method of supplementary variables. Performance Evaluation 20(1-3), 317–335 (1994)
16. Guet, C.C., Gupta, A., Henzinger, T.A., Mateescu, M., Sezgin, A.: Delayed continuous-time Markov chains for genetic regulatory circuits. In: Madhusudan, P., Seshia, S.A. (eds.) CAV 2012. LNCS, vol. 7358, pp. 294–309. Springer, Heidelberg (2012)
17. Haas, P.J.: Stochastic petri nets. Springer (2002)
18. Haddad, S., Mokdad, L., Moreaux, P.: Performance evaluation of non Markovian stochastic discrete event systems – a new approach. In: WODES 2004, p. 243. Elsevier (2005)
19. Hatefi, H., Hermanns, H.: Improving time bounded reachability computations in interactive Markov chains. In: Fundamentals of Soft. Engineering, pp. 250–266. Springer (2013)
20. Horváth, A., Paolieri, M., Ridi, L., Vicario, E.: Probabilistic model checking of non-Markovian models with concurrent generally distributed timers. In: QEST, pp. 131–140. IEEE (2011)
21. Horváth, A., Paolieri, M., Ridi, I., Vicario, E.: Transient analysis of non-markovian models using stochastic state classes. Performance Evaluation 69(7), 315–335 (2012)
22. Horváth, A., Ridi, L., Vicario, E.: Transient analysis of generalised semi-Markov processes using transient stochastic state classes. In: QEST, pp. 231–240. IEEE (2010)
23. Horváth, A., Telek, M.: Markovian modeling of real data traffic: Heuristic phase type and MAP fitting of heavy tailed and fractal like samples. In: Calzarossa, M.C., Tucci, S. (eds.) Performance 2002. LNCS, vol. 2459, pp. 405–434. Springer, Heidelberg (2002)
24. Horváth, A., Telek, M.: Phfit: A general phase-type fitting tool. In: Field, T., Harrison, P.G., Bradley, J., Harder, U. (eds.) TOOLS 2002. LNCS, vol. 2324, pp. 82–91. Springer, Heidelberg (2002)
25. Horváth, A., Telek, M.: Matching more than three moments with acyclic phase type distributions. Stochastic Models 23(2), 167–194 (2007)
26. Jensen, A.: Markoff chains as an aid in the study of Markoff processes. Scandinavian Actuarial Journal 1953(supp. 1), 87–91 (1953)
27. Jones, R., Ciardo, G.: On phased delay stochastic Petri nets: Definition and an application. In: Petri Nets and Performance Models, pp. 165–174. IEEE (2001)
28. Korenčiak, Ľ., Krčál, J., Řehák, V.: Dealing with zero density using piecewise phase-type approximation. CoRR, abs/1406.7527 (2014)
29. Kwiatkowska, M., Norman, G., Parker, D.: PRISM 4.0: Verification of probabilistic real-time systems. In: Gopalakrishnan, G., Qadeer, S. (eds.) CAV 2011. LNCS, vol. 6806, pp. 585–591. Springer, Heidelberg (2011)
30. Lindemann, C., Reuys, A., Thummler, A.: The DSPNexpress 2.000 performance and dependability modeling environment. In: Fault-Tolerant Comp., pp. 228–231. IEEE (1999)
31. Lindemann, C., Thümmler, A.: Transient analysis of deterministic and stochastic Petri nets with concurrent deterministic transitions. Performance Evaluation 36, 35–54 (1999)
32. Marsan, M.A., Chiola, G.: On Petri nets with deterministic and exponentially distributed firing times. In: Rozenberg, G. (ed.) APN 1987. LNCS, vol. 266, pp. 132–145. Springer, Heidelberg (1987)
33. Molloy, M.K.: Discrete time stochastic Petri nets. Software Eng. SE-11(4), 417–423 (1985)
34. Neuts, M.F.: Matrix-geometric solutions in stochastic models: An algorithmic approach. The Johns Hopkins University Press, Baltimore (1981)

35. O'Cinneide, C.A.: Phase type distributions: Open problems and a few properties. Communications in Statistics. Stochastic Models 15(4), 731–757 (1999)
36. Reinecke, P., Krauss, T., Wolter, K.: Hyperstar: Phase-type fitting made easy. In: QEST, pp. 201–202. IEEE (2012)
37. Sassoli, L., Vicario, E.: Close form derivation of state-density functions over DBM domains in the analysis of non-Markovian models. In: QEST, pp. 59–68. IEEE (2007)
38. Silverman, B.W.: Density estimation for statistics and data analysis. Monographs on Statistics and Applied Probability. Chapman and Hall (1986)
39. Thummler, A., Buchholz, P., Telek, M.: A novel approach for phase-type fitting with the EM algorithm. Dependable and Secure Computing 3(3), 245–258 (2006)
40. Zhang, L., Neuhäußer, M.R.: Model checking interactive Markov chains. In: Esparza, J., Majumdar, R. (eds.) TACAS 2010. LNCS, vol. 6015, pp. 53–68. Springer, Heidelberg (2010)

Uncertainty in On-The-Fly Epidemic Fitting

Roxana Danila, Marily Nika, Thomas Wilding, and William J. Knottenbelt

Department of Computing, Imperial College London
South Kensington Campus, London SW7 2AZ
{id210,marily,tw610,wjk}@imperial.ac.uk

Abstract. The modern world features a plethora of social, technological and biological epidemic phenomena. These epidemics now spread at unprecedented rates thanks to advances in industrialisation, transport and telecommunications. Effective real-time decision making and management of modern epidemic outbreaks depends on the two factors: the ability to determine epidemic parameters as the epidemic unfolds, and the ability to characterise rigorously the uncertainties inherent in these parameters. This paper presents a generic maximum-likelihood-based methodology for online epidemic fitting of SIR models from a single trace which yields confidence intervals on parameter values. The method is fully automated and avoids the laborious manual efforts traditionally deployed in the modelling of biological epidemics. We present case studies based on both synthetic and real data.

Keywords: Epidemics, Compartmental disease models, SIR models, Maximum likelihood estimation.

1 Introduction

> I have called the uncertainty that surrounds any response to a microbial outbreak the *Fog of Epidemics*, analogous to the *Fog of War* of which historians speak.
>
> Richard M. Krause

In this modern era, technological advances enable a deadly disease to spread across the globe in just a few days. If during the times of the Black Death people typically travelled less than 10 miles in a day, nowadays 14 000 miles can be covered in a day, resulting in unprecedented rates of infection spreading [19]. In addition to biological epidemics, phenomena such as social and technological epidemics have emerged due to the extensive coverage and penetration of the Internet and social media [3,14,23,4]. Online phenomena are characterised by a rapid, exponential spread through the population and are often triggered by seemingly inconsequential causes compared to the magnitude of their effects. The ability to predict and control such events is a topic of increasing interest.

Several formal quantitative approaches are available for making predictions about infectious disease. Although widely used in contingency planning, predictive modelling is still "the art of possible". The key requirement for a good

A. Horváth and K. Wolter (Eds.): EPEW 2014, LNCS 8721, pp. 135–148, 2014.
© Springer International Publishing Switzerland 2014

model is to provide accurate predictions, although it is already well established that such predictions cannot achieve perfect accuracy. This uncertainty arises due to two main factors: (i) the transmission of infection is stochastic in nature, making it very unlikely that one observes identical dynamic disease trajectories, even when the underlying epidemic processes are parameterically identical; (ii) models are an approximation, and rare or unforeseen behavioural patterns cannot be captured, but can have a significant impact on the disease dynamics [16]. Uncertainty can also result from assumptions made about the infectious agent and the environment, or even the technical details of the model.

The main contribution of this paper is a generic methodology for on-the-fly epidemic fitting of a classical compartmental epidemiological model, namely Susceptible-Infected-Recovered (SIR) [17], with inbuilt characterisation of parameter uncertainty. Given a single data trace of an evolving outbreak, a technique is developed for the fitting of SIR model parameters using an optimization method that employs a maximum-likelihood-based objective function. The output is a set of confidence intervals on key parameter values. In contrast with traditional approaches deployed in biological epidemics, which require laborious manual work for index case identification, lab testing and contact tracing, this method is fully automated.

A novel aspect of this research is connected to one of the major challenges: not knowing or being able to estimate from past data the initial number of susceptible and infected individuals. These initial conditions cause potentially large uncertainties in the estimation procedure. Our previous attempts to address this challenge using a least-squares fitting procedure yielded point estimates for parameters without any characterisation of related uncertainty [21].

The rest of this paper is organised as follows. Section 2 presents background information regarding infectious disease modelling. Section 3 describes the main optimisation methods used for fitting the models, estimating the parameters and setting confidence intervals to capture their uncertainty. Section 4 presents some example analyses on synthetic and real disease data. Section 5 concludes.

2 Background

Improved sanitation, antibiotics and vaccination programs created a confidence in the 1960s that infectious disease spreading would be eliminated. However, infectious disease agents adapt and evolve over time, so that new infectious diseases have emerged and some existing diseases have re-emerged. Mathematical models have become important tools in planning, implementing, evaluating and optimizing various detection, prevention, therapy and control programs. Epidemiology modelling can contribute to the design and analysis of epidemiological surveys, suggest crucial data that should be collected, identify trends, make forecasts and estimate the uncertainty in forecasts [15,13,6,9,11].

An epidemic is defined as a widespread occurrence of an infectious disease in a community at a particular time. Real-time forecasts of epidemic spread using data-driven models have been hindered by technical challenges posed by parameter estimation and validation [21]. Furthermore, traditional approaches rely on laborious and often infeasible approaches to initial estimates for parameters, such as studying in detail the index cases of the outbreak to infer, for example, the recovery rate as the reciprocal of the average infectious period [7].

In 1927 Kermack and McKendrick proposed one of the classical compartmental models most widely used in epidemiology, namely SIR [17]. Using Ordinary Differential Equations (ODEs), this models the evolution of an epidemic over time in terms of the number of Susceptible, Infected and Recovered individuals. Given a closed population of individuals, it defines

- $S(t)$ = individuals not yet infected at time t, but susceptible to infection
- $I(t)$ = individuals infected at time t by contact with susceptibles at a rate β
- $R(t)$ = individuals recovered at time t at a constant rate γ

We assume that the size of each compartment is a differentiable function of time. We ignore intricacies related to the pattern of contact between individuals, considering the instantaneous rate of new infections to be βSI. The recovery rate γ is proportional to the number of infected individuals, as each individual is assumed to recover at a constant rate γ.

These assumptions lead to the set of differential equations:

$$\frac{dS}{dt} = -\beta SI \tag{1}$$

$$\frac{dI}{dt} = \beta SI - \gamma I \tag{2}$$

$$\frac{dR}{dt} = \gamma I \tag{3}$$

The initial values of the SIR model must satisfy the following conditions:

$$S(0) = S_0 > 0 \tag{4}$$

$$I(0) = I_0 > 0 \tag{5}$$

$$R(0) = 0 \tag{6}$$

and at any time, t, $S(t) + I(t) + R(t) = N$, where N is the total population size.

Such compartmental models can forecast the disease spread between individuals, not only in one population but also in various subpopulations and across localities [20,2]. An outbreak originating in a seed subpopulation could potentially lead to a global-scale epidemic. A computational model called the Global Epidemic and Mobility model (GLEAM) is capable of integrating high-resolution data on human demography and mobility on a global scale in a metapopulation stochastic epidemic framework. GLEAM can simulate the global spread of influenza in order to provide insights on intervention strategies including vaccinations, antiviral treatment and travel restrictions [22].

A compelling interdisciplinary analysis of methods through which model uncertainty can be negotiated is presented in [9]. The study shows that many models provide only cursory reference to the uncertainties of the information and data, or the parameters used, concluding that a more careful consideration of the limitations and uncertainties present in modelling epidemic phenomena would drastically improve its value. It is therefore essential to implement a rigorous and transparent technique that can provide confidence intervals on parameters for a clear understanding of evolving scenarios.

3 Methodology

Given a data set, we estimate the parameters using a two-pass methodology that combines least squares (LS) and maximum likelihood (ML) based optimisation techniques. Uncertainty quantification is then performed using the profiles obtained from the ML estimates.

3.1 Model Fitting Procedure

Mathematical modelling of infectious disease dynamics relies on a series of assumptions regarding key parameters that cannot be measured directly. We discuss here the technique used to fit the parameters of our model as an outbreak unfolds over time. In particular, we consider the challenges of estimating the initial number of susceptible and infected individuals in the target population, when these values are unknown. Currently, there is no principled way of doing this, as traditionally they are either known or can be estimated from the context [21]. However, in an era of social and technological epidemics, we argue that time and speed of movement make it infeasible to obtain accurate manual estimates.

3.1.1 Online Model Fitting. We attempt to account for uncertainty as each outbreak unfolds, over time. To achieve this, we apply our fitting methodology on truncated data sets. We initially consider the first 10 observations from the outbreak. We then create new truncated datasets by adding each subsequent observation as the outbreak unfolds.

Using the SIR model, we propose two methodologies, one for estimating the parameters β, γ, S_0, and another for estimating β, γ, S_0 and I_0. By definition, all these quantities are positive, allowing us to apply a log transformation, yielding an unconstrained optimisation landscape with no possibility to explore infeasible values. Similarly a scaled logistic transformation can be applied to the initial number of susceptibles S_0 and infecteds I_0 when these are known to be bounded above by some constant C. The transformation function is:

$$\text{trans}(x) = \log(\frac{x}{C - x}) \tag{7}$$

and its corresponding inverse is:

$$\text{trans}^{-1}(y) = \frac{C}{1 + e^{-y}} \tag{8}$$

3.1.2   Parameter Estimation Using Maximum Likelihood. The Maximum Likelihood method is an analytic procedure for finding parameter vectors which maximise the likelihood of a dataset of iid observations. The likelihood function is defined as:

$$\mathcal{L}(\boldsymbol{\theta} \,|\, x_1, \ldots, x_n) = f(x_1, x_2, \ldots, x_n | \boldsymbol{\theta}) = \prod_{i=1}^{n} f(x_i | \boldsymbol{\theta}) \qquad (9)$$

where $f(x_1, x_2, \ldots, x_n | \boldsymbol{\theta})$ is the joint density function of the observations and $\boldsymbol{\theta}$ the vector of unknown parameters. The maximum likelihood estimator $\hat{\boldsymbol{\theta}}$ is then:

$$\hat{\boldsymbol{\theta}} = \arg\max_{\boldsymbol{\theta}} \mathcal{L}(\boldsymbol{\theta} \,|\, x_1, \ldots, x_n) \qquad (10)$$

Equivalently, one can minimise the negative log likelihood:

$$\hat{\boldsymbol{\theta}} = \arg\min_{\boldsymbol{\theta}} \left(-\log \mathcal{L}(\boldsymbol{\theta} \,|\, x_1, \ldots, x_n) \right) = \arg\min_{\boldsymbol{\theta}} \left(-\sum_{i=1}^{n} \log f(x_i | \boldsymbol{\theta}) \right) \qquad (11)$$

In the present work we assume the observations to be Poisson distributed. Typically, epidemiologists model variability in disease occurrence using either Binomial, Poisson or Exponential distributions. [12] argues that the three distributions have common attributes and underlying assumptions that tend to yield similar results. They also state that the Poisson distribution is widely used by epidemiologists when the data involves summary counts of cases. Moreover, since we deal with discrete observations, the variance is expected to scale with the number of infected individuals [5,10].

The estimates are computed using the *mle2* function in the *bbmle* R package, which requires a negative log-likelihood function and starting values for the initial parameters to be specified. A computational challenge arises through the calculation of confidence intervals within *mle2*. This requires calculating the covariance matrix for the parameters, which is done by inversion of the Hessian matrix at the optimum and can be unsuccessful depending on the initial parameters. To overcome this, we first applied a Least Square based fitting procedure and used the estimates provided as starting values in order to be able to successfully estimate the confidence intervals.

The set of parameters that gives the best Maximum Likelihood based fit to the data is found using the Nelder-Mead algorithm, a widely used gradient-free method for unconstrained multidimensional optimization [18]. The first-order ODEs are solved using the *lsoda* R package. For optimal results, it is important to specify a small threshold for the absolute error tolerance.

3.1.3   Confidence Intervals. We make use of profile confidence intervals to indicate how reliable the estimate for a parameter is. The level of confidence is taken to be the probability that the interval contains the true value of the parameter, given a distribution of samples.

Traditionally, Wald-type confidence intervals are used as an approximation to profile intervals. The standard procedure for computing such a confidence interval is:

$$\text{estimate} \pm (\text{percentile} \times \text{SE(estimate)}) \tag{12}$$

where SE is the standard error and the percentile represents the desired confidence level with respect to some reference distribution. Although easier to compute for complex models, it performs poorly when the likelihood surface is not quadratic.

A more robust technique for constructing confidence regions can be derived from the asymptotic χ^2 distribution of the likelihood ratio test statistic. Given a maximum likelihood estimate $\hat{\boldsymbol{\theta}}$ of a parameter vector $\boldsymbol{\theta}_0$, an approximate $(1 - \alpha)$ confidence interval for $\boldsymbol{\theta}_0$ is the set of values of satisfying:

$$\{\boldsymbol{\theta} : 2[l(\hat{\boldsymbol{\theta}}) - l(\boldsymbol{\theta}_0)] \leq c_{k;1-\alpha}\} \tag{13}$$

where $c_{k;1-\alpha}$ is the $(1 - \alpha)$th quantile of the χ^2 distribution with k degrees of freedom. Confidence intervals for individual parameters can be obtained by treating the others as "nuisance parameters" and maximising over them [24].

We compute two-sided confidence intervals using the *confint* function in the *bbmle* R package, at various confidence levels: 99%, 95%, 90%, 80% and 50%. In addition, we provide a 3D visualisation of the confidence intervals for the case when the unknown parameters vector is β, γ and S_0. This representation takes the shape of an ellipsoid, with each of the axis corresponding to one of the parameters to estimate. Note that the semi-axes may be unequal due to their asymmetric confidence intervals.

4   Results

In order to illustrate key aspects of the proposed approach we use both synthetic and a real-world datasets. The synthetic datasets were generated based on Gillespie's Stochastic Simulation Algorithm, using the *ssa* function in the *GillespieSSA* R package. The real dataset represents positive laboratory tests for influenza summed over all subtypes of the flu virus, as reported to the Centre of Disease Control (CDC) during the 2012/2013 flu season (starting in September 2012). The data were obtained via the FluView Web Portal[1].

4.1   Synthetic Data

The synthetic data set used in this section was generated by simulating an SIR epidemic with known parameters $\beta = 0.001$, $\gamma = 0.1$ and initial conditions $S_0 = 500$, $I_0 = 10$, $R_0 = 0$.

Synthetic Data with β, γ, S_0 Unknown. We fitted truncated datasets obtained of 25%, 50%, 75% and 100% of the data in order to analyse the uncertainty in the parameters as more data becomes available. As time progresses, we observe that our fits become more and more stable as illustrated in Figure 1.

---

[1] http://gis.cdc.gov/grasp/fluview/fluportaldashboard.html

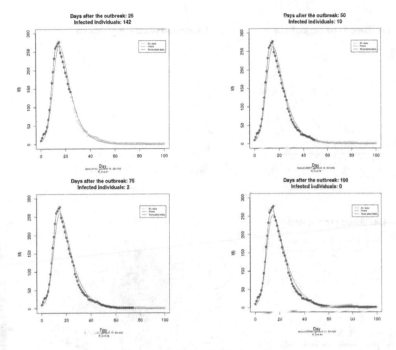

Fig. 1. Fitting of SIR model with β, γ, S_0 unknown to synthetic data

Figure 2 shows the profiles obtained from the ML estimate at various confidence levels for log-based transformations of each of the unknown parameters β, γ and S_0. For example we see that the 95% confidence interval for $\log(\beta)$ is (-7.083,-6.962), yielding a 95% confidence interval for β as (8.39e-04, 9.47e-04). As expected, the estimated range of possible values is wider as the confidence level increases. This is illustrated in the isosurface plot extended to three dimensions to visually represent the uncertainty inherent in the parameters.

Table 1 shows the lower and upper bounds on each parameter when the data is fitted over time. We observe the uncertainty of the parameters tends to decrease as more observations are considered.

Table 1. 95% Confidence Intervals for synthetic data

Data%	β		γ		$S0$	
	Lower	Upper	Lower	Upper	Lower	Upper
25%	5.66e-04	8.47e-04	1.08e-01	1.93e-01	569	962
50%	7.17e-04	8.36e-04	1.17e-01	1.35e-01	590	692
75%	7.62e-04	8.68e-04	1.13e-01	1.26e-01	568	646
100%	8.39e-04	9.47e-04	1.03e-01	1.14e-01	519	582

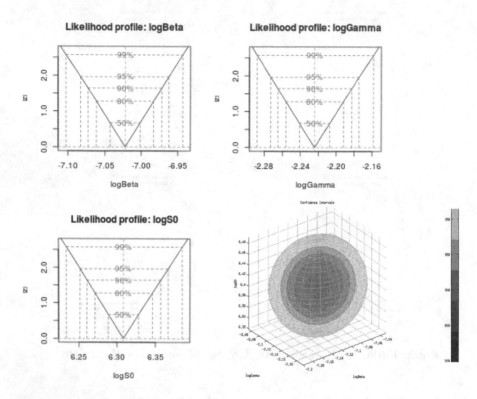

Fig. 2. Likelihood profile plots and corresponding isosurface plot for the estimated confidence intervals of transformed parameters when β, γ and S_0 are unknown (synthetic data)

Synthetic Data with β, γ, S_0, I_0 Unknown. Figure 3 captures the uncertainty characterised over the parameters β, γ, and the initial conditions S_0, I_0, where I_0 is bounded by S_0 using a logistic based transformation. The uncertainty ranges and estimated values are similar to the ones computed by the optimisation with known I_0, demonstrating the robustness of the optimisation.

True Value Recoverability Rate for Parameters. For a known set of ground truth parameters, we use Gillespie's stochastic simulation algorithm to generate 1 000 sample trajectories of the number of infected individuals over time. For each trajectory, we apply our methodology to obtain 95% confidence intervals for each parameter. We might have expected that 95% of the time, the true values of the parameters should lie within the 95% confidence interval. However, Table 2 shows that this is not the case. This emphasises how difficult it is to obtain accurate estimates of the uncertainty of the parameters from a single data trace. Such traces may be heavily affected by stochastic variation, especially in cases like our example where there are a relatively small number

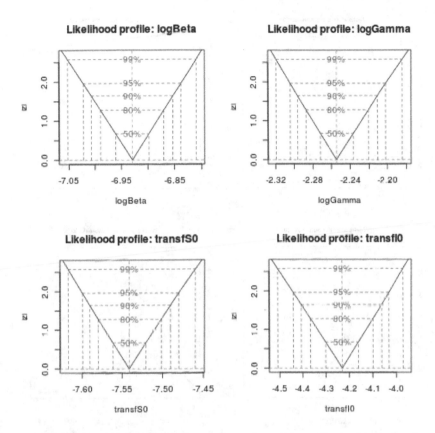

Fig. 3. Likelihood profile plots for the estimated confidence intervals of transformed parameters when β, γ, S_0 and I_0 are unknown (synthetic data)

Table 2. True value recoverability rate for unknown parameters β, γ and S_0 (left) and for β, γ, S_0 and I_0 (right)

Parameter	Recoverability rate
β	26.59%
γ	26.28%
S_0	31.82%
β, γ, S_0	8.86%

Parameter	Recoverability rate
β	41.99%
γ	26.28%
S_0	34.44%
I_0	48.04%
β, γ, S_0, I_0	9.46%

of initial susceptibles [1]. We also note the improvement in recovery rates for β and S_0 when I_0 is included as an unknown parameter, showing the benefits of maintaining flexibility with respect to this critical initial condition.

4.2 CDC Influenza Data

We used data regarding positive lab-based influenza tests reported to the Center of Disease Control and Prevention (CDC) during the 2012/2013 influenza season.

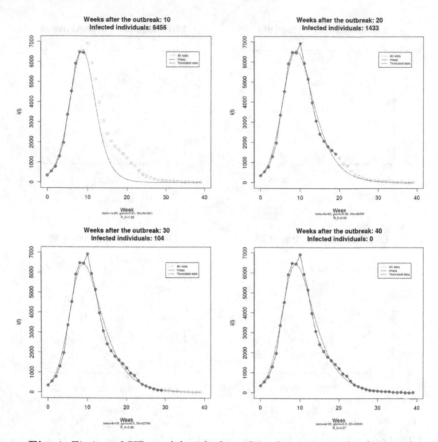

Fig. 4. Fitting of SIR model with β, γ, S_0 unknown to real influenza data

Figure 4 shows the fitting over time of truncated datasets, illustrating that the algorithm is robust enough to be applied to real data.

Figure 5, Figure 6 and Table 3 characterise the uncertainty of the parameters for the real data set. The similar behaviour to the synthetic data reinforces our results and methodology.

Table 3. 95% Confidence intervals for influenza data (* - non convergence)

Data%	β		γ		S_0	
	Lower	Upper	Lower	Upper	Lower	Upper
25%	*	*	*	*	*	*
50%	2.95e-05	3.22e-05	3.46e-01	3.81e-01	26769	30118
75%	3.50e-05	3.69e-05	2.90e-01	3.06e-01	22091	23515
100%	3.53e-05	3.70e-05	2.90e-01	3.03e-01	22031	23292

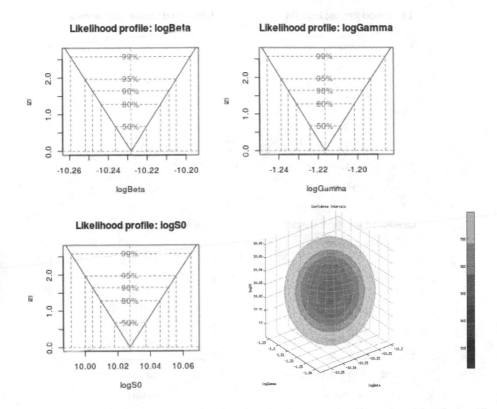

Fig. 5. Likelihood profile plots and corresponding isosurface plot for the estimated confidence intervals of transformed parameters when β, γ and S_0 are unknown (influenza data)

5   Conclusion

In this paper we provided a generic maximum-likelihood-based approach towards the on-the-fly epidemic fitting of SIR models from a single trace, which yields confidence intervals on parameter values. In contrast to traditional biological epidemiological modelling techniques, our approach is fully automated and the parameters to be estimated include the number of initial susceptibles and the initial number of infected in the population. Visualising the fitted parameters gives rise an isosurface plot of the feasible parameter ranges corresponding to each confidence level.

We generated multiple synthetic disease outbreak trajectories via stochastic simulation and fitted parameters to those trajectories. The "true" parameters were contained in the corresponding confidence bounds only for a relatively low proportion of the time, emphasising (a) the difficulty of obtaining accurate parameter estimations from a single epidemic trace and (b) the large potential impact of small random variations, especially those occurring early on in a trace.

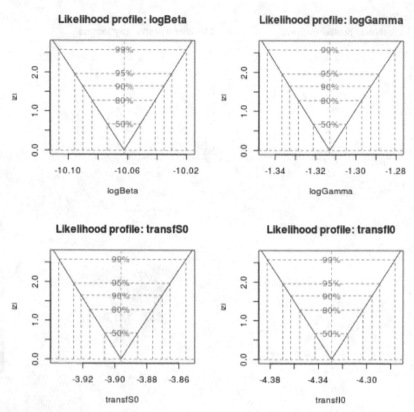

Fig. 6. Likelihood profile plots for the estimated confidence intervals of transformed parameters when β, γ, S_0 and I_0 are unknown (influenza data)

It is expected that real systems are likely to exhibit different characteristics than the ideal ones assumed by the classical SIR model; for example real systems may feature time-varying parameters and the homogeneous mixing assumption may not apply. Nevertheless, the models may have utility in predicting the stochastic impact of candidate interventions in real systems with bounds [8]; a simulation-based methodology for this will be the focus of our future work.

References

1. Anderson, H., Britton, T.: Stochastic Epidemic Models and their Statistical Analysis. Springer (2000)
2. Angulo, J., Yu, H., Langousis, A., Kolovos, A., Wang, J., Madrid, A., Christakos, G.: Spatiotemporal Infectious Disease Modeling: A BME-SIR Approach. PLoS ONE (2013)
3. Bakshy, E., Rosenn, I., Marlow, C., Adamic, L.: The Role of Social Networks in Information Diffusion. In: Proc. 21st International Conference on the World Wide Web, WWW 2012 (2012)

4. Bauer, F., Lizier, J.: Identifying Influential Spreaders. CoRR, abs/1203.0502 (2012)
5. Bolker, B., Ellner, S.: Likelihood and all that, for Disease Ecologists (2011), http://kinglab.eeb.lsa.umich.edu/EEID/eeid/2011_eco/mle_2011.pdf
6. Briggs, A., Weinstein, M., Fenwick, E., Karnon, J., Sculpher, M., Paltiel, A.: Model Parameter Estimation and Uncertainty: A Report of the ISPOR-SMDM Modeling Good Research Practices Task Force-6. Value in Health 15(6), 835–842 (2012)
7. Brooks-Pollock, E., Eames, K.: Pigs didn't Fly, but Swine Flu. Mathematics Today 47, 36–40 (2011)
8. Burr, T., Chowell, G.: Observation and Model Error Effects on Parameter Estimates in Susceptible-Infected-Recovered Epidemiological Models. Far East Journal of Theoretical Statistics 19(2), 163–183 (2013)
9. Christley, R., Mort, M., Wynne, B., Wastling, J., Heathwaite, A., Pickup, R., Austin, Z., Latham, S.: "Wrong, but Useful": Negotiating Uncertainty in Infectious Disease Modelling. PLoS One 8(10), e76277, 10 (2013)
10. Dolgoarshinnykh, R.: Epidemic Modelling Graduate Topics Course. Lecture Notes, http://www.stat.columbia.edu/~regina/research/
11. Elderd, B., Vanja, M., Dukic, V., Dwyer, G.: Uncertainty in Predictions of Disease Spread and Public Health Responses to Bioterrorism and Emerging Diseases. Proceedings of the National Academy of Sciences 103(42), 15693–15697 (2006)
12. Flanders, W., Kleinbaum, D.: Basic Models for Disease Occurrence in Epidemiology. International Journal of Epidemiology 24(1), 1–7 (1995)
13. Gilbert, J., Meyers, L., Galvani, A., Townsend, J.: Probabilistic Uncertainty Analysis of Epidemiological Modeling to Guide Public Health Intervention Policy. Epidemics 6, 37–45 (2014)
14. Hartmann, W., Manchanda, P., Nair, H., Bothner, M., Dodds, P., Godes, D., Hosanagar, K., Tucker, C.: Modeling Social Interactions: Identification, Empirical Methods and Policy Implications. Marketing Letters 19(3), 287–304 (2008)
15. Hethcote, H.: The Mathematics of Infectious Diseases. SIAM Review 42(4), 599–653 (2000)
16. Keeling, M.: State-Of-Science Review: Predictive and Real-time Epidemiological Modellig (2006), http://www.dti.gov.uk/assets/foresight/docs/infectious-diseases/s9.pdf
17. Kermack, W., McKendrick, A.: A Contribution to the Mathematical Theory of Epidemics. Proceedings of the Royal Society of London. Series A 115(772), 700–721 (1927)
18. Lagarias, J., Reeds, J., Wright, M., Wright, P.: Convergence Properties of the Nelder-Mead Simplex Method in Low Dimensions. SIAM Journal of Optimization 9, 112–147 (1998)
19. Lerner, R.: The Black Death and Western European Eschatological Mentalities. The American Historical Review 86, 533–552 (1981)
20. Nika, M., Fiems, D., Turck, K., Knottenbelt, W.J.: Modelling Interacting Epidemics in Overlapping Populations. In: Proc. 21st International Conference on Analytical & Stochastic Modelling Techniques & Applications (ASMTA 2014), Budapest, Hungary (2014)
21. Nika, M., Ivanova, G., Knottenbelt, W.J.: On Celebrity, Epidemiology and the Internet. In: Proc. 7th International Conference on Performance Evaluation Methodologies and Tools (VALUETOOLS), Turin, Italy (December 2013)
22. Tizzoni, M., Bajardi, P., Poletto, C., Ramasco, J., Balcan, D., Goncalves, B., Perra, N., Colizza, V., Vespignani, A.: Real-time Numerical Forecast of Global Epidemic Spreading: Case study of 2009 A/H1N1pdm. BMC Medicine 10(1), 165 (2012)

23. Tweedle, V., Smith, R.: A Mathematical Model of Bieber Fever: The most infectious disease of our time. In: Mushayabasa, S., Bhunu, C.P. (eds.) Understanding the Dynamics of Emerging and Re-Emerging Infectious Diseases using Mathematical Models, ch. 7, pp. 157–177. Transworld Research Network (2012)
24. Venzon, D., Moolgavkar, S.: A Method for Computing Profile-Likelihood-Based Confidence Intervals. Applied Statistics 37(1), 87–94 (1988)

Performance Modeling of Intelligent Car Parking Systems

Károly Farkas[1,3], Gábor Horváth[1,2], András Mészáros[1,3], and Miklós Telek[1,2]

[1] Budapest University of Technology and Economics
[2] MTA-BME Information Systems Research Group
[3] Inter-University Center of Telecommunications and Informatics Debrecen

Abstract. The performance analysis of the car parking process in a parking lot with various levels of assistance is considered in the paper. The input of the model is the description of the parking lot and a Markovian description for the driver behavior, the set of computable performance measures contains the average time necessary for the user to reach the desired destination, the amount of cars moving in the parking lot at the same time (thus, the environmental strain), etc. To overcome the state space expansion, that makes the direct analysis of the model computationally infeasible, we apply a mean field limit based approximation whose accuracy is investigated with discrete event simulation.

1 Introduction

Since its introduction at the beginning of the century [2], the smart city concept enjoys an enormous attention from the research and development community, e.g., real-time traffic monitoring based car routing (i.e., Personal Navigation Assistant (PNA) devices with Traffic Message Channel (TMC) support) are commonly used in several countries, as well as the real-time tracking of Global Positioning System (GPS) equipped public transport devices. Intelligent parking systems, which we investigate in this paper belong to this category as well. In an intelligent car parking system the driver selects its destination on a smartphone application when entering the parking garage, and the parking system selects the optimal parking field and provides navigation aid to it, considering the occupancy situation.

Several such systems have been introduced in the literature in the recent years, e.g. [3], [1] or [5]. All these papers are focusing on the technical and implementation aspects of the problem, while in this paper, we focus on the stochastic performance analysis of such a parking system and calculate performance measures like the mean time spent by searching for a parking place and by walking to the destination, and the mean number of cars rambling around to find a place. The results can be used to quantify the benefits of an intelligent parking system.

The direct analysis of the overall parking system including the state of each individual car is inhibitory complex and that is why we propose a mean field limit based efficient approximate analysis [4].

A. Horváth and K. Wolter (Eds.): EPEW 2014, LNCS 8721, pp. 149–163, 2014.

The rest of the paper is organized as follows. Section 2 introduces the considered model. Section 3 provides the description of the behavior of the drivers, both in the un-assisted and in the assisted cases. The mean field model is defined in Section 4, and some numerical examples are investigated in Section 5.

2   Model Description

2.1   The Floor Plan

The architectural plan of the studied parking lot (Figure 1) contains a grid divisions which we use for representing the position of the cars in the garage by a discrete variable. These rectangular fields are identified by their row and column position $p = (r, c)$, and are held by set $\mathcal{P} = \{p\}$.

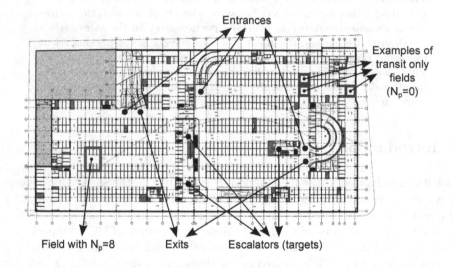

Fig. 1. Floor plan of the parking garage of the Allee mall in Budapest

Some fields contain a number of parking places, while others are serving as transit only. The maximal number of cars that can park at position p is denoted by N_p, from which the free ones are denoted by K_p. The "transit only" and the "transit+parking" type fields are distinguished by $N_p = 0$ and $N_p > 0$ respectively. Additionally, there are fields with special purposes (see Figure 1).

- Set $\mathcal{E}$ contains the positions of the entrances of the garage.
- Set $\mathcal{T}$ contains the positions of the possible targets of the drivers.
- Set $\mathcal{X}$ contains the exit positions (to leave the garage).

The roads in the garage can be either one- or two-ways. The set of neighboring positions where a car can proceed after leaving p is denoted by $\mathcal{N}(p)$, from which $succ(p)$ represents the one that can be reached without changing direction.

The shortest path between two fields will be used frequently in the sequel. When a tagged car is located at field $a \in \mathcal{P}$ we denote the next field along the shortest path towards field b by $\overrightarrow{d}_{a,b}$ if the driving directions on the roads are respected, and by $d_{a,b}$, if they are not respected. The corresponding distances along the shortest path are denoted by $\overrightarrow{c}_{a,b}$ and $c_{a,b}$, respectively, where $\overrightarrow{d}_{a,b}$ and $\overrightarrow{c}_{a,b}$ applies to cars, while $d_{a,b}$ and $c_{a,b}$ applies to pedestrians.

2.2   Behavior of the Cars

The life cycle of a car consists of the following four phases.

1. It *enters* the garage at one of the entrances;
2. it *searches* for a free parking field, preferably as close to the target destination as possible;
3. it selects a free parking field and *parks*. The driver leaves the car behind proceeds the trip on foot.
4. Finally, after some time, the driver returns to the car and *leaves* the garage.

We assume that the selection of the entrance and the target is done by a random choice. The behavior of the drivers can be very different depending on their experiences. how much information the drivers have on the floor plan of the parking garage. Completely uninformed drivers have no information on the floor plan, and no clue on their target destination, whereas returning drivers know the floor plan and are able to take the distance information into consideration. In case of intelligent parking systems the occupancy information of the parking fields is available, too. The measurement based car behavior description is not available yet, thus our motion models are based on intuition. We are using the following four different motion models.

– *Uninformed drivers.*
  In this case the motion of the cars is similar to a random walk. The actions the driver can take are:
  - Stop and park. The more free places are in the current position, the easier is to park the car, thus the higher is the probability of parking.
  - Change direction. The probability of choosing a direction is proportional with the number of parking places available in that direction.
  - Go forward, without changing direction. It is not typical that a car changes direction at every possible places, thus this action has a higher probability than changing direction.
– *Returning (distance aware) drivers.*
  If the driver has prior knowledge on the parking garage, apart of the available positions he/she can utilize the distance from the selected target destination. Thus, the probabilities associated with the three actions above will decrease with the distance from the destination.
  If the utilization of the garage is high and parking fields close to the target are all occupied, distance aware drivers have a hard time to find an appropriate

place to park. To avoid unrealistically long search phase, we introduce the *patience* of the drivers, which is a random variable. Once a driver becomes impatient, it gives up the distance preference and acts like an uninformed driver.

– *Intelligent parking assist systems.*
  If an intelligent assist system is installed, the number and the distribution of the free parking places can be used in proposed guidance. When entering the garage, the driver selects the desired destination and the system directions to the best possible parking field. Due to the dynamic evolution of the occupancy of parking fields the parking assist system must re-calculate the optimal parking field and the corresponding directions continuously. We are considering two possible assist strategies:
  • solely based on the distance from the desired destination, to minimize the walking distance,
  • based on the distance from the desired destination and the driving distance, to minimize the driving plus walking time.

Regarding the departure phase of the life-cycle of a car, we assume that informative traffic signs guide the cars towards the exits on the shortest path.

3   Formalizing the Car Motion Models

The entrance, target and exit selections are assumed to be random choice with the following distributions.

– $q_i^{(E)}$ is the probability that the car enters at entrance $i \in \mathcal{E}$,
– $q_j^{(T)}$ is the probability that the driver intends to visit target $j \in \mathcal{T}$,
– $q_k^{(X)}$ is the probability that the driver chooses exit $k \in \mathcal{X}$ to leave.

3.1   Parking Preference Functions

We introduce two functions to characterize the parking preference of the drivers. Function $f_p^{(F)}$ represents the parking preference based on K_p, the number of free parking fields at position p. Obviously, if $K_p = 0$ (all fields are occupied at p) we have that $f_p^{(F)} = 0$, and $f_p^{(F)}$ is monotonously increasing with K_p, reflecting the observation that drivers prefer to park at positions where more free fields are available. If $f_p^{(F)} = 1$, position p is ideal for parking and every driver wants to take it. This function is approximated by a complementary Gaussian curve according to

$$f_p^{(F)} = 1 - e^{-K_p^2/\sigma_F^2}, \tag{1}$$

where the variance parameter σ_F^2 will be used to tune the shape of this function. The choice of Gaussian curve is motivated by its frequent appearance in

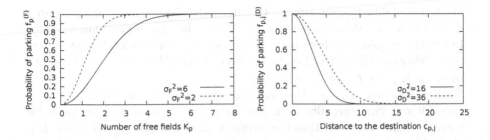

Fig. 2. Parking preference functions

stochastic models and its simplicity. We do not have experimental data (at all) to validate this choice.

Function $f_{p,j}^{(D)}$ represents the parking preference at position p based on the distance of p from the desired destination j. If the distance is 0 ($p = j$), we have that $f_{p,j}^{(D)} = 1$. Furthermore, $f_{p,j}^{(D)}$ decreases monotonously to zero with the distance $c_{p,j}$, since every driver wants to find a place as close to its destination as possible. Function $f_{p,j}^{(D)}$ is approximated by a Gaussian function (with the same lack of experimental data)

$$f_{p,j}^{(D)} = c^{-c_{p,j}^2/\sigma_D^2}, \tag{2}$$

where the variance parameter σ_D^2 controls how important the distance is to the drivers. The shapes of the functions with the parameters used in the upcoming numerical example are depicted in Figure 2.

3.2 Motion Model for Uninformed Drivers

Our model for the uninformed driver is based on intuition, and might seem to be a bit artificial. We hope to improve it in the future based on empirical measurements. The reason to include it in this paper is to have a less sophisticated driver behavior in the comparison with the assisted case, for which we have an accurate model.

In our simple model, an uninformed driver can take one of the following actions while being at position $p \in \mathcal{P}$:

- It can park with probability q_p, where $q_p = f_p^{(F)}$. The σ_F^2 parameter (corresponding to $f_p^{(F)}$, see (1)) depends on the average occupancy of the parking fields at position p, which the driver can estimate by looking around. The higher the occupancy of the parking lot is, the smaller σ_F^2 is, thus it gets more likely that the driver takes every single empty parking place it encounters. $\sigma_F^2 \to 0$ corresponds to the case where the driver selects the first free place it encounters. With this setting the garage is filled up along concentric circles around the entrances. We use $\sigma_F^2 >> 0$, thus uninformed drivers just drive for a while to choose a place.

– With probability $r_{p,i}, i \neq succ(p)$ the driver decides to move on and change direction to a neighboring position $i \in \mathcal{N}(p)$. Choosing a direction where no free field is visible (e.g., because that field is for traversal only) has a non-zero probability as well (denoted by $1-\gamma$). Accordingly, the weight corresponding to $r_{p,i}$ is $r_{p,i} = \gamma f_i^{(F)} + 1 - \gamma$.

– With probability $r_{p,succ(p)}$ the driver decides to move on to the next position without changing direction. Not changing direction is more probable than changing direction, thus the weight assigned to this case is larger, it is $W(\gamma f_{succ(p)}^{(F)} + 1 - \gamma)$, $W > 1$.

Thus, the probability of moving from position p to position $i \in \mathcal{N}(p)$ is

$$r_{p,i} = \begin{cases} \frac{1}{\varphi_1}(\gamma f_i^{(F)} + 1 - \gamma), & i \neq succ(p), \\ \frac{1}{\varphi_1} W(\gamma f_{succ(p)}^{(F)} + 1 - \gamma), & i = succ(p), \\ 0 & \text{otherwise}, \end{cases} \tag{3}$$

where $\varphi_1 = \sum\limits_{j \in \mathcal{N}(p), j \neq succ(p)} \gamma f_j^{(F)} + 1 - \gamma + W(\gamma f_{succ(p)}^{(F)} + 1 - \gamma)$. Hereafter we assume that a car does not turn over 180° and go back to its preceding position unless this is the only possibility to proceed.

3.3 Motion Model for Distance Aware Drivers

In case of returning drivers the distance information is also available when making the parking decision. The probability that the driver stops and parks at position p depends both on the number of free parking fields K_p and on the distance to the destination j as $q_p = f_p^{(F)} \cdot f_{p,j}^{(D)}$. The variance parameters σ_F^2 and σ_D^2 of distance aware drivers are less than the ones in the uninformed case, as returning drivers are more determined.

If the driver decides not to park, the distance information plays a role in the distribution of the next position of the car, too, according to

$$r_{p,i} = \begin{cases} \frac{1}{\varphi_2}(\gamma f_i^{(F)} + 1 - \gamma)f_{i,j}^{(D)}, & i \neq succ(p), \\ \frac{1}{\varphi_2} W(\gamma f_i^{(F)} + 1 - \gamma)f_{succ(p),j}^{(D)}, & i = succ(p), \\ 0 & \text{otherwise}, \end{cases} \tag{4}$$

where $\varphi_2 = \sum\limits_{j \in \mathcal{N}(p), m \neq succ(p)} (\gamma f_m^{(F)} + 1 - \gamma)f_{m,j}^{(D)} + W(\gamma f_{succ(p)}^{(F)} + 1 - \gamma)f_{succ(p),j}^{(D)}$.

To avoid endless rambling the patience of the drivers has to be included in the model as well. We assume that the patience follows a discrete phase-type (DPH) distribution defined by initial probability vector α and transient transition probability matrix A, yielding $P(\text{patience} = k) = \alpha A^{k-1}(I - A)\mathbb{1}$. Once the patience of the driver is over, it gives up optimizing on the distance and switches to the uninformed strategy defined in Section 3.2.

3.4   Motion Model in the Presence of Intelligent Parking Systems

The intelligent parking system continuously sends the suggested driving direction towards the most appealing parking field to the driver. Given that the car is currently at position p and the desired destination is target j, the appeal of a field at position i (denoted by $a_{p,j,i}$) is derived from the parking preference function $f_i^{(F)}$ and $f_{p,j,i}^{(D)'}$, which depends on the distance to be driven from p to i and the walking distance from i to the destination j, that is $f_{p,j,i}^{(D)'} = e^{-(\xi_d \overrightarrow{c}_{p,i} + \xi_w c_{i,j})^2/\sigma_D^2}$, where ξ_d and ξ_w determine the weights of driving and walking distance in the decision, respectively. If the parking system optimizes on solely the walking distance, we have that $\xi_d = 0$ and $\xi_w = 1$.

The appeal $a_{p,j,i}$ is then modeled by $a_{p,j,i} = f_i^{(F)} \cdot f_{p,j,i}^{(D)'}$, and the best position p^{opt} is selected according to $p^{opt} = \arg\min_{i \in P} a_{p,j,i}$. We also take into account that there are drivers that do not follow the guidance of the parking system till the end, but with probability $U \cdot f_p^{(F)} \cdot f_{p,j}^{(D)}$ can park to a parking field even if it is not the recommended one (parameter U stands for drivers independence). All in all, an assisted car at position p heading to target j can take the following actions.

- A car stops and park with probability

$$q_p = \begin{cases} 1, & \text{if } p = p^{opt}, \\ U \cdot f_p^{(F)} \cdot f_{p,j}^{(D)}, & \text{otherwise.} \end{cases} \qquad (5)$$

- Given that the car did not park, it moves forward to position i with probability

$$r_{p,i} = \begin{cases} 1, & \text{if } i = \overrightarrow{d}_{p,p^{opt}}, \\ 0, & \text{otherwise.} \end{cases} \qquad (6)$$

3.5   Distribution of the Parking Time

The duration of parking is modeled by a discrete phase type distribution with initial probability vector and transient transition probability matrix denoted by β and B, respectively, that is $P(\text{parking time} = k) = \beta B^{k-1}(I - B)\mathbb{1}$.

After parking the cars leave the garage along the shortest path to the selected exit. In the considered closed system scenario when a car leaves the garage, a new car enters immediately to keep the number of cars in the garage constant. We apply the closed system scenario to eliminate the randomness caused by the randomly changing load (number of cars) and to amplify the performance implications of the considered driver behavior.

4   The Mean-Field Model

We introduce a discrete time model in this section to characterize the state of the parking garage. Each car being in search or leaving phase moves exactly one field along the grid in each time step.

4.1   The State Space of the System

Cars in the garage are either search for a parking field, or parked, or leaving. For cars in the *search* phase we need to keep track of

- the current position p,
- the desired target destination j,
- the current orientation of the car o (to avoid complete turn-overs),
- and, in case of the distance-aware strategy,
    - the phase of the DPH distribution representing the patience n.
    - a flag f indicating that the car lost patience and gave up optimizing on distance

For *parked* cars we have to follow the

- the position of the car p,
- the phase of the DPH corresponding to the parking time m.

Finally, for *leaving* cars we have to include into the state space

- the current position of the car p,
- and the selected exit where the car is heading to x.

Thus, the state of a car at time k can be represented by

$$\mathcal{S}_k \in \{(search, p, j, o, n, f)\} \cup \{(parked, p, m)\} \cup \{(leaving, p, x)\}.$$

if the cars are using a distance aware strategy, otherwise it can be represented by

$$\mathcal{S}_k \in \{(search, p, j, o)\} \cup \{(parked, p, m)\} \cup \{(leaving, p, x)\}.$$

4.2   State Transitions

The possible state transitions of cars being in the search phase at time n are the following in case of the uninformed and the assisted case.

$$P(\mathcal{S}_{k+1} = (parked, p, m) | \mathcal{S}_k = (search, p, j, o, n, f)) = q_{p_{eff}} \beta_m,$$

$$P(\mathcal{S}_{k+1} = (search, i, j, o', n, f) | \mathcal{S}_k = (search, p, j, o, n, f)) = r_{p,i}$$

where q_p and $r_{p,i}$ depends on the car motion model (see Section 3), and o' is the orientation of the car after moving from p to i. Note that instead of the parking probability q_p, a modified quantity, the effective parking probability $q_{p_{eff}}$ is used in the formula to ensure that the parking capacity N_p of field p is never exceeded (in the mean sense). If there are S_p cars on the same position, and each individual car wants to park with probability q_p, the mean number of parking cars in the next step would be $S_p q_p$, that can be larger than the capacity of p. Therefore the effective parking probability is given by

$$
q_{p_{eff}} = \begin{cases} q_p, & \text{if } S_p q_p \leq N_p, \\ q_p K_p / N_p & \text{otherwise,} \end{cases} = \begin{cases} q_p, & \text{if } s_p q_p \leq n_p, \\ q_p k_p / n_p & \text{otherwise,} \end{cases} \tag{7}
$$

where S_p denotes the number of cars being in the search phase at position p and $s_p = S_p/N$, $n_p = N_p/N$, $k_p = K_p/N$ are the population normalized parameters which equally define the transition probability.

If the cars are moving according to the distance aware strategy, the patience information needs to be taken into account as well, hence we have

$$
P(\mathcal{S}_{k+1} = (parked, p, m) | \mathcal{S}_k = (search, p, j, o, n, false)) = q_{p_{eff}}^{(DA)} \beta_m,
$$

$$
P(\mathcal{S}_{k+1} = (parked, p, m) | \mathcal{S}_k = (search, p, j, o, n, true)) = q_{p_{eff}}^{(UI)} \beta_m,
$$

$$
P(\mathcal{S}_{k+1} = (search, i, j, o', n, true) | \mathcal{S}_k = (search, p, j, o, n, true)) = r_{p,i}^{(UI)},
$$

$$
P(\mathcal{S}_{k+1} = (search, i, j, o', n', false) | \mathcal{S}_k = (search, p, j, o, n, false)) = r_{p,i}^{(DA)} A_{n,n'},
$$

$$
P(\mathcal{S}_{k+1} = (search, i, j, o', n, true) | \mathcal{S}_k = (search, p, j, o, n, false))
$$

$$
= r_{p,i}^{(DA)} \left(1 - \sum_{n'} A_{n,n'} \right)
$$

where $q_{p_{eff}}^{(UI)}$ and $q_{p_{eff}}^{(DA)}$ are the effective parking probabilities associated with the uninformed and distance aware case, respectively, and the routing probabilities $r_{p,i}^{(UI)}$ and $r_{p,i}^{(DA)}$ are distinguished in the same way.

Cars being in the parked phase can either stay parked or enter to the leaving state according to

$$
P(\mathcal{S}_{k+1} = (parked, p, m') | \mathcal{S}_k = (parked, p, m)) = B_{m,m'},
$$

$$
P(\mathcal{S}_{k+1} = (leaving, p, x) | \mathcal{S}_k = (parked, p, m)) = \left(1 - \sum_{m'} B_{m,m'} \right) \cdot q_x^{(X)}.
$$

Finally, the state transition probabilities corresponding to leaving cars are

$$
P(\mathcal{S}_{k+1} = (leaving, \overrightarrow{d}_{p,x}, x) | \mathcal{S}_k = (leaving, p, x)) = 1, \qquad \text{if } p \neq x,
$$

$$
P(\mathcal{S}_{k+1} = (search, e, j, o_e, n, false) | \mathcal{S}_k = (leaving, p, x)) = q_e^{(E)} q_j^{(T)} \alpha_n, \text{ if } p = x,
$$

given that the orientation of the cars entering at entrance e is o_e.

4.3   The Occupancy Vector and Its Mean Field Limit

Let $X_n^N(k)$, $n = 1, \ldots, N$ denote the state of car n at time step k, where N is the (constant and finite) number of cars in the garage. The state transition probabilities given in Section 4.2 define a discrete time Markov chain (DTMC) for a given number of cars, where the DTMC keeps track of the states of all the cars in the garage. In this Markov chain the stochastic behavior of cars are interdependent because the movement (the state transition probabilities) of a given car depends on the positions of the other cars, but it is important to note that the state transition probabilities do not depend on the position of particular cars $X_n^N(k)$, but only on the number of cars which stay in the grids (i.e., vector M^N defined below). We refer to this property as density dependence.

When N is large ($\geq$ several tens), which is the common case in practice, the state space of the DTMC gets extremely large and the analysis of this large DTMC becomes infeasible. To overcome this limitations we can approximate the DTMC which describe the behavior of N (finite) cars in the parking lot with its mean field limit, which is obtained when the number of cars increases to infinity, due to the density dependence of the model.

In order to evaluate the mean field limit we introduce the occupancy measure ([4]) which is a row vector $M^N(k) = [M_i^N(k), i \in \mathcal{S}]$ where $M_i^N(k)$ is the proportion of cars being in state i at time step k which is $M_i^N(k) = \frac{1}{N} \sum_{n=1}^N I_{\{X_n^N(k)=i\}}$, with $I_{\{\}}$ being the indicator function, and introduce the normalized versions of the transition probability functions.

The state transitions probabilities given in Section 4.2 depend on the number of cars which stay in the grids. E.g., $q_{p_{eff}}$ depends on K_p which is the number of free parking fields at position p, and on S_p which is the number of cars searching a parking field at p. The population normalized versions of these quantities can be expressed from the occupancy vector as $k_p(k) = n_p - \sum_{i \in \{(parked,p,m)\}} M_i^N(k)$, $s_p(k) = \sum_{i \in \{(search,p,j,o,n,f)\}} M_i^N(k)$. Let $\bar{\Pi}(M^N(k))$ denote the state transition probability matrix containing the state transitions probabilities given in Section 4.2 as a function of the population normalized number of cars in different states. The mean field limit of the occupancy vector $M^N(k)$, denoted by $M(k)$, satisfies the difference equation $M(k+1) = M(k)\bar{\Pi}(M(k))$ [4]. This difference equation is much less expensive to compute than the direct analysis of the large DTMC model representing the product space of the states of all cars in the garage.

The mean field limit is obtained when the number of cars and the capacity of the parking lot increases proportionally (i.e., the "size" of the cars decreases to zero). However, the mean field model gives an approximate analysis of the system, since the number of cars being in state i calculated by $N \cdot M_i(k)$ is a real number, while it is an integer variable in the reality.

4.4   Performance Measures

We evaluate the following performance measures based on the occupancy measure $M(k)$, the mean driving time to parking, L_S, the mean walking distance from the selected parking field to the target destination, L_W, the mean of the total latency including the driving and walking time. L_T, and the ratio of cars moving in the garage at the same time (either in search or leaving phase), C. At time k C is obtained as $C(k) = \sum_{i \in \{(search,p,j,o,n,f)\} \cup \{(leaving,p,x)\}} M_i(k)$.

To compute the rest of the performance measures we introduce $\ell_{j,p}(t)$, which is the probability that a car heading to target j finds a parking field at time t and position p. For $\ell_{j,p}(t)$ we have $\ell_{j,p}(t) = a \cdot (\bar{\Pi}_{search,search})^{t-1} \cdot b_{j,p}$, where row vector a reflects the starting state of an individual car. The entries of a are given by

$$a_i = \begin{cases} q_e^{(E)} q_j^{(T)}, & \text{if } i = (search, e, j, o_e, n, f), \\ 0, & \text{otherwise.} \end{cases}$$

Furthermore, $\bar{\Pi}_{search,search}$ is derived from $\lim_{k\to\infty} \bar{\Pi}(M(k))$ by setting entries but the ones belonging to the $search$ phases to zero. Thus, the entries of $(\bar{\Pi}_{search,search})^t$ are the probabilities that the car is still in the search phase after time t with the corresponding state transitions. Finally, $b_{j,p}$ is a column vector whose entry i is the probability that a car in a search phase heading to target j being at position p stops and selects a parking field. It can be obtained as

$$(b_{j,p})_i = \begin{cases} \sum_{\forall m} P(\mathcal{S}_{k+1} = (parked, p, m) | \mathcal{S}_k = i), & \text{if } i = (search, p, j, o, n, f), \\ 0, & \text{otherwise.} \end{cases}$$

With $\ell_{j,p}(t)$ the mean search time is $L_S = \sum_{t=0}^{\infty} t \cdot \sum_{\forall j \in \mathcal{T}, p \in \mathcal{P}} \ell_{j,p}(t)$, and the mean walking distance to the target is $L_W = \sum_{\forall j \in \mathcal{T}, p \in \mathcal{P}} c_{p,j} \sum_{t=0}^{\infty} \ell_{j,p}(t)$.

To calculate L_T we have to take into consideration that walking is slower than driving. By denoting the time required to walk through a field in the grid relative to the driving time by R the total time to target is given by $L_T = \sum_{\forall j \in \mathcal{T}, p \in \mathcal{P}} \sum_{t=0}^{\infty} (t + c_{p,j} R) \cdot \ell_{j,p}(t)$.

5 Numerical Experiments

We implemented the mean field method in C++ environment and compared the performance of the parking strategies discussed in the paper[1]. The floor plan corresponds to the "Allee" shopping mall in Budapest. The parameters of the model have been determined by intuition due to the lack of real data according to Table 1. The DPH distribution generating the parking time has a mean of 4000 seconds and the squared coefficient of variation is 1.5. The corresponding parameters are

$$\beta = \begin{bmatrix} 0.16 & 0.84 & 0 \end{bmatrix}, B = \begin{bmatrix} 0.99988 & 0.00012 & 0 \\ 0 & 0.99925 & 0.00075 \\ 0 & 0 & 0.99925 \end{bmatrix}. \tag{8}$$

---

[1] The software is open source and can be downloaded from
http://www.hit.bme.hu/~ghorvath/software

Table 1. Values of the parameters used in the examples

Strategy	Parameters
Uninformed	$\sigma_F^2 = 6$; $\gamma = 0.3$; $W = 3$
Distance aware	$\sigma_F^2 = 2$; $\gamma = 0.2$; $W = 1.5$; $\sigma_D^2 = 16$; $\alpha = \begin{bmatrix} 0.321 & 0.379 & 0 \end{bmatrix}$; $A = \begin{bmatrix} 0.99679 & 0.00321 & 0 \\ 0 & 0.99 & 0.01 \\ 0 & 0 & 0.99 \end{bmatrix}$ (Mean patience=300 seconds, squared coefficient of variation=0.8)
Assisted, case 1.	$\sigma_F^2 = 2$; $\sigma_D^2 = 36$; $U = 0.05$; $\xi_d = 0$; $\xi_w = 1$
Assisted, case 2.	$\sigma_F^2 = 2$; $\sigma_D^2 = 36$; $U = 0.05$; $\xi_d = 0.2$; $\xi_w = 0.8$

5.1 Distribution of the Occupied Parking Fields

Figure 3 compares the mean occupancy of the parking fields under light load ($N = 250$). When the drivers are uninformed, they choose lightly occupied fields for parking along the main roads on the garage, instead of the ones located close to the targets.

In the distance aware case it is visible that the cars are parking around the three escalator entrances of the garage. Observe, however, that the garage can be divided into two main parts, and there is only a single possibility to move to the left part from the right one. Therefore there are drivers who choose sub-optimal parking field at the right part because they do not find the way to the less loaded left part. This situation is reflected on the heat map as well, the circular region corresponding to the more occupied parking fields is asymmetric, there are much more cars parking at the right side.

The occupancy of the parking fields is better distributed around the targets in the first assisted case (where only the walking distance is the subject of optimization), since the intelligent parking system is able to guide the cars to the left part if there are more free parking fields there.

The heat map belonging to the second assisted case (where the driving time and the walking time are both optimized) does not differ too much in this particular example.

The execution times were 2 ms, 8 ms, 2 ms, 21 ms per iteration in the uninformed, distance aware, the first and the second assisted case, respectively on an average PC with a 3.4 GHz CPU and 4 GB of memory.

5.2 Comparing the Uninformed Model with Simulation

To assess the precision of the mean field approximation we implemented the uninformed driver case in OMNeT++ [6], which is a C++ based framework for discrete event simulation.

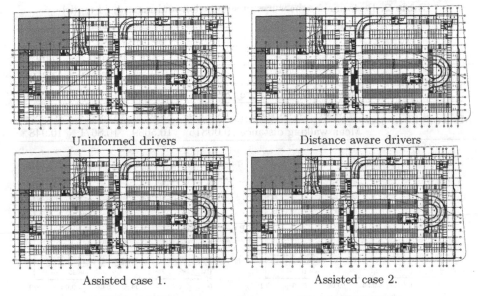

Uninformed drivers   Distance aware drivers

Assisted case 1.   Assisted case 2.

Fig. 3. The distribution of the occupied parking fields

We examined the mean occupancy of the parking fields as well as the average searching times and walking distances of the drivers. The simulations were run for a total of 10^6 parking events, of which the first 10^5 were considered a warm-up period thus they were not taken into account in the statistics calculations. Between several test runs the difference in mean occupancy was less than 0.005 for more than 90% of the parking fields (with an average less than 0.002), while the maximum difference was around 0.02. The relative differences between the searching and walking times were both less than 0.2% for every test run.

In the comparison we made the inspection for four different loads ($N = 100, 300, 400, 500$). To demonstrate the transition between the simulation and the mean field model we introduce the ω scaling parameter ($1 \leq \omega < \infty$). The meaning of this parameter in the simulation is the following. If $\omega > 1$, the number of cars in the system and the parking fields at each position are multiplied by ω. Furthermore, to compensate the scaling $f_p^{(F)}$ becomes $f_{p,\omega}^{(F)} = 1 - e^{-K_p^2/(\omega\sigma_F)^2}$. For $\omega = 1$ the model corresponds to the original physical realization. When $\omega \to \infty$ the behavior converges to the mean field model. Figure 4 shows the cumulative distribution functions of the errors of the mean occupancy for light and heavy loads. In accordance with the expectations the errors decrease when increasing ω. It can also be seen, that the precision of the mean field model decreases with the increasing load. Table 2 shows the performance measures of interest. The same tendency can be observed in this case as well, however, the error is small except for the mean search time in the $N = 500$ case, therefore the mean field approximation proved to be quite precise for low and medium load.

Table 2. Performance measures of simulation and mean field model

		Number of cars in the garage			
		100	300	400	500
Simulation	Mean search time	7.57176	12.1356	16.411	37.8561
	Mean walking distance	8.97576	10.3427	10.8522	10.7663
Mean field model	Mean search time	7.61508	12.5713	16.9241	47.2762
	Mean walking distance	9.00815	10.5659	10.8671	10.948

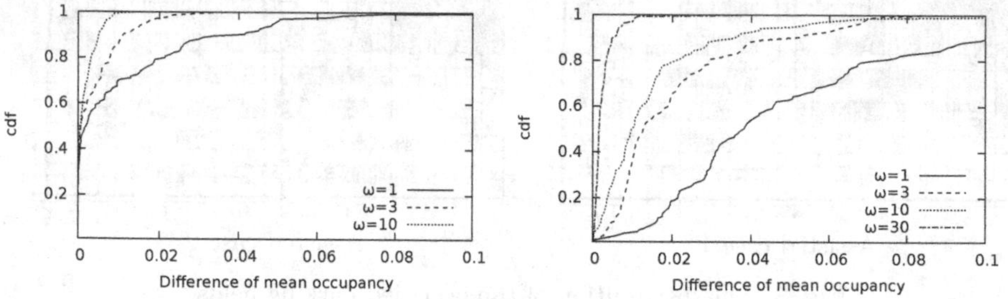

Fig. 4. The effect of scaling to the error of mean occupancy

5.3   The Effect of the Load of the Garage

Figure 5 depicts the performance measures as the function of the load. To record the results $5 \cdot 10^5$ iterations were executed, which was enough to get high precision results in the cases where the iteration converged. In the uninformed and the distance aware cases the convergence was fast and found a (supposedly) global at attractor from any random initial states. However, with the assisted strategies the mean field iteration did not converge, it had an oscillating behavior. The corresponding plots on the figures show the average of the results from 10 different random initial points and also have error bars indicating the minimum and maximum values obtained.

With the increase of the load, the performance of both the informed and distance aware strategies drop sharply. Cars spend too much time rambling around to find a position where the number of free spaces and the distance from the target is appropriate. To be fair, our model did not include an important factor: in the reality the behavior of the drivers (in particular the σ_F and σ_D parameters) depends on the utilization of the garage.

Nevertheless, the results make it clear that the intelligent parking systems can be really efficient. Both variants were able to reduce the time to reach the target, including the search phase and the walking time. Especially the "assisted case 2.", that takes the total time to target into account, was successful in reducing the number of moving cars in the garage even at high load, which is beneficial from the environmental protection point of view as well.

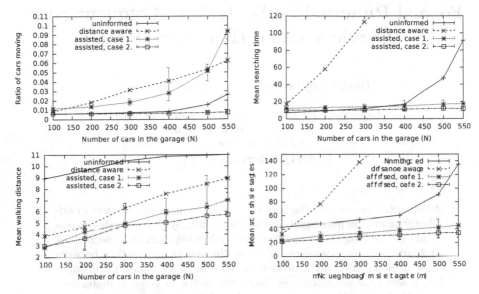

Fig. 5. The performance measures as the function of the load

Acknowledgment. This work was supported by the Hungarian research project OTKA K101150, by the European Union (co-financed by the European Social Fund) through the TAMOP-4.2.2C-11/1/KONV-2012-0001 project, and by the János Bolyai Research Scholarship of the Hungarian Academy of Sciences.

References

1. Bojja, J., Kirkko-Jaakkola, M., Collin, J., Takala, J.: Indoor 3D navigation and positioning of vehicles in multi-storey parking garages. In: Acoustics, Speech and Signal Processing (ICASSP), pp. 2548–2552. IEEE (2013)
2. Caragliu, A., Bo, C.D., Nijkamp, P.: Smart cities in Europe. Technical report (2009)
3. Gódor, G., Huszák, Á., Farkas, K.: Intelligent indoor parking. In: Proc. of Global Virtual Conference, GV-CONF (2013)
4. Le Boudec, J.-Y., McDonald, D., Mundinger, J.: A generic mean field convergence result for systems of interacting objects. In: Quantitative Evaluation of Systems, QEST, pp. 3–18. IEEE (2007)
5. Liu, J., Chen, R., Chen, Y., Pei, L., Chen, L.: iParking: An intelligent indoor location-based smartphone parking service. Sensors 12(11), 14612–14629 (2012)
6. Varga, A.: The OMNeT++ discrete event simulation system. In: Proceedings of the European Simulation Multiconference (ESM 2001), vol. 9, p. 185. sn (2001)

Formal Punctuality Analysis of Frequent Bus Services Using Headway Data

Daniël Reijsbergen* and Stephen Gilmore

Laboratory for Foundations of Computer Science
The University of Edinburgh
Edinburgh, Scotland
dreijsbe@inf.ed.ac.uk

Abstract. We evaluate the performance of frequent bus services in Edinburgh using the punctuality metrics identified by the Scottish Government. We describe a methodology for evaluating each of these metrics that only requires measurements of bus 'headways' — the time between subsequent bus arrivals. Our methodology includes Monte Carlo simulation and time series analysis. Since one metric is given in ambiguous language, we provide a formal description of the two most plausible interpretations. The automated nature of our method allows public transport operators to continuously assess whether the performance of their network meets the targets set by government regulators. We carry out a case study using Automatic Vehicle Location (AVL) data involving two frequent services, including the AirLink service to and from Edinburgh airport.

Keywords: Public transportation, punctuality, headways.

1 Introduction

A key feature of a sustainable city is a well-run public transportation network. This is witnessed, among other reasons, by the fact that satisfaction with public transport quality is included as an indicator for a 'smart' city [4]. One important measure for the performance of a public transport network is its *punctuality*, as this has been observed to be a major factor in passenger satisfaction and perceived service quality [3]. However, a formal definition of punctuality is not straightforward to give, partially because passenger perception of punctuality may depend on the nature of the service. In particular, for a non-frequent service (e.g., one bus departure every 30 minutes) strict timetable adherence is the main factor for punctuality. However, strict timetable adherence is less relevant for frequent services, which are defined as those with one bus departure every ten minutes or less. Punctuality metrics for frequent services are primarily dependent on the probability distribution of the times between departures — the so-called 'headways'. In general, less headway variance means better punctuality.

---

* This work is supported by the EU project QUANTICOL, 600708. Corresponding author.

A. Horváth and K. Wolter (Eds.): EPEW 2014, LNCS 8721, pp. 164–178, 2014.

Several punctuality metrics have been proposed in the scientific literature; [9] and [11] are two recent papers that present an overview. In this paper, we focus on the three punctuality metrics for frequent services identified in the guidance document on Bus Punctuality Improvement Partnerships by the Scottish Government [12]; all of these depend on the headways. Two of these metrics coincide with the metrics identified in [9] and [11].[1] The third metric does not; furthermore, it is ambiguously worded, so we formalise the two most plausible interpretations, resulting in a total of four metrics. We then provide a formal methodology for the evaluation of the four metrics that only requires headway measurements. The methodology is statistical in nature, so we particularly focus on providing approximate confidence intervals for the estimates of the metrics. This is a challenge because the probability distributions of some of the quantities under consideration are unclear. The evaluation of the two new metrics in particular is non-trivial, and we apply a range of statistical techniques including time series analysis, bootstrapping [6] and Monte Carlo simulation. Finally, we apply our methodology to a real-world set of headway measurements obtained using low-frequency Automatic Vehicle Location (AVL) data provided to us by the Lothian Buses company, based in Scotland and operating an extensive bus network in Edinburgh.

The outline of the paper is as follows. In Section 2, we discuss the routes considered and the datasets used. In Section 3, we formally define the three bus punctuality metrics used by the Scottish Government. We discuss a time series model for sequences of headway measurements in Section 4, and discuss the bootstrapping method for constructing approximate confidence intervals in Section 5. In Section 6, we evaluate the performance of two services operated by Lothian Buses using the punctuality metrics of Section 3. Section 7 concludes the paper.

2 Description and Visualisation of Routes and Data

In this section we explain the data processing that was applied to the raw AVL data before using it to compute the punctuality measures of interest. We had six datasets available: three for Route 100 (the AirLink service) and three for Route 31. For Route 100, three bus stops are of interest: the airport, the zoo, and George Street in Edinburgh city centre. For Route 31, the bus stops of interest are East Craigs, the zoo, and the Scott Monument on Princes Street in the city centre. The number of observations in each dataset is specified in Table 1.

The AVL data records the position of each bus in the fleet. Each bus has a unique identifier called a fleet number, and the assignment of buses to routes is captured in a schedule which is drawn up before the bus service begins for the day. The bus schedule maps buses by fleet number to routes but it can

---

[1] In particular, the metric of Section 6.1 is also mentioned in [9], while the related notion of the headway coefficient of variation is preferred in [11]. The metric of Section 6.2 is related to what is identified as an *"Extreme-Value based"* waiting time measure in [9]; it is also related to the Earliness Index of [11].

Table 1. Overview of the dataset sizes $n_1, \ldots, n_6$. Measurements were collected between 28th January 2014 at 11:31:14 and 30th January 2014 at 12:38:31.

Route	100 (AirLink)			31		
Stop	Airport	Zoo	George St	East Craigs	Zoo	Princes St
# measurements	127	126	128	102	102	105

change dynamically during the day in response to unpredictable problems such as mechanical failures of vehicles, or unexpectedly high or low levels of passenger demand. Thus the bus schedule serves as a guide for interpreting the AVL data but is not always accurate because it has not always been updated to record all unexpected events which occurred during the day. To address this problem, we use our custom visualisation tool [16] to plot buses on a map of Edinburgh. This allows us to check that they are serving the routes which we believe they are. If this was not done, incorrectly assigned buses would invalidate the computation of headway on routes. One Route 31 bus had to be identified manually. Furthermore, we suspect that one or two of the measurements in the Princes Street dataset correspond to wrongly assigned buses, but we have no evidence of this.

The schedule changes which make headway computation more difficult tend to occur at the start and the end of the day, when bus services have low frequency and the same bus is being used to serve several different routes. To eliminate this potential source of error in our interpretation of the data we restricted our observations to lie only between 9:00 and 17:00, when buses are frequent and rarely subject to route reassignments.

We linearly interpolate the AVL measurement data down to a granularity of one second between data points, and we detect departures from stops by dividing bus routes into zones and counting a departure as occurring when a bus moves from a zone containing a stop to the subsequent zone, using interpolated data. The bus stop zones were chosen such that they did not contain traffic lights.

3   Punctuality Measures

As mentioned in the introduction, we focus on the punctuality metrics set out by the Scottish government in [12]; we formalise these metrics in this section. Since buses are subject to a variety of unpredictable influences such as the number of passengers at bus stops and road congestion, the requirements are inherently *stochastic*. The randomness of the system is modelled through the headway, denoted by a random variable Y which takes values from $\mathbb{R}^+$ and is measured in seconds. The kth dataset, $k = 1, \ldots, 6$, is then a sequence $(y_{k1}, y_{k2}, \ldots, y_{kn_k})$ of realisations of Y_k, where n_k is as given in Table 1 (in the paper, we often leave out the dataset index k for brevity). Let $\mu = \mathbb{E}(Y)$ and $\sigma^2 = \mathrm{Var}(Y)$. The requirements are then as follows.

In §2.13 of [12], which is under the header *"Starting point of the journey"*, we find the following. *"For frequent services it is expected that on at least 95% of occasions:*

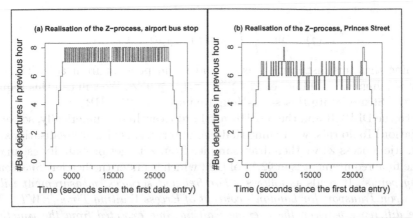

Fig. 1. Realisations of the process Z on 29th January 2014 for the airport bus stop dataset (left) and the Princes Street dataset (right)

- *Six or more buses will depart within any period of 60 minutes; and*
- *The interval between consecutive buses will not exceed 15 minutes"*

Using seconds as the granularity, the latter requirement can be expressed as $\mathbb{P}(Y \leq 900) \geq 95\%$. We will call the probability $\mathbb{P}(Y \leq 900)$ the Extreme-Value Waiting Time (EVWT).

The former requirement is more intricate: given a sequence of headway measurements $(y_1, y_2, \ldots, y_{n_k})$, define for $t \in \mathbb{R}^+$

$$u(t) = \max\left\{ j : \sum_{i=1}^{j} y_i \leq t \right\} \quad \text{and} \quad d(t) = \max\left\{ j : \sum_{i=1}^{j} y_i \leq t - 3600 \right\}.$$

Let $z(t) = u(t) - d(t)$, then $z(t)$ denotes the number of buses that departed in the hour prior to t. By construction, $u(t) \geq d(t)$ for all t so z is defined on $\mathbb{R}^+$. Figure 1 depicts the evolution of z for two of the six datasets.

The requirement that on 95% of "occasions" there must be six or more buses departures "within any period of 60 minutes" is slightly ambiguous; we will consider two interpretations. First, if we focus on the word "any", we could say that an "occasion" represents a time interval $[a, b]$ (a reasonable assumption would be that a denotes an hour after the departure of the first bus and b the departure of the last bus in a single day), and that $z(t)$ needs to be at least 6 at "any" point $t \in [a, b]$. The full requirement can then be expressed as: $\mathbb{P}(\forall t \in [a, b] : z(t) \geq 6) \geq 0.95$. We will call this requirement the Day-Long Buses-per-Hour Requirement (DLBHR).

The second interpretation is in terms of steady-state probabilities: for any measurement (assumed to be conducted when z is in steady-state) the probability that z is at least 6 needs to be at least 95%. To express this formally, define for any Boolean expression A

$$\mathbf{1}(A) = \begin{cases} 1 & \text{if } A, \\ 0 & \text{otherwise,} \end{cases}$$

and

$$\pi_z(j) = \lim_{t \to \infty} \frac{1}{t} \int_0^t \mathbf{1}(z(\tau) = j)d\tau, \quad \forall j \in \mathbb{N},$$

assuming that this distribution exists and is independent from $z(0)$. Then the latter requirement could be read as $\sum_{j=6}^{\infty} \pi_z(j) \geq 95\%$. We will call this requirement the Steady-State Buses-per-Hour Requirement (SSBHR).

Both the DLBHR and the SSBHR are hard to evaluate numerically, so we use simulation. To do this, we assume that the observations of z are realisations of a stochastic process Z; we then draw samples from Z to get probability estimates.

The final requirement is in §2.14 of [12], which is under the header '*Subsequent timing points*'. It reads as follows. "*For frequent services, measurement will be based upon Transport for London's concept of Excess Waiting Time (EWT). This is the difference between the average waiting time expected from the timetable, and what is actually experienced by passengers on the street. TC standards specify that EWT should not exceed 1.25 minutes.*"

The "*average waiting time expected from the timetable*" is assumed to be $\frac{1}{2}\mu$,[2] while the average waiting that "*is actually experienced by passengers on the street*" is given by $\frac{1}{2}(\mu^2 + \sigma^2)/\mu$ (see [7] or [14]). Hence, the Excess Waiting Time equals $\frac{1}{2}\sigma^2/\mu$. The maximum, according to the standards of the Traffic Commissioner (TC), is 1.25 minutes, or, equivalently, 75 seconds.

In summary, the EVWT, SSBHR and DLBHR are relevant only at journey starting points (which in our case study refers to the airport for Route 100 and to East Craigs for Route 31), while the EWT is relevant at all subsequent timing points. However, the former three metrics are also evaluated at the other timing points in Section 6 ; this is done for illustrative purposes.

4   Time Series Modelling

As we discussed in the previous section, the stochasticity of the headway between frequent buses is modelled using the random variable Y. To investigate whether the four requirements are satisfied, varying degrees of knowledge of the distribution of Y are needed. To calculate the EWT, we only need to know the expectation and variance of Y. To calculate whether the requirement on the time between subsequent bus departures is valid, we need to know the 95th percentile (i.e., the value x such that $\mathbb{P}(Y < x) = 95\%$; the requirement is satisfied if x is below 900). This is known if we know the quantiles and, hence, the entire probability distribution. To evaluate the requirements on the Z-process, we need to know the distribution of vectors $(Y_1, Y_2, \ldots)$ of measurements. This would be as hard as knowing the distribution of individual samples from Y if the samples that make up the vector were mutually independent. We will argue later in this section that they are not.

---

[2] There is no official scheduled headway for the AirLink service because the timetable for this service only says: "*at least every 10 minutes*". For Route 31, the difference between the timetabled and average observed headway is negligible (600 vs. roughly 580 seconds).

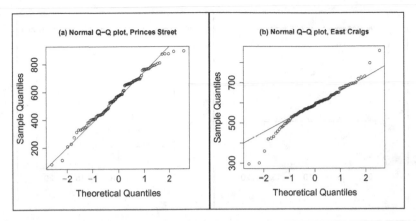

Fig. 2. Normal distribution Q-Q plots for the Princes Street dataset (left) and the East Craigs dataset (right)

In this paper, we assume that individual samples drawn from Y have a normal (Gaussian) distribution. To visualise whether the normality assumption is valid, we use normal Q-Q plots, i.e., plots of the quantiles of the empirical distribution against those of the normal distribution. This is done in Figure 2 for two datasets. A formal measure of the resemblance of an empirical distribution to the normal distribution is the Shapiro-Wilk test statistic, which assigns a value between 0 and 1 to an empirical distribution such that high values of the statistic represent close resemblance to the normal distribution. As the name suggests, it is used for the Shapiro-Wilk test [5] in which the null hypothesis that the sample has been drawn from the normal distribution is evaluated. The key result of the test is its p-value, which is the confidence in the validity of the null hypothesis. Values below 5% imply that the null hypothesis can be rejected at the 95% level. As we can see in Table 2, this only happens for the East Craigs dataset because of its fat (compared to the normal distribution) tails, especially on the left. Assuming that Y is normally distributed, the probability of interest can be computed using the normal cumulative distribution function, which is implemented in the statistical package R [13].

To generate a sample $(y_1, y_2, \ldots)$, we need to incorporate the correlation between measurements. Figure 3(a) depicts an Autocorrelation Function (ACF) plot for the dataset for the airport bus stop. The lag one autocorrelation is especially visible. The correlation is due to at least three sources:

1. Dependence by construction. Consider three buses with μ time units between departures; if the second is ϵ time units late, then the first headway will be $\mu + \epsilon$ and the second headway $\mu - \epsilon$ (negative correlation). This affects the lag one AC.
2. If a bus is late, then the number of passengers at the stop will be greater than normal, causing an additional delay. The next bus will need to pick up fewer passengers and may start to run early (negative correlation). This phenomenon is also mentioned in [1], and affects the lag one AC.

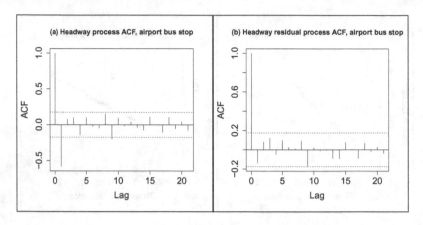

Fig. 3. ACF plots for the headway process $(y_1, \ldots, y_n)$ of the airport bus stop dataset (left) and its residual process $(\hat{\epsilon}_1, \ldots, \hat{\epsilon}_n)$ after MA(1) fitting (right). The blue lines represent the levels at which the ACs are significant at the 5% level. Note that with 20 ACs plotted, the expected number of false positives at the 5% level is 1 (this could explain the seemingly significant lag 9 AC in both graphs).

3. Factors that cause headway variation may persist: if one bus is late due to heavy traffic, then this traffic may have an influence on the next bus as well (positive correlation). This affects all ACs, in a decreasing fashion as the lag becomes bigger.

The lag one autocorrelation between the headways can be modelled using a Moving Average MA(1) time series model:

$$Y_i = \mu + \epsilon_i + \theta\epsilon_{i-1}, \quad \text{where } \epsilon_i \sim N(0, \sigma_\epsilon^2) \;\; \forall i = 1, \ldots, n.$$

The parameter θ captures the three forms of lag one correlation. The estimates of θ — denoted by $\hat{\theta}$ — for each of the six datasets are displayed in Table 2. Since the values $\hat{\theta}$ are estimates, we include the confidence in the null hypothesis that the true value θ equals 0. In each case, this is below 5%. Low values indicate that the time series model has a low explanatory power.

The ϵ's are commonly termed the *error terms*; the estimates $\hat{\epsilon}_1, \ldots, \hat{\epsilon}_n$ of these error terms based on a time series fit are called the *residuals*. We display an ACF plot for the residuals of the MA(1) model in Figure 3(b); as we can see, the lag one autocorrelation observed in Figure 3(a) is not present here. MA(1) models are part of the wider class of Autoregressive Moving Average ARMA(p,q) models. In general, these can be expressed as

$$Y_i = \mu + \sum_{j=1}^{p} \phi_j Y_{i-j} + \sum_{j=0}^{q} \theta_j \epsilon_{i-j}, \quad \text{where } \epsilon_i \sim N(0, \sigma_\epsilon^2) \;\; \forall i = 1, \ldots, n,$$

with $\theta_0 = 1$. The autoregressive terms (i.e., the ones that depend on the ϕ_j's) are well-suited to capture the effect of the third source of correlation on the higher lag ACs. However, if the higher lag ACs are small then little explanatory power is added while the effect of the MA-terms becomes harder to isolate, so the net effect of adding these terms need not be positive. We have observed that MA(1) provides the best fit for all datasets, except the Princes Street dataset for Route 31. In this case the Autoregressive AR(1) process is best, although the difference is rather small (in terms of the Akaike Infomation Criterion). The MA(1) parameters θ and σ_ϵ can be estimated using methods of the tseries package in R. Given estimates of θ and σ_ϵ, we can draw realisations of $(Y_1, Y_2, \ldots)$ by drawing realisations of ϵ; methods for drawing standard normal random variables are implemented in R and the SSJ package in Java.[3] Realisations of Z can then be drawn analogously.

Table 2. Estimates of θ. Note that the values for $\hat{\theta}$ are negative because the first two sources of correlation mentioned in Section 4 apparently outweigh the third. The p-values of the t-test for $\theta = 0$ and the Shapiro-Wilk normality test for Y are given in the final two columns.

#	Stop	$\hat{\theta}$	t-test p-value	SW p-value
	Airport	-0.74903	$< 2 \cdot 10^{-16}$	0.1615
100	Zoo	-0.68416	$< 2 \cdot 10^{-16}$	0.0912
	George St	-0.56441	$2.22 \cdot 10^{-16}$	0.4436
	East Craigs	-0.21045	$2.07 \cdot 10^{-2}$	0.0047
31	Zoo	-0.36521	$6.24 \cdot 10^{-5}$	0.3439
	Princes St	-0.24497	$2.62 \cdot 10^{-3}$	0.1783

5   The Bootstrapping Method for Confidence Intervals

In the previous section, we described how to estimate the punctuality metrics used by the Scottish government. The estimates are based on realisations of $(Y_1, \ldots, Y_n)$, and will typically be different when the experiment is repeated. To account for the variation in the estimates, the estimates are given in the form of an interval estimate called the *confidence interval*. The interpretation of a $(1 - \alpha)$ confidence interval for a statistic is as follows: if the experiment is repeated N times, then the number of confidence intervals that do not contain the true value of the statistic is expected to be αN. Throughout this paper we use $\alpha = 5\%$.

Whether a confidence interval for a statistic can be computed analytically (or approximated numerically) depends on how easy it is to express the probability distribution of the statistic. For some commonly-used test statistics their distribution is known explicitly, which means that confidence intervals can be constructed using methods implemented in common statistical tools such as R. However, even when nothing is known about the probability distribution of the test statistic, one can construct approximate confidence intervals using the *bootstrapping* method. A broad variety of bootstrapping methods exist; we use two

---

[3] www.iro.umontreal.ca/~simardr/ssj/indexe.html

of them, namely the non-parametric method of *case resampling* (through Monte Carlo simulation) and *parametric* bootstrapping.

5.1 Case Resampling (Monte Carlo)

Case resampling is one of the most general forms of bootstrapping; given a sample $x = (x_1, \ldots, x_n)$ of independent and identically distributed realisations of some random variable X, and a test statistic f that is a function of x, the approach is as follows. Let b be some positive integer. For each $j \in \{1, \ldots, b\}$, randomly draw n elements of x with replacement; let the new sample be x_j and $f_j = f(x_j)$. Let $f^{(i)}$ be the ith smallest element obtained this way; an $(1 - \alpha)$ confidence interval is then given by

$$[f^{(b\alpha/2)+1}, f^{(b(1-\alpha/2))}], \tag{1}$$

assuming that $b\alpha/2$ and $b(1 - \alpha/2)$ are integers (we would use the floor and ceiling functions otherwise). The approximation gets better when the sample (empirical) distribution more closely resembles the true distribution of X.

5.2 Parametric Bootstrap

Parametric bootstrapping works the same way as case resampling, with the exception that we now have a stochastic model that allows us to draw random samples from X. The bootstrapping samples x_j, $j = 1, \ldots, n$, are then obtained directly from the distribution of X. We still use (1) as the confidence interval.

6 Results

In this section, we discuss the results for the four performance metrics and requirements discussed in the previous sections: the Excess Waiting Time (EWT), the Extreme-Value Waiting Time (EVWT), the Steady-State Buses-per-Hour Requirement (SSBHR) and the Day-Long Buses-per-Hour Requirement (DLBHR). Each of these has its own subsection.

6.1 Excess Waiting Time

The Excess Waiting Time is relatively easy to evaluate: its computation only requires knowledge of μ and σ. These basic statistics are displayed for each of the six datasets in Table 3. Note that AirLink buses depart roughly every eight minutes, whereas Route 31 buses depart roughly every 10 minutes. We note that despite difference in means, the headway variance in related stops is very close. Consequently, the EWT is higher for the AirLink service than for Route 31. We further observe that the variance increases as the buses complete a larger part of the route, and with it the excess waiting time, which is what one would expect (as a bus completes its route it is increasingly subjected to sources of journey time variation, e.g., passenger numbers at stops).

Table 3. EWT for each dataset, together with estimates of μ and σ. Confidence intervals were obtained using a parametric bootstrap with $b = 10\,000$.

#	Stop	$\hat{\mu}$	$\hat{\sigma}$	EWT	EWT c.i.
	Airport	477.2047	87.2564	7.9773	[5.182, 9.417]
100	Zoo	475.2381	129.0810	17.5301	[12.261, 22.226]
	George St	473.5781	183.5627	35.5752	[26.087, 46.722]
	East Craigs	585.7255	88.2538	6.6488	[5.261, 9.350]
31	Zoo	588.7157	119.5863	12.1458	[9.231, 16.954]
	Princes St	568.8952	170.6736	25.6018	[19.156, 34.388]

In each dataset, the EWT is below the 75 second threshold. However, as mentioned in Section 5, the EWT estimates are subject to uncertainty because they are based on random samples. The empirical EWT can be computed from the sample variance and sample mean of a set of headway measurements $(Y_1, \ldots, Y_n)$. However, to generate a bootstrapped confidence interval for this statistic, we cannot use case resampling, as the measurements are correlated (although the effect of the correlation would vanish in larger samples). To remedy this, we conduct a parametric bootstrap using the time series model. The error terms ϵ are assumed not to have autocorrelation, so we could either use case resampling using the empirical dataset or draw samples directly from the normal distribution. The confidence intervals in Table 3 were obtained using the latter approach. In all datasets, the EWT is well below the 75 second mark with 95% confidence.

6.2 Extreme-Value Waiting Time

Table 4 summarises the results for the EVWT, i.e., the probability of a headway of over 900 seconds. As with the EWT, an estimate for the EVWT is easy to obtain; we only have to count the number of times this event occurred in the empirical dataset. Since this number is approximately binomially distributed, we can construct Clopper-Pearson confidence intervals [6] for the true probability. These intervals are very broad, owing to the small number of samples. For example, the upper bound of the confidence interval for the airport dataset is 2.863%, even though one would expect a much smaller probability based on the fact that the variance of Y is very small (as can be seen in Table 3). Using the assumption that Y is normally distributed, we can construct an estimate of the EVWT by using the normal distribution function combined with the estimates for μ and σ of Table 3. The results are in the '$\mathbb{P}(Y > 900)$' column of Table 4. Note that for East Craigs, the probability will be underestimated because this dataset differs so much from one with a normal distribution because it contains many more extreme values than one would expect.

Based on the binomial confidence intervals, we can conclude that for all datasets except George Street, the requirement on the EVWT is met with more than 95% confidence. Based on the assumption of normality, the requirement is met for all datasets.

Table 4. Probability of over 15 minute headway for each dataset. The first two numerical columns contain empirical estimates of these probabilities and exact (binomial / Clopper-Pearson) confidence intervals. In the final column we display the exact probabilities based on the normal distribution.

#	Stop	EVWT	EVWT c.i.		$\mathbb{P}(Y > 900)$
	Airport	0	[0,	2.863] $\cdot 10^{-2}$	$6.3166 \cdot 10^{-7}$
100	Zoo	0	[0,	2.885] $\cdot 10^{-2}$	$4.9976 \cdot 10^{-4}$
	George St	$2.344 \cdot 10^{-2}$	[0.486,	6.697] $\cdot 10^{-2}$	$1.0089 \cdot 10^{-2}$
	East Craigs	0	[0,	3.552] $\cdot 10^{-2}$	$1.8470 \cdot 10^{-4}$
31	Zoo	0	[0,	3.552] $\cdot 10^{-2}$	$4.6205 \cdot 10^{-3}$
	Princes St	0	[0,	3.452] $\cdot 10^{-2}$	$2.6191 \cdot 10^{-2}$

6.3 Steady-State Buses-per-Hour Requirement

An empirical estimate of the SSBHR for a given service is easy to obtain; in the realisation of z in this dataset (as is visualised in Figure 1), count the amount of time that z is lower than 6 and divide this by the total time. Formally, given a realisation of z on $[0, t]$, this means that the estimate $\hat{\pi}_z(k)$ for the steady-state probability of being in state k can be computed as

$$\hat{\pi}_z(k) = \frac{1}{t} \int_0^t \mathbf{1}(z(\tau) = k) d\tau.$$ (2)

The results are given in Table 5; instead of just the percentage of time that z spends below 6, the entire empirical steady-state distribution is given. For a given day, we start observing z one hour after the first bus departure (we assume that the process has approximately reached steady-state by then), and stop at the final bus departure.

Table 5. Empirical steady-state distributions of z for each of the six datasets

#	Stop	$\hat{\pi}_z(5)$	$\hat{\pi}_z(6)$	$\hat{\pi}_z(7)$	$\hat{\pi}_z(8)$	$\hat{\pi}_z(9)$
	Airport	0	$3.431 \cdot 10^{-4}$	0.460	0.540	0
100	Zoo	0	$1.292 \cdot 10^{-3}$	0.472	0.525	$1.702 \cdot 10^{-3}$
	George St	0	$3.336 \cdot 10^{-2}$	0.483	0.454	$2.927 \cdot 10^{-2}$
	East Craigs	$3.001 \cdot 10^{-2}$	0.753	0.215	$1.573 \cdot 10^{-3}$	0
31	Zoo	$5.540 \cdot 10^{-2}$	0.705	0.240	0	0
	Princes St	$8.765 \cdot 10^{-2}$	0.675	0.233	$5.167 \cdot 10^{-3}$	0

Again, the empirical steady-state distribution is subject to variation, so we want to construct confidence intervals for these values. We have two options. First, we can use the time series model to generate long-run realisations of Z and use this to construct a parametric bootstrapping interval. Unfortunately, we have found that the parametric model is not well-suited for estimating the relatively small steady-state probabilities of low values of Z. Reaching the lower values of Z is particularly influenced by tail behaviour that is not captured well by the time series model, which tries to capture the dependence between subsequent headways in a single parameter (even though this dependence may vary throughout the day).

Therefore, we aim to use case resampling to construct a bootstrapping confidence interval. The question is what values to resample; obviously, sampling from the realisations $(y_1, \ldots, y_n)$ directly cannot be expected to work well because this would ignore the correlation between these realisations and the behaviour of Z depends on the behaviour of sequences of realisations of Y. Hence, we apply a renewal-like argument to estimate $\pi_z(j)$. Given a realisation of Z on $[0, t]$ and $m, l \in \mathbb{N}$, we partition $[0, t]$ into intervals $(I_{j1}, \ldots, I_{jm}) = ([\underline{I}_{j1}, \overline{I}_{j1}], \ldots, [\underline{I}_{jm}, \overline{I}_{jm}])$, when z equals j, and intervals $[\underline{J}_{j1}, \overline{J}_{j1}], \ldots, [\underline{J}_{jl}, \overline{J}_{jl}]$, when z does not equal j. It follows trivially from (2) that

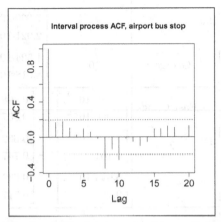

Fig. 4. ACF plot (see also Figure 3) for the series $(I_{j1}, \ldots, I_{jm})$ for $j = 7$ for the airport bus stop

$$\hat{\pi}_z(j) = \frac{1}{t} \sum_{i=1}^{m} (\overline{I}_{ji} - \underline{I}_{ji}).$$

The idea is then to resample from the vector $(I_{j1}, \ldots, I_{jm})$ to construct a bootstrapping confidence interval. The key observation is that the intervals $(I_{j1}, \ldots, I_{jm})$ do have autocorrelation, but 1) that this is significantly less than for the headways and that 2) the effect of the correlation vanishes in large samples while the autocorrelation between the headways has an impact on the probability distribution of Z. The autocorrelation between the intervals for $j = 7$ for the airport dataset is displayed in Figure 4.

The results of the bootstrapping procedure are displayed in Table 6. Note that we have also resampled the values for $[\underline{J}_{j1}, \overline{J}_{j1}], \ldots, [\underline{J}_{jl}, \overline{J}_{jl}]$, to obtain the total times; again the correlation between measurements of I and J vanishes asymptotically. The confidence intervals for different values of j are not independent because they are implicitly based on the same samples. We observe that the AirLink seems to always satisfy the SSBHR. However, based on the confidence intervals for $z = 5$, we cannot conclude with 95% confidence for the Route 31 that the requirement will be satisfied. In fact, for Princes Street we can conclude with 95% confidence that the requirement will *not* be satisfied. We note that this is not necessarily a problem, as the SSBHR is only imposed on the starting points of the routes.

Table 6. Confidence intervals for steady-state distributions of z for each of the six datasets, generated using bootstrapping with case resampling with $b = 1\,000\,000$

#	Stop	$\hat{\pi}_z(5)$	$\hat{\pi}_z(6)$	$\hat{\pi}_z(7)$	$\hat{\pi}_z(8)$	$\hat{\pi}_z(9)$
	Airport	0	$\begin{bmatrix} 2.875 \cdot 10^{-4}, \\ 4.253 \cdot 10^{-4} \end{bmatrix}$	$\begin{bmatrix} 0.434, \\ 0.485 \end{bmatrix}$	$\begin{bmatrix} 0.513, \\ 0.566 \end{bmatrix}$	0
100	Zoo	0	$\begin{bmatrix} 7.660 \cdot 10^{-4}, \\ 2.921 \cdot 10^{-3} \end{bmatrix}$	$\begin{bmatrix} 0.439, \\ 0.505 \end{bmatrix}$	$\begin{bmatrix} 0.493, \\ 0.557 \end{bmatrix}$	$\begin{bmatrix} 6.200 \cdot 10^{-4}, \\ 4.298 \cdot 10^{-3} \end{bmatrix}$
	George St	0	$\begin{bmatrix} 1.524 \cdot 10^{-2}, \\ 8.018 \cdot 10^{-2} \end{bmatrix}$	$\begin{bmatrix} 0.437, \\ 0.53 \end{bmatrix}$	$\begin{bmatrix} 0.374, \\ 0.528 \end{bmatrix}$	$\begin{bmatrix} 1.450 \cdot 10^{-2}, \\ 9.070 \cdot 10^{-2} \end{bmatrix}$
	East Craigs	$\begin{bmatrix} 1.819 \cdot 10^{-2}, \\ 5.060 \cdot 10^{-2} \end{bmatrix}$	$\begin{bmatrix} 0.698, \\ 0.807 \end{bmatrix}$	$\begin{bmatrix} 0.157, \\ 0.288 \end{bmatrix}$	$\begin{bmatrix} 4.511 \cdot 10^{-4}, \\ 3.502 \cdot 10^{-3} \end{bmatrix}$	0
31	Zoo	$\begin{bmatrix} 3.294 \cdot 10^{-2}, \\ 0.101 \end{bmatrix}$	$\begin{bmatrix} 0.662, \\ 0.747 \end{bmatrix}$	$\begin{bmatrix} 0.186, \\ 0.32 \end{bmatrix}$	0	0
	Princes St	$\begin{bmatrix} 6.044 \cdot 10^{-2}, \\ 0.13 \end{bmatrix}$	$\begin{bmatrix} 0.631, \\ 0.717 \end{bmatrix}$	$\begin{bmatrix} 0.184, \\ 0.291 \end{bmatrix}$	$\begin{bmatrix} 1.547 \cdot 10^{-3}, \\ 1.580 \cdot 10^{-2} \end{bmatrix}$	0

6.4 Day-Long Buses-Per-Hour Requirement

Table 7 summarises the results for the DLBHR. Note that it is impossible to estimate this measure from our current dataset without assuming an underlying time series model, as we only have a single measurement of an entire day (29th January). Hence, we draw realisations from Z by simulating the underlying time series model despite the weaknesses of this approach discussed in Section 6.3.

Confidence intervals are easy to generate because in a sample of day-long executions of Z, the number of days in which the process did not drop below 6 is binomially distributed. Hence, we can construct Clopper-Pearson confidence intervals using R. We conclude that, assuming that our time series model is correct, the requirement is met for all the AirLink stops with over 95% confidence (although it just barely holds for George Street), and the requirement is met at none of the Route 31 stops. This is not surprising; because the Route 31 service operates slightly over six buses per hour, even small deviations from the schedule cause a violation. The AirLink service, which runs about 7.5 buses per hour, is much more robust in terms of the DLBHR.

To construct the Clopper-Pearson confidence intervals for Table 7, it is necessary to fix a sample size beforehand. We have used 100 000 samples for the results in Table 7, but this choice is typically non-trivial; to evaluate whether the true probability is smaller than or greater than 95%, larger sample sizes are needed when the true probability is closer to 95%. A solution is to use the conceptual framework of hypothesis testing for *statistical model checking* [8]. In particular, we can use sequential tests that are able to terminate the simulation procedure as soon as enough evidence has been collected to make a statement

Table 7. Whole day probability estimates, based on 100 000 samples. We also include Clopper-Pearson confidence intervals and the sample sizes N needed by the SPRT to reach a conclusion (which was correct in all of our experiments).

#	Stop	$\hat{p}$	95% C.I.			SPRT N
	Airport	1	[0.9999631,	1]		1399
100	Zoo	0.99999	[0.9999443,	0.9999997]		1399
	George St	0.95256	[0.9512243,	0.9538694]		12701
	East Craigs	0.00159	[1.352615,	1.857001]	$\cdot 10^{-3}$	75
31	Zoo	0.00022	[1.378778,	3.330638]	$\cdot 10^{-4}$	74
	Princes St	0.00229	[2.003263,	2.606186]	$\cdot 10^{-3}$	75

about whether the requirement has been satisfied. We use the Sequential Probability Ratio Test (SPRT) [15] with indifference level $\delta = 0.001$ and $\beta = \alpha = 5\%$.[4] As we can see in the table, the requirement is the hardest to check for George Street; in all other cases, fewer than 10 000 samples were needed.

7 Conclusions

In this paper we have formalised the bus punctuality metrics used by the Scottish government. We investigated the performance of two services operated by Lothian Buses using these metrics. To do this, we have applied a number of statistical techniques such as time series modelling, bootstrapping and the sequential testing framework that is also employed in statistical model checking. Route 31, which operates six buses per hour, does better in terms of Excess Waiting Time, while the AirLink service, which runs over seven buses per hour, does better in terms of the Steady-State and Day-Long Buses-per-Hour Requirement.

A key feature of our methodology is its automated nature. Currently, when bus networks are subjected to a formal review by traffic regulators, the headway data is gathered manually by inspectors who are physically sent to the selected bus stops. Since AVL data is gathered systematically to aid live bus arrival time prediction at bus stops, an automated methodology for using the data to evaluate punctuality allows the traffic operator to detect potential shortcomings prior to the review, meaning that a fine can be avoided. This is of particular interest to Lothian Buses, which has been fined by regulators in the past [2].

As part of further research, we aim to improve our model by incorporating the non-Gaussian tail behaviour of the East Craigs dataset. We also hope to investigate possible time dependence (within the day) of θ and the error terms.

Acknowledgements. This work is supported by the EU project *QUANTICOL*, 600708. The authors thank Bill Johnston of Lothian Buses and Stuart Lowrie of

---

[4] Note that the SPRT's assumptions with $\delta = 0.001$ are only just valid for the George Street dataset; for a discussion of the effect of the parameter choice on the test's output, see [10].

the City of Edinburgh council for providing access to the data which was used for the case study. We would also like to thank Allan Clark and Mirco Tribastone for their helpful comments on a draft version of this paper.

References

1. Berkow, M., Chee, J., Bertini, R.L., Monsere, C.: Transit performance measurement and arterial travel time estimation using archived AVL data. ITE District 6 Annual Meeting (2007)
2. Lothian Buses faced fine for bad service. Edinburgh Evening News, Web (June 21, 2010)
3. Friman, M.: Implementing quality improvements in public transport. Journal of Public Transportation 7(4) (2004)
4. Giffinger, R., Fertner, C., Kramar, H., Kalasek, R., Pichler-Milanović, N., Meijers, E.: Smart cities: Ranking of European medium-sized cities. Technical report, Vienna University of Technology (2007)
5. Hill, T., Lewicki, P.: Statistics: methods and applications: a comprehensive reference for science, industry, and data mining. StatSoft (2006)
6. Hollander, M., Wolfe, D.A., Chicken, E.: Nonparametric statistical methods. John Wiley & Sons (2013)
7. Holroyd, E.M., Scraggs, D.A.: Waiting times for buses in central London. Printerhall (1966)
8. Legay, A., Delahaye, B., Bensalem, S.: Statistical model checking: An overview. In: Barringer, H., et al. (eds.) RV 2010. LNCS, vol. 6418, pp. 122–135. Springer, Heidelberg (2010)
9. Ma, Z., Ferreira, L., Mesbah, M.: A framework for the development of bus service reliability measures. In: Proceedings of the 36th Australasian Transport Research Forum (ATRF), Brisbane, Australia (2013)
10. Reijsbergen, D.: Efficient simulation techniques for stochastic model checking. PhD thesis, University of Twente, Enschede (December 2013)
11. Saberi, M., Zockaie, A., Feng, W., El-Geneidy, A.: Definition and properties of alternative bus service reliability measures at the stop level. Journal of Public Transportation 16(1), 97–122 (2013)
12. The Scottish government. Bus Punctuality Improvement Partnerships (BPIPs) guidance (2009)
13. R Development Core Team. R: A language and environment for statistical computing (2005)
14. Yang, Y., Gerstle, D., Widhalm, P., Bauer, D., Gonzalez, M.: The potential of low-frequency AVL data for the monitoring and control of bus performance. In: Proceedings of the 92nd Annual Transportation Research Board Meeting (2013)
15. Younes, H.L.S., Simmons, R.G.: Statistical probabilistic model checking with a focus on time-bounded properties. Information and Computation 204(9), 1368–1409 (2006)
16. Yuan, S.: Simulating Edinburgh buses. Master's thesis, The University of Edinburgh (2013)

Markov Decision Process and Linear Programming Based Control of MAP/MAP/N Queues

András Mészáros[1,3] and Miklós Telek[1,2]

[1] Budapest University of Technology and Economics
[2] MTA-BME Information Systems Research Group
[3] Inter-University Center for Telecommunications and Informatics Debrecen
{meszarosa,telek}@hit.bme.hu

Abstract. We investigate the control problem of the optimal choice of idle server (if any) for arriving customer in order to minimize the mean system time (waiting time + service time). The considered MAP/MAP/N queue consists of a common infinite buffer and multiple identical servers with MAP service processes whose phases (internal states) are known. Customers arrive according to a MAP (whose phase is also known) and are served with work conserving policy. Idle servers preserve their phases.

We transform the obtained infinite state optimization problem to a finite state one and apply two optimization procedures, policy iteration of finite state MDP and linear programming.

Keywords: Markov arrival process, Markov decision process, MAP/MAP/N queue.

1 Introduction

Suboptimal control of multi-server systems may result in lower utilisation and, consequently, higher system time, therefore finding the optimal control scheme in these systems may be critical. Different types of queueing systems have been analysed from the point of view of optimal control. Earlier works typically consider Poisson arrival process and exponential service time, see e.g. [11] for a survey. The work dealing with problems closest to our topic is probably that of Efrosinin [5], which analyses several types of queueing systems including the MAP/PH/K/B-K structure. Efrosinin uses Markov decision processes (MDPs) to investigate various multi-server systems with a common finite queue and independent service times. In our work we consider an infinite queue with correlated arrival and service times, characterised by Markov arrival processes (MAPs).

The direct Markov chain description of the infinite queueing system contains infinite states, for which the classical MDP and linear programming (LP) solution techniques cannot be applied. With the use of the matrix analytic methodology, however, we have derived two finite state formalizations of the optimization problem which can be used to find the optimal policy of the infinite system employing

A. Horváth and K. Wolter (Eds.): EPEW 2014, LNCS 8721, pp. 179–193, 2014.
© Springer International Publishing Switzerland 2014

finite MDP solvers and LP, respectively. According to [6], the finite MDP formalization of the problem ensures that a pure stationary optimal control policy exists. The LP formalization is an alternative description of the optimization problem which is more efficient in certain cases.

The rest of the paper is organised as follows. In Section 2 we present the necessary theoretical background on MAPs and MDPs. In Section 3 we provide the matrix analytic description of the MAP/MAP/N system. Based on this description we give the finite state MDP model and the LP description of the problem in Section 4 and 5. Section 6 presents some numerical results, finally Section 7 concludes the paper.

2   Background

In the following we will use the form M to denote matrices without and M_k with an index. For their elements in position (i, j) we will use notations $M_{i,j}$ and $M_{k i,j}$ respectively. Furthermore we will use $\mathbb{1}$ to denote a column vector of 1s and $\underline{e_i}$ to denote a column vector for which $\underline{e_{i_j}} = \delta_{i,j}$, where $\delta_{i,j}$ is the Kronecker delta.

2.1   Markov Arrival Processes

The standard description of a MAP is given using the square matrix pair (D_0, D_1), where $D_0 + D_1$ is the infinitesimal generator of the background CTMC [8]. D_0 describes state transitions without arrival and D_1 with an arrival. The average arrival intensity of a MAP is $\mu = \frac{1}{\underline{v}(-D_0)^{-1}\mathbb{1}}$, where $\underline{v}$ is the solution of the system of linear equation $\underline{v}\left((-D_0)^{-1}D_1 - I\right) = 0$ and $\underline{v}\mathbb{1} = 1$.

2.2   Markov Decision Processes

Definition 1. *Let us consider a process $X(k)$ on a discrete time Markov chain with state space S, a set of decisions A, a set of transition probability matrices $P = \{P_a, \ a \in A\}$ such that $P_{a i,j} = Pr(X(k+1) = j | X(k) = i, a_k = a)$, $\forall i, j \in S$, $a \in A$, $k \in \mathbb{N}$ and a set of cost functions $C = \{C_a(s), \ a \in A, \ s \in S\}$. We say that the tuple (S, A, P, C) is a Markov decision process.*

MDPs are powerful tools for optimal control of Markovian systems [11]. The previous definition stands only for discrete time homogeneous MDPs and can be generalised to continuous time and heterogeneous cases, but the above definition is sufficient in the current discussion. We also note that S can be finite or infinite, however the common algorithms are only applicable for the finite case.

Any function $\pi(s)$ that assigns an $a \in A$ to every $s \in S$ is called a strategy. The standard problem of MDPs is to find an optimal strategy, i.e. a $\pi(s)^*$ that minimizes a given objective function. The objective function used in this paper is the average cost per step in steady state, thus the optimal policy is

$$\pi^* = \arg\min_{\pi} \mathrm{E}_{\pi}\left[\lim_{k \to \infty} \frac{1}{k} \sum_{i=1}^{k} C_{\pi(X(i))}(X(i))\right], \tag{1}$$

or equivalently

$$\pi^* = \arg\min_\pi \sum_{s \in S} \alpha_\pi(s) C_{\pi(s)}(s), \tag{2}$$

where $\alpha_\pi(s)$ is the steady state probability of being in state s for policy π.

The previous description stands for pure strategies (i.e. we always make the same decision in a state with 1 probability). In general a convex combination of pure strategies (called mixed strategies) can be considered as well, however, as shown in [6], there always exists a pure strategy that gives the optimum for the average cost per step problem.

3   Infinite State Description of the Queueing System

The service processes of the N servers of a MAP/MAP/N queue are stochastically identical, but in the considered control problem apart of the pair of matrices characterizing the MAP service process the phase of the individual servers are also known and the available idle servers (if any) are distinguished based on this information at customer arrival. The goal is to find the policy for assigning the arriving customer with the optimal idle server.

The natural structured representation of the MDP characterizing the MAP/MAP/N queue with this control option is presented below for $N = 2$. Extension to more servers is quite straightforward, but would needlessly complicate the description. We refer to the totality of states that have k customers in the system as level k and denote state i of level k as (k, i). Levels for which the number of customers in the system is higher than the number of servers are called regular the others are called irregular. Let the MAP describing the interarrival times be of size n_a and defined by (A_0, A_1) and the MAP describing the service times be of size n_s and defined by (S_0, S_1). We recall again that we consider only work conserving schemes. Thus the MAP/MAP/2 queue can be described as a continuous time Markov chain that has the standard structure of a quasi birth-death (QBD) process [8] with infinitesimal generator

$$Q = \begin{pmatrix} L_0 & F_0 & 0 & \cdots & & \\ B_1 & L_1 & F_1 & 0 & \cdots & \\ 0 & B_2 & L & F & 0 & \\ \vdots & 0 & B & L & F & \ddots \\ & & \ddots & \ddots & \ddots & \ddots \end{pmatrix}, \tag{3}$$

where

$$L_0 = A_0 \otimes I(n_s^2), \qquad F_0 = \left(\Delta(A_1 \otimes I(n_s^2)) \mid (I(n_a) - \Delta)(S_1 \otimes I(n_s^2)) \right),$$

$$F_1 = \begin{pmatrix} A_1 \otimes I(n_s^2) \\ \hline A_1 \otimes I(n_s^2) \end{pmatrix}, \qquad B_1 = \begin{pmatrix} I(n_a n_s) \otimes S_1 \\ \hline I(n_a) \otimes S_1 \otimes I(n_s) \end{pmatrix},$$

$$L_1 = \begin{pmatrix} A_0 \otimes I(n_s^2) + I(n_a n_s) \otimes S_0 \mid & 0 \\ 0 & \mid A_0 \otimes I(n_s^2) + I(n_a) \otimes S_0 \otimes I(n_s) \end{pmatrix},$$

$$B_2 = \left(I(n_a n_s) \otimes S_1 \mid I(n_a) \otimes S_1 \otimes I(n_s) \right), \qquad L = A_0 \oplus S_0 \oplus S_0,$$

$$F = A_1 \otimes I(n_s^2), \qquad B = I(n_a) \otimes S_1 \otimes I(n_s) + I(n_a n_s) \otimes S_1,$$

and $I(x)$ is the identity matrix of size x.

As the system is work conserving (i.e. servers can only be idle, if the queue is empty), decisions have to be made at the arrival of a new customer on levels $i = 0, \ldots, N - 2$, that is, when 2 or more servers are empty at the time of an arrival. For the considered $N = 2$ case this means only the 0th level. The specific control here is determined by diagonal matrix Δ. Assuming that the size of the arrival MAP is n_a and the size of the service MAP is n_s Δ has the following special structure (for N=2):

$$\Delta_{i,i} = \begin{cases} p_{j,k,k} = 0.5, & \text{if } i = (j-1) * n_s^2 + (k-1) * n_s + k, \\ p_{j,k,l}, & \text{if } i = (j-1) * n_s^2 + (k-1) * n_s + l, \\ 1 - p_{j,k,l}, & \text{if } i = (j-1) * n_s^2 + (l-1) * n_s + k, \end{cases}$$

where $j = 1, \ldots, n_a$ and $k, l = 1 \ldots, n_s$, with $k < l$. Parameter $p_{j,k,l}$ is the probability, that we choose the first server if the MAP of the arrival is in phase j, the MAP of the first server is in phase k, and the MAP of second server is in phase l. From this $0 \le p_{j,k,l} \le 1$. If both servers are in the same state we choose both with the same 0.5 probability. Otherwise, the only constraint is that $p_{j,k,l} = 1 - p_{j,l,k}$ for any given j, k, l set. This constraint corresponds to the assumption that the probability of choosing the server in phase k does not depend on whether it is labeled first or second.

The steady state solution of the system is partitioned according to the levels as $\alpha = (\alpha_0 \ \alpha_1 \ \alpha_2 \ \ldots)$. Due to the level independent behaviour of (3) for $i \ge 2$ we have

$$\alpha_i = \alpha_2 R^{i-2}, \tag{4a}$$

where R is the minimal non-negative solution of the quadratic matrix equation [8]

$$0 = F + RL + R^2 B.$$

Matrix R can be determined using efficient numerical methods [8]. Based on (3) and using matrix R, the irregular part of the steady state distribution is the solution of the linear system

$$(\alpha_0 \ \alpha_1 \ \alpha_2) \begin{pmatrix} L_0 & F_0 & 0 \\ B_1 & L_1 & F_1 \\ 0 & B_2 & L + RB \end{pmatrix} = 0, \tag{4b}$$

with normalization condition

$$\alpha_0 \mathbb{1} + \alpha_1 \mathbb{1} + \alpha_2 (I - R)^{-1} \mathbb{1} = 1. \tag{4c}$$

Using the steady state distribution (4c), the mean number of customers in the system can be expressed as

$$E(n) = \sum_{i=0}^{\infty} i\alpha_i \mathbb{1} = \alpha_1 \mathbb{1} + \sum_{i=2}^{\infty} i\alpha_2 R^{i-2}\mathbb{1} \tag{5}$$

$$= \alpha_1 \mathbb{1} + 2\alpha_2 (I - R)^{-1} \mathbb{1} + \alpha_2 R (I - R)^{-2} \mathbb{1},$$

and, applying Little's law, the mean system time can be calculated as

$$T = \frac{E(n)}{\lambda}, \tag{6}$$

where λ is the expected value of the inter-arrival time. Based on the connection between T and $E(n)$ it is clear that optimizing one is equivalent with optimizing the other. In the following we will use $E(n)$ as objective function in the optimization.

These equations can be easily extended for the $N > 2$ case. Doing so we get

$$E(n) = \sum_{i=1}^{N-1} i\alpha_i \mathbb{1} + N\alpha_N (I - R)^{-1} \mathbb{1} + \alpha_N R (I - R)^{-2} \mathbb{1}. \tag{7}$$

Equation (5) and (6) is relatively simple, however in the expression of the α_i vectors terms including $p_{j,k,l}^{-1}$ and $(1 - p_{j,k,l})^{-1}$ will appear. This makes the straightforward optimisation a non-linear problem.

4 Finite State MDP Formalization of the Problem

In this section we present a finite state MDP formalization of the queueing system control problem. This formalization is based on the following observations:

- Decisions have to be made only on levels $0, \ldots, N - 2$.
- The objective function of the optimization is (7), which has a similar form to the objective function of the MDP (2) and contains only α_i, $i = 1, \ldots N$.

Using these our goal is to make an MDP for which (2) (the objective function of the MDP) is identical to (7) (the objective function of the optimization problem). To achieve this we use the following method.

In the first step we apply the simple transformation: $P = \frac{1}{\gamma}Q + I$, where $\gamma = \max_{i,j} |Q_{i,j}|$, i.e., the absolute value of the element of Q with the largest absolute value. This ensures that P is a valid DTMC transition matrix. Furthermore $\alpha Q = 0$ (where α is the steady state probability vector of Q), thus $\alpha P = \alpha \frac{1}{\gamma}Q + \alpha I = \alpha$, consequently α is the steady state probability vector of the new DTMC as well. Note that this transformation is the same as the one used in randomization [7]. It is easy to see that this DTMC defines the S, A, P sets of an infinite state MDP, where decisions correspond to possible server choices upon arrival. For example in the two server example: $A = \{1,2\}$, and if $i = (j-1) * n_s^2 + (k-1) * n_s + l$, then $p_{j,k,l} = 1$ if $\pi(i) = 1$ and $p_{j,k,l} = 0$ if $\pi(i) = 0$, in other words P_1 is P with $p_{j,k,l} = 1$, $\forall j, k, l$ and P_2 is P with $p_{j,k,l} = 0$, $\forall j, k, l$. Note that the resulting DTMC is a discrete time QBD with $B_i' = \frac{1}{\gamma} B_i$, $F_i' = \frac{1}{\gamma} F_i$ and $L_i' = \frac{1}{\gamma} L_i + I$ for all i, where M_i' denotes the discrete time pair of the continuous time QBD's M_i matrix.

In the second step we change the infinite state regular part to a finite set of states and transitions while keeping the steady state probabilities of the irregular part unaffected. This step will be discussed in more detail shortly.

In the third step we assign the costs to the states based on objective function of the original optimization problem (7). This assignment is fairly straightforward. The cost of state i of level k is

$$
C_{(k,i)} = \begin{cases} k, & \text{for } k = 0, 1, \ldots, N-1, \\ \underline{e_i}^T \left(N \left(I - R \right)^{-1} + R \left(I - R \right)^{-2} \right) \mathbb{1}, & \text{for } k = N, \\ 0, & \text{otherwise.} \end{cases} \tag{8}
$$

We would like to stress that the dynamic behaviour of the MDP, i.e., the cost collected in a state of the MDP, does not have to be proportional to the waiting time accumulated in the corresponding state of the CTMC. The only important thing is that the irregular part of the MDP has the same steady state probabilities as the irregular part of original CTMC. Consequently, using the appropriate costs, the MDP has the same objective function - thus the same optimum - as the original problem.

The main question of the above procedure is how to carry out step two, i.e., how to substitute the infinite state regular part of the DTMC so that the steady state probabilities of the irregular part remain the same. For this we need to introduce two new matrices, G and H. $G_{i,j}$ shows the probability that, starting from (k,i), $k > N$ we reach $k-1$ and the first time this occurs we arrive in $(k-1,j)$, i.e.,

$$
G_{i,j} = Pr(\tau < \infty, \; X(\tau) = (k-1,j)|X(0) = (k,i)), \tag{9}
$$

where τ is the time of the first arrival to level $k-1$. $H_{i,j}$ is the the expected time (number of steps) of reaching level $k-1$ ($k > N$) if we start in (k,i) supposing we arrive in phase $(k-1,j)$, multiplied by $G_{i,j}$, i.e.,

$$
H_{i,j} = \mathrm{E}[\tau I_{X(\tau)=(k-1,j)}|X(0) = (k,i)], \tag{10}
$$

where again τ is the time of the first arrival to level $k-1$ and I is the indicator function. It can be shown that G and H are the solutions of

$$
B' + (L' - I)G + F'G^2 = 0 \tag{11}
$$

and

$$
G + (L' - I)H + F'GH + F'HG = 0, \tag{12}
$$

respectively. Equation (11) can be solved numerically using efficient numerical methods, while (12) is a Sylvester equation, which is linear in the elements of H thus can be solved analytically if G is known. More details, including the derivation of the equations and the applicable numerical procedures for (11) can be found in [8].

Using G and H a finite state equivalent of the infinite QBD can be given. The irregular part of the DTMC consists of the 0th to Nth level. These are left unchanged during the transformation process. The regular part is substituted by M^2 states, where M is the size of one level. The probability of the event that the process, starting from (N, i) goes up to any phase of level $N + 1$ in the next step and reaches level N again in phase j for the first time (after possibly multiple transitions on higher levels) is

$$b_{i,j} = \sum_{k=1}^{M} F'_{i,k} G_{k,j}. \tag{13}$$

Let the random variable $\tau_{i,j}$ be $\tau_{i,j} = \left(\tau I_{X(\tau)=(k-1,j)} | X(0) = (k,i)\right) / G_{i,j}$. Note that $\mathrm{E}[\tau_{i,j}] = \dfrac{H_{i,j}}{G_{i,j}}$. For solving the optimization problem we only have to know the steady state probabilities of states on the irregular levels, therefore we do not need to distinguish states of the regular levels, thus we can modify the system the following way. From level $0, \ldots, N$ to level $0, \ldots, N$ transitions happen as before. We substitute the regular part (levels $N+1$ and above) of the DTMC with a level of M^2 states denoted by $s_{i,j}$, $i, j = 1, \ldots, M$. From state (N, i) transition to state $s_{i,j}$ happens with $b_{i,j}$ probability. If the process reaches $s_{i,j}$ it transitions to (N, j) after staying in $s_{i,j}$ for $\tau_{i,j}$ time. This structure has the following interpretation. The instant the process would enter the regular part of the DTMC we determine the first state it arrives to upon first reaching the irregular part again. Instead of moving on to a state of the regular part the process moves to an intermediate state where it stays for the random time which is the same as the time needed in the original DTMC to go back to level N conditional on the fact that it arrives to state (N, j). It is clear, that the substitution does not make a difference from the irregular part's point of view. This new structure, however, is not Markovian as the distributions of transition times from $s_{i,j}$ are not memoryless. Processes where transition probabilities are according to a transition matrix, but transition times may have a general distribution are called semi-Markov processes. For semi-Markov processes the steady state distribution depends only on the expected value of the transition times. For proof see e.g. [3]. Thus, without affecting the steady state probabilities of the irregular part we can change $\tau_{i,j}$ to the geometrically distributed $\tau'_{i,j}$ if $\mathrm{E}[\tau'_{i,j}] = \mathrm{E}[\tau_{i,j}] = \dfrac{H_{i,j}}{G_{i,j}}$. This geometrical distribution can be achieved using a feedback in $s_{i,j}$ with probability $q_{i,j}$ and transition to (N, j) with probability $1 - q_{i,j}$, where $q_{i,j} = \dfrac{H_{i,j} - G_{i,j}}{H_{i,j}}$. Now the modified system is a DTMC and its irregular part has the same steady state probability distribution as the original CTMC. In the following we will denote the transition matrix of this DTMC by P^*. Matrix P^* coupled with the previously defined costs and actions form a finite state MDP that can be optimized using standard methods (e.g. linear programming, value iteration, policy iteration) and has the same optimal strategy and optimal cost as the original infinite continuous time system.

Using the description of Section 3, the size of level $0, \ldots, N$ is $n_a n_s^N$ each, while the size of level $N + 1$ is $(n_a n_s^N)^2$ consequently the size of $\boldsymbol{P}^*$ is $N n_a n_s^N + (n_a n_s^N)^2$. To improve the speed of the optimization this size has to be reduced, which can be done using two methods. The first one is the reduction of the original QBD, the second one is the reduction of the part used for substituting the regular part of the QBD.

We will not discuss the first method in detail just present its basic idea. It is easy to see, that the labelling of the servers is arbitrary, i.e., while in the description in Section 3 the phase of every server is followed individually, it is enough to keep track of the number of empty and busy servers at each point (and the phase of the arrival process of course). Consequently the irregular levels (where the buffer is empty) can be described using a set of $2n_s + 1$ numbers. The first element of the set indicates the phase of the arrival and can be between 1 and n_s, the next n_s elements show the number of empty servers in each of the service phases, the final n_s elements show the number of occupied servers in each of the service phases. As there are a total of N servers, the sum of the last $2n_s$ elements of the set is N. E.g. if the queueing system has three servers ($N = 3$), the service process is described by a size 3 MAP and the arrival process by a size 4 MAP, then $(4|0, 1, 1|0, 0, 1)$ denotes the state where the arrival process is in phase 4, there are two empty servers, one in phase 2 one in phase 3, and there is one working server in phase 3. It can be easily seen that there are a total of $n_a \left(\left(\binom{2n_s}{N} \right) \right)$ different configurations for the set, where $\left(\binom{n}{k} \right) = \binom{n+k-1}{k}$ is the k combination of n with repetition. Using the above idea a more efficient QBD description can be constructed where each state of the irregular part corresponds to a specific configuration of the set. This construction is done by combining multiple equivalent states into one, which is called lumping and is a standard method for state space reduction. The previously described method of making a finite state MDP from the infinite QBD can still be applied without any changes. Using the same thought process the finite state substitute of the regular part can be reduced to $\left(n_a \left(\left(\binom{n_s}{N} \right) \right) \right)^2$ states.

The second improvement can be made by realizing that there are a few constraints that the regular part of the original CTMC and its substitute has to satisfy. If these requirements are met, the size of the substituted part and its exact structure are not important. These constraints come from $\alpha \boldsymbol{Q} = 0$, $\sum_{i=0}^{N-1} \alpha_i \mathbb{1} + \alpha_N \left(\boldsymbol{I} - \boldsymbol{R} \right)^- \mathbb{1} = 1$ and the general Markovian constraints of a CTMC. For simplicity's sake we discuss the $N = 2$ case, from which the general case can be easily derived. Let us consider the modified CTMC with generator $\hat{\boldsymbol{Q}}$ that has the following structure

$$\hat{Q} = \begin{pmatrix} L_0 & F_0 & 0 & 0 \\ B_1 & L_1 & F_1 & 0 \\ 0 & B_2 & L & F^* \\ 0 & 0 & B^* & L^* \end{pmatrix}. \tag{14}$$

We have to keep the steady state probabilities of level $0, 1, 2$ the same, but the new part can have a different α_3^*, $\alpha_3^* \neq \alpha_3$ steady state probability vector. First of all, from $\alpha^* \mathbb{1} = \alpha \mathbb{1} = 1$ using $\alpha_3 = \alpha_2 R$ we get

$$\alpha_3^* \mathbb{1} = \alpha_2 R (I - R)^{-1} \mathbb{1}. \tag{15}$$

From $\alpha Q = 0$ we have $\alpha_1 F_1 + \alpha_2 L + \alpha_3 B = 0$ and from $\alpha^* \hat{Q} = 0$ we have $\alpha_1 F_1 + \alpha_2 L + \alpha_3^* B^* = 0$. By subtracting these equations from each other and using $\alpha_i = \alpha_2 R^{i-2}$ we get

$$\alpha_3^* B^* = \alpha_2 R B. \tag{16}$$

Furthermore from $\alpha^* \hat{Q} = 0$ and $\hat{Q} \mathbb{1} = 0$ we get

$$\alpha_2 F^* + \alpha_3^* L^* = 0, \tag{17}$$

$$(B_2 L + F^*) \mathbb{1} = 0, \tag{18}$$

$$(B^* + L^*) \mathbb{1} = 0. \tag{19}$$

Finally, we have the standard sign constraints, i.e. all elements of B^*, L^*, F^* are non-negative, except for the diagonal of L^* which is strictly negative.

These constraints give a constrained linear equation system if we first tie α_3^*. We can assign values to α_3^* randomly or e.g. make all the elements equal. This system can be solved using linear programming if a solution does exist. In general it is not guaranteed that a constrained linear equation system has a solution, however the method presented previously in this section gives one where the size of L^* is $M^2 \times M^2$ if M is the size of the last irregular level. Thus one possible method is to start by trying to find a solution to the above equations with an L^* of size $n \times n$, $n = 1$ (the size of B^* and F^* are $n \times 1$ and $1 \times n$ respectively) and increase the size until the system can be solved or $n = M^2$ is reached. After this the optimization problem can be solved (after transforming the CTMC with generator $\hat{Q}$ into a DTMC) using standard MDP methods.

5   Linear Programming Solution of the Problem

In this section we give a method for solving the queueing system control optimization problem using linear programming (LP). First we mention that linear programming is one of the classical ways to optimize a finite state MDP. As such, the last step of the optimization after the transformation of the infinite state MDP to a finite state one could be using LP to find the optimal strategy. In this section, however, we make use of the flexibility of LP to describe the problem without forming a finite MDP, although the description will be based on the LP formalization of MDP optimization. Consequently we first present the LP formalization of the general average cost MDP problem. We will follow the same thought process as in [1] with different notations. After that we make the necessary customization for the problem at hand.

Let us take an MDP with the previously introduced (S, A, P, C) notation, but consider also mixed strategies. In case of a mixed strategy decision $a \in A$ is made at state $i \in S$ with probability $u_{i,a}$, i.e., $u(i,a) = Pr(a_k = a | X(k) = i)$, with $\sum_{a \in A} u(i,a) = 1$, $\forall i \in S$. The goal is to optimize $u(i,a)$ for all $i \in S$, $a \in A$ according to the given objective function. Let us define U such that $U_{i,a} = u(i,a)$, $\forall i \in S$ and $a \in A$ and P_u such that $P_{u_{i,j}} = \sum_{a \in A} u(i,a) P_{a_{i,j}}$. Now the objective function in (2) changes to $\sum_{i \in S} \alpha(u)_i \sum_{a \in A} U_{i,a} C_a(i)$. From these the optimization problem can be given in a form that is similar to the standard LP form as

$$\min \sum_{i \in S} \alpha_i \sum_{a \in A} U_{i,a} C_a(i),$$
$$\text{s.t. } \alpha(P_u - I) = 0,$$
$$U \mathbb{1} = \mathbb{1}, \tag{20}$$

where the variables are the elements of α and U. This problem is non-linear because P_u depends on the elements of U, thus the products of the elements of α and U appear in the constraints, however, it can be linearised by introducing new variables $x_{i,a} = \alpha_i u_{i,a}$. That is, $x_{i,a}$ is the steady state probability that the process is in state i and decision a is made. Let us introduce matrix X with $X_{i,a} = x_{i,a}$, $\forall i \in S$ and $a \in A$ and denote its ith column by $X_{*,i}$. Notice that $(X\mathbb{1})^T = \alpha$. As such the optimization problem can now be defined as an LP as

$$\min \sum_{i \in S} \sum_{a \in A} X_{i,a} C_a(i),$$
$$\text{s.t. } \sum_{a \in A} X_{*,a}^T P_a - (X\mathbb{1})^T = \underline{0},$$
$$\sum_{i \in S} \sum_{a \in A} X_{i,a} = 1,$$
$$X_{i,a} \geq 0, \quad \forall i \in S, \forall a \in A. \tag{21}$$

Here $X_{i,a} P_{a_{i,j}}$ is the steady state probability that the process is in state i, decision a is made, and as a result the process transitions to state j.

As mentioned before, the above introduced LP could be used to solve the finite state MDP of Section 3. Instead we make use of the fact that the LP optimization is a general purpose tool unlike methods that are particularly developed to solve MDPs, e.g. value and policy iteration. The MAP/MAP/N optimization problem can be formalized using LP by noticing that equations (4b) and (4c) (more precisely their N server generalizations) are sufficient constraints to solve the optimization problem. To be consistent with the previous discrete time description we use the discrete time counterpart of (4c), i.e., $P = \frac{1}{\gamma}Q + I$ just as in Section 3. Thus, for the $N = 2$ case for example

$$\alpha = \begin{pmatrix} \alpha_0 & \alpha_1 & \alpha_2 \end{pmatrix}, \tag{22}$$

$$P = \frac{1}{\gamma} \begin{pmatrix} L_0 & F_0 & 0 \\ B_1' & L_1' & F_1' \\ 0 & B_2' & L' + RB' \end{pmatrix} + I. \tag{23}$$

These can be substituted into (21). The only difference is in the last normalization constraint. This changes to $\alpha_0 \mathbb{1} + \alpha_1 \mathbb{1} + \alpha_2 (I - R)^{-1} \mathbb{1} = 1$ according to (4c). Finally the cost vector is the same as in Section 3, thus, when the index of the last state of level i is k_i, the LP problem is

$$\min \sum_{i \in S} \sum_{a \in A} X_{i,a} C_a(i),$$

$$\text{s.t. } \sum_{a \in A} X_{*,a}^T P_a - (X\mathbb{1})^T = \underline{0},$$

$$\sum_{i=1}^{k_1} \sum_{a \in A} X_{i,a} + \sum_{i=k_1+1}^{k_2} \sum_{j=k_1+1}^{k_2} \sum_{a \in A} X_{i,a}(I - R)^{-1}_{i-k_1, j-k_1} = 1,$$

$$X_{i,a} \geq 0, \quad \forall i \in S, \forall a \in A. \tag{24}$$

As in the previous section P_1 and P_2 can be obtained by using substitution $p_{j,k,l} = 1$ and $p_{j,k,l} = 0$ respectively $\forall j, k, l$. If, for a given $i \in S$ and a_1, a_2, $a_1 \neq a_2$ we get $u_{i,a_1}, u_{i,a_2} > 0$, then the optimization gives a mixed strategy as optimum. In that case choosing either of the decisions with 1 probability gives the same optimum. The significance of enabling mixed strategies is that it makes the optimization an ordinary LP problem instead of the integer LP problem that results from considering only pure strategies.

6 Numerical Experiments

6.1 Computational Complexity

The computational complexity of building up the QBD, finding its G and H matrices and transforming the problem to finite state is negligible compared to the one of the solution of the resulting MDP or LP problem, therefore we only consider complexity of the solution the finite MDP and LP.

The complexity of the basic MDP solution methods is summarized in [9]. Let $|S|$ be the number of states and $|A|$ be the number of possible decisions. We upper bound the number of decisions by $|A| = n_s$, which means that there is an empty server in every phase for all the states of the MDP. Policy iteration has a complexity $O(|S|^3)$ per step and requires $O(|A|)$ steps in the average case. As seen in Section 4 $|S| = n_a \left(\binom{2n_s}{N} \right) + \left(n_a \left(\binom{n_s}{N} \right) \right)^2$ - or less, if the substitute of the regular part can be further reduced. The second term of the expression is usually higher than the first one, for the computationally tractable cases the difference is $0 - 2$ orders of magnitude.

For the LP approach the number of variables is $|X| = n_a \left(\binom{2n_s}{N} \right) n_s$ and the number of constraints (from (24)) is $n_c = |X| + n_a \left(\binom{2n_s}{N} \right) + 1$. There is a vast number of algorithms for solving LP problems. The simplest and most frequently used are probably the simplex and the revised simplex methods. For the latter the average computational cost is $O(|X|^3)$ (see e.g. [10]). Using more involved methods this cost can be reduced. See [4] for example for a comprehensive summary.

In general the two methods have similar computational costs. The MDP approach is better if the substitute of the regular part can be reduced, while the LP approach is more efficient if this is not possible and a more involved LP solver is used.

6.2   Numerical Examples

M/MAP(2)/2 Systems. First we discuss the simplest interesting case, the M/MAP(2)/2 queue, in which arrivals happen according to a Poisson process, and the service is carried out by two servers that have the same order 2 MAP service time. The first part of this segment is the reiteration of the results of the corresponding section in [2].

In the M/MAP(2)/2 queue there is one simple question to be answered: If both servers are idle, one of them is in phase 1 and the other one is in phase 2, which server has to process the next arriving customer to have a minimal average system time? In other words, what is the optimal value of $p_{1,1,2}$? It seems natural, and from the first part of the paper we already know, that a pure strategy is optimal, i.e., $p_{1,1,2}$ is either 1 or 0. The intuitive answer is to choose the server which can serve the customer faster. This means that we compare the mean service time starting from phase 1 and phase 2, i.e., $e_1^T(-S_0)^{-1}\mathbb{1}$ and $e_2^T(-S_0)^{-1}\mathbb{1}$, and if the first expression is smaller, we choose the server in phase 1 ($p_{1,1,2} = 1$), otherwise the one in phase 2 ($p_{1,1,2} = 0$).

This greedy decision can be motivated by the fact that we would like to serve the customer as fast as possible to have an idle queue as soon as possible. For the examined system, however, the numerical results show that the opposite choice is better, i.e., it is better to choose the server which serves the customer slower. This counter-intuitive result can be interpreted the following way. If we use the faster server for the first customer, the probability of finishing the service before a new arrival is high, as the mean service time of the faster state is smaller than the mean inter-arrival time of a new customer. Upon service there is a chance that the server moves to the slower state, leaving the system with two servers in the phase with higher service time. In this state there is a higher chance that more than 2 consecutive customers arrive before the first customer can be served, which leads to a higher average system time. In other words, assigning the customer with the faster server leads to a more deteriorated state after service completion, while assigning the customer with the server in the slower phase, there is a chance that the server will move to the faster state upon service, thus the state of the system improves. One can think of this effect as the repair of the server at the cost of a slower service. Our extensive numerical investigations suggest that choosing the server with higher service time is optimal for MAP(2) servers regardless of their other characteristics and the intensity of arrivals.

A natural question to ask is what does the possible gain, i.e., the magnitude of the difference between the worst and best strategies, depend on. In other words, how can we characterize systems where the best strategy is significantly better then the worst. In the following we denote by g_a the absolute value of the maximum gain, and by g_r its ratio to the cost of the best strategy. That is, $g_a = \mathrm{E}_\pi^-(n) - \mathrm{E}_\pi^+(n)$, $g_r = \frac{\mathrm{E}_\pi^-(n) - \mathrm{E}_\pi^+(n)}{\mathrm{E}_\pi^+(n)}$, where $\mathrm{E}_\pi(n)$ is the mean number of customers for policy π, and π^+ and π^- are the optimal and worst strategy respectively. According to our experiments, classic statistical measures of the service MAP such as its autocorrelation or the moments of its marginal distribution cannot be used to characterize the gain in the general case. This can be understood by knowing that different MAPs can have the same exact statistical properties. We have found that it is best to consider the following simple characteristics:

- The ratio of the mean arrival (λ) and service rate (μ (see Section 2.1 for computation)), $r_1 = \frac{\lambda}{\mu}$. (To have a stable queue $r_1 < N$ has to hold.)
- The ratio of mean service times of the server starting from the different phases of the MAP, i.e. $r_2 = \frac{\max(m_1, m_2)}{\min(m_1, m_2)}$, with $m_1 = \underline{e_1}^T(-\boldsymbol{S_0})^{-1}\mathbb{1}$, $m_2 = \underline{e_2}^T(-\boldsymbol{S_0})^{-1}\mathbb{1}$.
- The steady state probability vector of the phases embedded to the arrivals of the service MAP, $\underline{v} = \{v_1, v_2\}$ (see Section 2.1).

Intuitively the higher r_1, the bigger the difference is. This is true for the absolute gain g_a but not for the relative gain g_r. Figure 1 shows the relative gain versus the arrival intensity for the M/MAP(2)/2 queue with service MAP

$$\boldsymbol{S_0} = \begin{pmatrix} -1/10 & 1/20 \\ 0 & -100 \end{pmatrix}, \quad \boldsymbol{S_1} = \begin{pmatrix} 1/20 & 0 \\ 5 & 95 \end{pmatrix}, \tag{25}$$

It can be seen that g_r has a maximum around $\lambda \approx 0.44$. Based on our experiments this behaviour is typical, i.e., the relative gain is the highest for medium load. Intuitively, the higher r_2, the bigger the g_r is, as there is a more significant difference between the possible decisions. This time the intuition is correct. Finally we found that g_r can get higher if the MAP is more "balanced", i.e. both elements of the $\underline{v}$ embedded probability vector are high enough.

The above observations for the M/MAP(2)/2 case are reflected in Figure 2, where 1000 service MAPs with a given mean and completely random elements were taken and their relative gains g_r were plotted against flexibility f that represents the randomness due to the embedded stationary vector and the difference of the mean service times starting from different initial states as $f(r_2, \underline{v}) = \sqrt{(r_2 - 1) * v_1 * v_2}$. The correlation between f and g_r was ≈ 0.98.

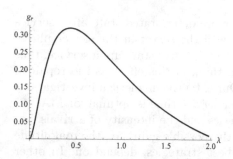

Fig. 1. Relative gain as the function of arrival intensity for a given service process

Fig. 2. Relative gain in mean system time as the function of flexibility f

We examined more special types of random MAPs as well, the correlation between f and g_r was always > 0.9.

More Complex Systems. Intuitive understanding of the optimal control of more complex systems becomes increasingly hard. Again we refer back to [2], where it was demonstrated that for M/MAP(3)/2 queues already the optimal strategy cannot be explained as simply as for the M/MAP(2)/2 case. For example let us take the service MAP with

$$S_0 = \begin{pmatrix} -1 & 0 & 0 \\ 0 & -2.3 & 0 \\ 0 & 0 & -100 \end{pmatrix}, \quad S_1 = \begin{pmatrix} 0 & 1 & 0 \\ 0 & 0 & 2.3 \\ 100 & 0 & 0 \end{pmatrix}.$$

For $\lambda = 1.5$ the optimal strategy is to always prioritize the server in phase 1 and choose the server in phase 3 over the one in phase 2. For $\lambda = 1.2$ the priority of phase 2 and phase 3 are swapped. The $r_1, r_2, \underline{v}$ factors introduced for the M/MAP(2)/2 queue can still be used to roughly evaluate the system.

For $N > 2$ using simple intuitive rules gets even harder. In these cases the optimal server choice can even depend on the phase of the servers that are occupied at the time of a new arrival. For example let us consider an M/MAP(3)/3 queue with arrival intensity $\lambda = 3.6$ and service MAP

$$S_0 = \begin{pmatrix} -0.76 & 0 & 0 \\ 0 & -10 & 0 \\ 0 & 0 & -1 \end{pmatrix}, \quad S_1 = \begin{pmatrix} 0 & 0.76 & 0 \\ 0 & 0 & 10 \\ 1 & 0 & 0 \end{pmatrix}.$$

If the queue is on level 1, the empty servers are in phase 1 and 3 and the occupied server is in phase 1, the optimal decision is to process the new customer by the server in phase 3. However if the occupied server is in phase 3, the optimal choice is the server in phase 1.

Finally, we have to stress that the intuitive explanations of the optimal control are only conjectures based on numerical experiments, and that in spite of the relatively efficient optimization techniques the proposed methods can only be used for relatively small systems (depending on the size service and arrival MAPs for $N = 2, \ldots, 10$ servers) due to the multiplicative increase of the state space.

7 Conclusion

In this paper we presented two procedures for finding the optimal policy in MAP/MAP/N queues. Both procedures are based on the matrix analytic methods, which make an efficient treatment possible. Using these procedures we demonstrated some of the characteristics of the MAP/MAP/N systems. We showed that even the simplest queues have counter intuitive behaviour and illustrated the lack of simple intuitive rules in case of more complex systems (e.g. $N \geq 3$), which makes the use of a computational approach is necessary.

Aconwledgement. The authors gratefully acknowledge the support of the TÁMOP-4.2.2C-11/1/KONV-2012-0001 and the OTKA K101150 projects.

References

1. Bello, D., Riano, G.: Linear programming solvers for Markov decision processes. In: IEEE Systems and Information Engineering Symposium, pp. 90–95 (2006)
2. Bodrog, L., Gribaudo, M., Horváth, G., Mészáros, A., Telek, M.: Control of queues with map servers: experimental results. In: 8th International Conference on Matrix-Analytic Methods in Stochastic Models (2014)
3. Çinlar, E.: Exceptional paper-Markov renewal theory: A survey. Management Science 21(7), 727–752 (1975)
4. Dantzig, G.B., Thapa, M.N.: Linear programming 2: Theory and Extensions. Springer (2003)
5. Efrosinin, D.: Controlled Queueing Systems with Heterogeneous Servers. PhD thesis, University of Trier (2004)
6. Filar, J., Vrieze, K.: Competitive Markov Decision Processes. Springer (1997)
7. Gross, D., Miller, D.R.: The randomization technique as a modeling tool and solution procedure for transient Markov processes. Operations Research 32(2), 343–361 (1984)
8. Latouche, G., Ramaswami, V.: Introduction to Matrix Analytic Methods in Stochastic Modeling. ASA-SIAM Series on Statistics and Applied Probability. Society for Industrial and Applied Mathematics (1999)
9. Littman, M.L., Dean, T.L., Kaelbling, L.P.: On the complexity of solving Markov decision problems. In: Proc. of 11th Conf. on Uncertainty in Artificial Intelligence, pp. 394–402. Morgan Kaufmann Publishers Inc. (1995)
10. Pan, V.: On the complexity of a pivot step of the revised simplex algorithm. Computers & Mathematics with Applications 11(11), 1127–1140 (1985)
11. Stidham Jr., S., Weber, R.: A survey of Markov decision models for control of networks of queues. Queueing Systems 13(1-3), 291–314 (1993)

A Decision Making Model of Influencing Behavior in Information Security

Iryna Yevseyeva, Charles Morisset, Thomas Groß, and Aad van Moorsel

Centre for Cybercrime and Computer Security
School of Computing Science, Newcastle University
Newcastle upon Tyne NE1 7RU, UK
firstname.lastname@newcastle.ac.uk

Abstract. Information security decisions typically involve a trade-off between security and productivity. In practical settings, it is often the human user who is best positioned to make this trade-off decision, or in fact has a right to make its own decision (such as in the case of 'bring your own device'), although it may be responsibility of a company security manager to influence employees choices. One of the practical ways to model human decision making is with multi-criteria decision analysis, which we use here for modeling security choices. The proposed decision making model facilitates quantitative analysis of influencing information security behavior by capturing the criteria affecting the choice and their importance to the decision maker. Within this model, we will characterize the optimal modification of the criteria values, taking into account that not all criteria can be changed. We show how subtle defaults influence the choice of the decision maker and calculate their impact. We apply our model to derive optimal policies for the case study of a public Wi-Fi network selection, in which the graphical user interface aims to influence the user to a particular security behavior.

1 Introduction

People continuously make information security decisions: should I use this wireless, should I put this person's USB stick in my laptop, how do I choose and memorize passwords? Almost always, the decision involves a trade-off between security and other concerns, such as being able to complete an important task or being able to easily do something that otherwise could be cumbersome. The decisions are often complex, with several objectives to be considered simultaneously, and the optimal decision may very much depend on the specific situation: while using a stranger's USB stick is not advisable, the importance of the job to be completed and/or knowledge about the owner of the USB stick may make it advisable to put the USB stick in one's laptop, despite the associated information security risks.

In situations such as above, a simple compliance policy (such as, not to allow USB sticks at all) would be suboptimal. Instead, one would want to allow some freedom for the owner of the laptop to decide the best course of action.

A. Horváth and K. Wolter (Eds.): EPEW 2014, LNCS 8721, pp. 194–208, 2014.

In general terms, unless one can specify a compliance policy that is optimal under all possible circumstances (which is a rare real-world case), there is room for improvement by allowing the user to make the final decision. There exist other situations, in which the user should play a role in the security decision making. For instance, in case of BYOD (bring your own device) [7], where the device owner uses their own device for work-related activities, the fact that the user owns the device puts certain restrictions on what the employer can decide without the owner's input. However, an employer might still want to influence the decisions of its employees, since the employer is impacted by these decisions.

In all these situations the end user is involved in the information security decision making, and is in fact responsible for the final choice. Then, and this is key for this paper, it may be advisable that service providers (telecoms, online banks), device vendors, employers, or other parties are able to *influence* the decision making, without restricting the end user. In the literature, this is often referred to as nudging [22] implemented widely in healthcare and social policies, see e.g. [21]. Nudging leaves the choice with the user, but aims to influence the decision so that the user is more likely to make a beneficial decision, e.g., by presenting choices in a particular manner that aims to impact the choice a person ends up making. There are many aspects to nudging that deserve discussion, but, in this paper, we do not debate the specific approach, but aim to derive results for influencing in general.

In [17] a first formalization of the concept of influencing was provided assuring it is as general as possible, but at the same time is intuitive and useful. Earlier, Heilmann [11] presented schematically the nudge success conditions from perspective of influencing autonomous system, also called System 1 (and not reflective, also called System 2) [13], and showed the difference of these conditions for different types of nudges with respect to taxonomy of Bovens [4].

In this paper, we provide a *model for influencing human decision making in security contexts*. A model aim to analyze users' decisions and behavior in order to be able to define better security policies and procedures from both an employer and its employees points of view. In particular, it gives an opportunity to an employer to influence decisions of its employees; however, leaving the final choice and responsibility for the decision to the employee who made it.

We believe such a model is necessary to enable a solid quantitative evaluation of influence. In particular, we want to be able to apply mathematical optimization to decision making as well as to the decision on how to influence, and for that we need a rigorous underpinning and understanding of the problem at hand.

Finally, we want to be able to evaluate the level of success of influencing behaviors, be it experimentally or theoretically–again, a formal model allows us to define the experimental or theoretical setting under which we carry out the evaluation. This paper will not reach all these goals, but provide the underlying quantitative model for human decision making evaluation for security decisions.

Our model is based on a well-known practical approach to modeling human decision making, *multicriteria decision analysis*, see e.g. [2], in particular, on *multiattribute utility theory* [15] presented in Section 2. We assume that such a model can be used both for the *decision maker* (e.g., the employee of a company), and for the *stakeholder* (e.g., the company). Given a set of alternatives evaluated on a set of criteria, we can define a policy that represents the choice of optimal decisions by the decision maker, and we can calculate the optimal modification of these criteria with respect to the stakeholder. A particular contribution of this work is to model the freedom of choice left by the stakeholder to the decision maker by considering that only a subset of all criteria are modifiable. We illustrate in Section 4 the case, where a stakeholder is effectively unable to influence the decision maker.

We illustrate each stage of our model and its merits using a public Wi-Fi selection scenario taken from [23] in each section. In the Wi-Fi example, a device user decides between networks and the device presents choices so as to influence the decision of the device user. In this case, the decision maker represents the device user and the stakeholder represents the company of the user. We show how changing presentation of some Wi-Fi's may alter the choice of decision maker. However, the approach is designed generally enough to be applied to other case studies, e.g. for choosing among access control policies [18]. Finally, Section 5 discusses possible extensions of the framework, in particular, considering influencing populations.

2   Decision Making

In order to model human decision making and to evaluate the different alternatives for a decision maker, we consider Multi-Criteria Decision Analysis (MCDA). MCDA is particularly useful in situations, where alternatives are evaluated on multiple, often conflicting, criteria, and in search of solutions that represent the best trade-off(s) between these criteria. In information security, this trade-off is usually between security and productivity/usability, for instance, a decision maker has to select between a more secure network and a faster one.

When compared to other approaches to model security decision making, e.g. through Markov Decision Process and reward models [3] or using the experience, e.g. by reinforcement learning [20], MCDA provides transparency to the process of making decisions and illustrates explicitly how trading-off between criteria is obtained. Transparency of the decision making process is desirable by both decision makers and stakeholders, who are interested in seeing how their preferences with respect to criteria are considered within a model. Moreover, MCDA allows for possible behavioral biases to be taken into account within a model, e.g. in a similar way as in [14].

For selecting a set of criteria that influence security decisions, it may be advisable to look at attributes related to technology, to management, to economy, to culture and to personal preferences. However, the general MCDA recommendation is to consider a set of criteria most relevant to a particular problem to be solved [10] from [15].

Here, by making a decision, we assume choice of an alternative among available ones. A decision maker is responsible for selecting an alternative a. We write $\mathcal{A}$ for the set of alternatives available to the decision maker.

In MCDA, alternatives are evaluated and compared using a set of criteria $\mathcal{G}$, such that each criterion should be either minimized or maximized (the direction of optimization). Each criterion comes with a scale, in which alternatives can be compared. Typical scales include real numbers, intervals, ratios, binary or verbal values (qualitative descriptions), which are ordered with respect to the optimization direction. Each criterion $g \in \mathcal{G}$ is, therefore, associated with a scale $\mathcal{K}_g$, and we write $g^{\min} \in \mathcal{K}_g$ and $g^{\max} \in \mathcal{K}_g$ for the minimal and maximal values of g, respectively. We write $\mathcal{K} = \bigcup_{g \in \mathcal{G}} \mathcal{K}_g$ for the set of all possible scales, and without any loss of generality, we assume that all criteria are maximized (minimized criteria can simply be multiplied by -1).

Each alternative is evaluated on each criterion $g \in \mathcal{G}$ by means of an evaluation function $\sigma_g : \mathcal{A} \to \mathcal{K}_g$. We write Σ_g for all possible σ_g functions, and $\Sigma_{\mathcal{G}} = \prod_{g \in \mathcal{G}} \Sigma_g$ for the cartesian product of all criteria evaluation functions. When no confusion can arise, we write $\sigma = (\sigma_{g_1}, \dots, \sigma_{g_n})$ for a vector of evaluation functions, and $\sigma[g]$ for the evaluation function of σ corresponding to the criterion g.

We now present the basics of MCDA and Multi-Attribute Utility Theory, in particular. We then detail how to define the policy of a decision maker, and we illustrate this approach for selection of a public Wi-Fi case study.

2.1 Multi-Attribute Utility Theory

Multi-Attribute Utility Theory (MAUT) [15] is an MCDA approach, which assumes that decision makers aim to maximize their implicit utility function. MAUT is a compensatory technique, since it allows smaller values on a subset of criteria to be compensated by a large value on a single criterion, and is based on expected utility theory with some strong technical assumptions related to comparability, transitivity, continuity, and independence of outcomes (that assumes independence of criteria). MAUT is attractive because of its sound theoretical foundations (based on expected utility theory), its non-monetary nature and its applicability to be used as a basis for comparison of new, not yet considered alternatives with the same utility function constructed for the same decision maker. In addition, its natural approach to modeling risk behavior is particularly attractive for designing security decisions, where risk attitude of decision makers plays crucial role in their decision patterns.

The global utility of an alternative is obtained by aggregating individual criteria values amplified by criteria weights for all criteria. However, before aggregation, these criteria values must be *normalized*, in order to provide a fair basis for comparison. A normalization function, which in MAUT corresponds to *marginal utility function*, is a function $n_g : \mathcal{K}_g \to [0, 1]$. This function can change from one decision maker to another, thus, encoding some notion of preference/meaning interpretation.

In addition, preferences can be encoded using criteria weights, which in MAUT represent trade-offs between criteria. Here, a weight shows the relative importance of the criterion, when compared to other criteria. In particular, it defines how many units of one criterion can be traded-off for a unit of another criterion.

Here, we assume deterministic criteria weights defined by the decision maker with a criteria function $w : \mathcal{G} \to [0, 1]$ such that $\sum_{w(g) \in W} w(g) = 1$.

Determining weights explicitly may be difficult for decision makers. It may be cognitively hard to quantify weights also due to the meaning of weights may not be straightforward and even differ in different MCDA methods [2].

We can now define the notion of MAUT model.

Definition 1. *A MAUT model is a tuple* $M = (\mathcal{A}, \mathcal{G}, \Sigma_{\mathcal{G}}, n, w)$, *where* $\mathcal{A}$ *is a set of alternatives,* $\mathcal{G}$ *is a set of criteria,* $\Sigma_{\mathcal{G}}$ *is a set of criteria evaluation functions,* n *is a set of normalization functions with* $n_g : \mathcal{K}_g \to [0, 1]$ *for each* g, *and* $w : \mathcal{G} \to [0, 1]$ *is a weights function, such that* $\sum_{w(g) \in W} w(g) = 1$.

After mapping all criteria utilities to their scales, normalizing them and defining weights, the alternatives can be evaluated. For aggregating marginal criteria utilities for each alternative some form of aggregation function should be used, e.g. multiplicative, additive or some combination of both is usually applied. When compared to additive aggregation function, which allows some criteria for alternatives to be of zero value, multiplicative aggregation function requires presence of non-zero values for all criteria to make alternative useful.

For now, we introduce one of simplest forms of aggregating evaluations on all criteria values for each alternative, weighted sum, which we will also use for the Wi-Fi case study:

Definition 2 (Utility function). *Given a model* $M = (\mathcal{A}, \mathcal{G}, \Sigma_{\mathcal{G}}, n, w)$, *the utility of an alternative* $a \in \mathcal{A}$, *is defined as:*

$$v(a, w, \sigma) = \sum_{g \in \mathcal{G}} w(g) \cdot n_g(\sigma_g(a)).$$

Note that for the sake of simplicity, we consider that the normalization function is unique for all criteria, and therefore we do not pass it as an argument of v.

We now assume that decision makers base their decision making process using a MAUT model[1]. Given a vector of evaluation functions σ, and a weights function w, the policy of a decision maker is defined as:

$$\pi(w, \sigma) = \arg \max_{a \in \mathcal{A}} v(a, w, \sigma).$$

Note that in order for a decision maker to be deterministic, we assume the existence of an arbitrary ordering over alternatives, so that if there are several alternatives maximizing the utility function, the decision maker selects the highest one according to that ordering.

---

[1] We are aware of strong assumptions of MAUT and biases from rational behavior of the decision makers studied, e.g. by Kahneman and Tversky [13], [14], and here establish a basic model for influencing human decision making in security context also to initiate investigation of these biases.

2.2   Case Study: Selection of a Wi-Fi Network

As a case study, let us consider an example of influencing a choice of a publicly available wireless network (Wi-Fi). The dangers of choosing non-secure Wi-Fi are well documented: it exposes device and transmitted data to increased chances of spoofing and man-in-the-middle attacks [1], [5], and new attacks appear regularly.

For instance, recently, it was reported that the penetration testing tool BDF-Proxy (BackdoorFactory Proxy), which acts as a proxy for network communication, has the capability to infect any binary executable download the user makes with a Metasploit malware [16]. Thus, it can compromise the user's device with malicious software and gain control over the device. This attack is particularly problematic for untrusted Wi-Fi's as the Wi-Fi router can engage in ARP (Address Resolution Protocol) spoofing (i.e., by manipulating the lowest-level address resolution to make the user's client go through the proxy without the user's knowledge: The Wi-Fi can make the client fall for the trap without the user noticing anything.

Ability to influence choice of trusted Wi-Fi is of special interest in the context of recent consumerization of IT trend [24] and BYOD, in particular, since employees work on their own devices and define security protection of their devices by themselves, thus, potentially, exposing sensitive information [19]. In general, over one billion workers will work remotely by 2015, over a third of the total worldwide workforce [12]. A company that allows BYOD may want to influence the employee so that the trade-off decision between security and productivity is done in the company's interest. Alternatively, users may want to have influencing software on their phone to assist in making the information security decisions for work as well as home use.

Influencing was earlier introduced in the security context of Wi-Fi selection in [23]. There, the focus is on introducing the user interface design nudges, and on evaluating them with a user group. Here, we want to model human behavior further when modifying context of decision making by introducing an affect influencing factor, color, and by computing impact of such or another modification of the context on the decision to be made. In particular, a traffic light effect [8] is used with red-green colors (and associated emotions and meanings), which was also applied for framing choices to nudge individuals away from privacy-invasive applications in [6].

Let us consider a user in a coffee-shop having to choose between two different public wireless networks $\mathcal{A} = \{s, f\}$: s is a secure Wi-Fi with weak signal; f is a Wi-Fi of the coffee shop, with strong signal, but not necessarily safe. We want to illustrate with this example the trade-off between security and productivity/usability, and therefore consider the set of criteria $\mathcal{G} = \{t, r, l\}$, indicating the trust of the network, its strength and color in which its name is drawn, respectively. For the sake of simplicity, we assume that the scales for the trust and strength criteria are defined as $\mathcal{K}_t = \mathcal{K}_r = \{0, 1, 2\}$ (the higher the better). For the color criterion, a scale is defined as $\mathcal{K}_l = \{R, N, G\}$, corresponding to red, neutral and green colors of paint used for drawing names of networks,

Table 1. Decision matrix for $\sigma = [s \mapsto (1,1,N), f \mapsto (0,2,N)]$

	criteria		
	trust	strength	color
	$\{0,1,2\}$	$\{0,1,2\}$	$\{R,N,G\}$ (scale)
	$t \to \max$	$r \to \max$	$l \to \max$ (direction)
	0.5	0.3	0.2 (weights)
alternative			
s	1	1	N
f	0	2	N
$\pi(w,\sigma)$		f	

respectively. This categorical scale can be mapped into a scale of quantitative values $\mathcal{K}_l = \{0, 0.5, 1\}$, taking into account traffic light similar effect, with red color associated with danger, green color – with no danger, and neutral color, e.g. white, with no special affect on users (when compared to a standard amber color with attention bringing effect).

Note that here, we consider a simple and abstract notion of trust, and, in practice, this notion can be defined using the presence of Wi-Fi network providers in a white list predefined by security officer or system administrator of the company or by the employee him-/herself. More sophisticated evaluation of 'trust' criterion may take into account other aspects, e.g. current location of an employee [9].

Finally, the decision maker has to define the criteria weights $w = (0.5; 0.3; 0.2)$, meaning that connecting to a trusted Wi-Fi is more important for the decision maker than choosing a Wi-Fi with strong signal. The color of the presented name of a Wi-Fi is less significant for the decision maker than the two other criteria.

In the following, for the sake of compactness, we write $\sigma = [s \mapsto (v_1, v_2, v_3), f \mapsto (v_4, v_5, v_6)]$, associating s Wi-Fi with a trust of v_1, a strength of v_2 and a color of v_3 values, respectively; and f Wi-Fi with a trust of v_4, a strength of v_5 and a color of v_6 values, respectively. Table 1 represents the traditional decision matrix [2] for a decision maker, evaluation of a set of alternatives on a set of criteria. Assuming that a decision maker uses a linear normalization function of the following form:

$$n_g(\sigma) = \frac{g - g^{min}}{g^{max} - g^{min}}, \tag{1}$$

we can calculate the utility for each alternative as follows:

$$v(s, w, \sigma) = 0.5 * 0.5 + 0.3 * 0.5 + 0.2 * 0.5 = 0.5$$
$$v(f, w, \sigma) = 0.5 * 0 + 0.3 * 1 + 0.2 * 0.5 = 0.4.$$

From these calculations it follows that the decision maker selects $\pi(w, \sigma) = s$.

3   Decision Evaluation

3.1   Impact

To be able to measure the efficiency of an alternative, we introduce an *impact function* such that, given an alternative a, a weight function w and evaluation functions σ, $\rho(a, w, \sigma)$ represents the impact of criteria weights and criteria evaluations of alternative on selection of that alternative as the final choice of the decision maker. In the rest of the paper, we consider that the impact function intuitively represents a benefit for the system, and as such, the aim of a stakeholder is to maximize the impact, i.e., a higher impact is 'better'. Note, the impact function should be seen as an ideal valuation of the possible alternatives, and as a way to evaluate the behavior of the decision makers, rather than as a way to define the behavior of the decision makers.

In general, this function can be defined in many different ways (for instance, through an access control policy stating which alternatives are secure [18]). We propose here to define it using a MAUT model, which is however slightly different from the one defined above. The impact function can be defined directly as the utility function v of M. However, in the context of information security, we want to clearly distinguish the alternatives, so that there are 'good' and 'bad' alternatives. Hence, given an alternative a, a weight function w and a set of criteria evaluation functions σ, we define the impact function as:

$$\rho(a, w, \sigma) = \begin{cases} 1 & \text{if } v(a, w, \sigma) \geq v(a', w_i, \sigma) \text{ for any } a', \\ 0 & \text{otherwise.} \end{cases}$$

In other words, an alternative has an impact if, and only if, it is maximal according to the utility function. Note that more complex impact functions can be considered, for instance, when different levels of security can be defined.

3.2   Utility Function Parameters

To evaluate the efficiency of a decision made by a decision maker, a stakeholder may compare it to his own choice or a choice of an 'ideal' decision maker from a company perspective in the same situation. Four cases are possible here: In ideal case, the stakeholder would wish the decision maker behaving in an optimal way from the stakeholder's point of view. This would mean the decision maker (user / employee) having the same with stakeholder (or company) preferences, or the same weight function $w_c = w_u$, where w_c is a stakeholder's weight function and w_u is a user's weight function, and at the same time having the same set of criteria evaluation functions: $\sigma_c = \sigma_u$, where σ_c is a stakeholder's set of criteria functions and σ_u is a set of user's criteria evaluation functions.

However, the reality might be different with the most general case with both sets of evaluation functions $\sigma_c \neq \sigma_u$ and weighting functions $w_c \neq w_u$ being different for a stakeholder and a decision maker. The two special cases with either different weights or different criteria evaluation functions will be considered below.

3.3   Evaluation in Case Study: Selection of a Wi-Fi Network

To illustrate decision evaluation scenarios for the case of public Wi-Fi selection, let us consider the stakeholder and the decision maker having the same weight functions $w_c = w_u = (0.5; 0.3; 0.2)$. However, they have different evaluation functions: $\sigma_c = [s \mapsto (1, 1, N), f \mapsto (0, 2, N)]$ for stakeholder and $\sigma_u = [s \mapsto (1, 1, N), f \mapsto (1, 2, N)]$ for the decision maker. Indeed, the decision maker considers the alternative f as being more trusted, with $\sigma_u[t](f) = 1$, when compared to the company, which assigns to it a smaller trust value with $\sigma_c[t](f) = 0$. This small difference results in the different utilities of the alternatives $v(s, w_u, \sigma_u) = 0.5$ and $v(f, w_u, \sigma_u) = 0.65$, and leads to the decision maker choosing $f = \pi(w_u, \sigma_u)$, while $\rho(f, w_c, \sigma_c) = 0$, meaning that the decision maker selects an alternative that is suboptimal for the stakeholder.

We may also consider another case of the company and the user having the same set of criteria evaluation functions $\sigma_c = \sigma_u$, but different preferences with respect to criteria weights. For instance, the stakeholder considers trust being more important $w_c(t) = 0.5$, when compared to the decision maker $w_u(t) = 0.3$. They may also have different opinions about importance of the strength of the Wi-Fi signal: the stakeholder assumes it is as less important $w_c(r) = 0.4$, when compared to the decision maker $w_u(r) = 0.6$. But they agree on color being not very important $w_c(l) = w_u(l) = 0.1$. Here, again $\pi(w_u, \sigma_u) = f$, while $\rho(f, w_c, \sigma_c) = 0$.

4   Influencing Decisions

As said in the introduction section, in the BYOD context, companies allow their employees to use personal devices for work (or company devices for personal purposes), and the border between personal and company data becomes blurred. In such situations, companies may try to take some control over personal devices for better protection of their data. Applying strong security policies for such personal devices may meet opposition reaction from their employees, since employees ownership perception of devices will be disturbed, which may push employees towards overriding such security policies. Therefore, companies must search for 'softer' ways of influencing their employees behavior.

In this work, we suggest a 'soft' strategy for stakeholders to assist in security decisions by their employees with limited changes to the information taken into account by the decision maker, based on the idea that even small changes can influence final choices of decision makers [13], [14]. Such an approach was considered widely for health and social solutions, see e.g. [21], [22], and recently studied in the context of security and privacy decision making [6].

Next, we examine an example of a company adopted BYOD strategy or stakeholder, which wants to protect its employees, users of devices, from non secure behavior and emphasize safer choices for them. Note, that we assume here a 'good' stakeholder, who wants to help and protect a user in a paternalistic way, and exclude a 'bad' influencer, for instance, aiming to attack users and manipulate their choices motivated by 'bad' incentives, leaving this special case as a future work. As our working case study for demonstrating influencing effect, we keep selection of a Wi-Fi to connect to in a public place among several available ones. We consider when influencing may be beneficial to both a stakeholder and a user, and how it may be performed, assuming MAUT model as a basis for human decision making.

4.1 Influence

Given a MAUT model $M = (\mathcal{A}, \mathcal{G}, \Sigma_\mathcal{G}, n, w)$, we can consider ways, in which a stakeholder may influence choices of a decision maker. By definition of the model, there are two ways to affect the result of the model evaluation: either by affecting a weighting function, and corresponding set of weights, or by affecting a set of criteria evaluation functions, and corresponding set of criteria values for alternatives.

Influencing weighting of criteria means influencing implicit trade-off preferences of decision makers with respect to different criteria. In principle, this approach may be efficient, but, in practice, it is time-consuming, since it requires training and education of users and their subsequent conscious reflection on the issues they were taught. For instance, for security decisions, it would require training sessions on the security policy of the company to increase employees awareness of risks; their education on the security issues related to the policy of their company and on possible consequences of such decisions for them and their company; and promoting a security culture, e.g., with rewards for secure behavior. These are efficient, but long-term approaches, which require time and involve user awareness and conscious decision making. Moreover, while users may be aware and intend to behave securely, these intentions do not always translate into actual behavior.

Therefore, an alternative and/or complementary approach would be to try to influence the behavior of decision makers directly at the moment of the decision making. This approach would involve changing values for some criteria. Having possibility to change all criteria would be ideal for a stakeholder. However, there are different reasons why it may not be possible in most cases; to name a few: it may not be legal or ethical to change values for some criteria or too costly for the company to do it. However, a stakeholder may still be able to change values for some 'modifiable' criteria via a set of evaluation functions, assuming the values for the rest of criteria are non-changeable.

Given a set of criteria $\mathcal{G}$, we consider a subset of 'modifiable' criteria $\mathcal{M} \subseteq \mathcal{G}$, for which stakeholder can change criteria values. The exact definition of this subset depends of course on the context, but intuitively, it corresponds to the aspects taken into account by the decision maker that are controlled by the stakeholder. Given a vector of evaluation functions σ, we define the set of possible modified functions as:

$$P_\mathcal{M}(\sigma) = \{\sigma' \mid \forall g \in (\mathcal{G} \setminus \mathcal{M}) \; \sigma'[g] = \sigma[g]\}.$$

Table 2. Impact of all modifications to the color criterion for initial alternatives evaluations $\sigma_u = [s \mapsto (1, 1, N), f \mapsto (1, 2, N)]$ and criteria weights $w_u = (0.3; 0.5; 0.2)$ of the decision maker with $\rho(f, w_c, \sigma_c) = 1$ and $\rho(s, w_c, \sigma_c) = 0$

σ_{xy}	$v(s, w_u, \sigma_{xy})$	$v(f, w_u, \sigma_{xy})$	$a = \pi(w_u, \sigma_{xy})$	$\rho(a, w_c, \sigma_c)$
σ_{NN}	0.5	0.6	f	0
σ_{NR}	0.5	0.5	f	0
σ_{NG}	0.5	0.7	f	0
σ_{GN}	0.6	0.6	f	0
σ_{RN}	0.4	0.6	f	0
$\sigma_{\mathbf{GR}}$	**0.6**	**0.5**	s	1
σ_{RG}	0.4	0.7	f	0
σ_{RR}	0.4	0.5	f	0
σ_{GG}	0.6	0.7	f	0

In general, more complex restrictions on $P_{\mathcal{M}}$ can be defined, for instance, reflecting an incremental change in the values of criteria (e.g., the value of a criterion can only be incremented or decremented by a given factor).

Hence, we assume that there is an *influence*, if and only if, the decision maker would behave differently without being influenced. The raw impact of an influence can be measured in a differential way: given a vector of evaluation functions σ and a weight function w, we say that the decision maker was influenced whenever $\pi(w, \sigma) \neq \pi(w, \sigma')$. Note that the set of alternatives $\mathcal{A}$ does not change with the application of criteria evaluation modifications. In other words, influencing a decision maker does not change the set of alternatives available to the decision maker.

We are now in position to define the optimal modification possible by a stakeholder over a decision maker.

Definition 3. *Given a stakeholder with a weights function w_c and a vector of evaluation functions σ_c, and a decision maker with a weights function w_u and a vector of evaluation functions σ_u, the optimal vector of modified evaluation functions for the decision maker is given by:*

$$\mathsf{opt}(w_u, w_c, \sigma_u, \sigma_c) = \arg \max_{\sigma'_u \in P_{\mathcal{M}}(\sigma_u)} \rho(\pi(w_u, \sigma'_u), w_c, \sigma_c).$$

4.2 Influence in Case Study: Selection of a Wi-Fi Network

Let us consider a subset of modifiable criteria $\mathcal{M} = \{l\}$, i.e., only the color, in which a network is displayed, can be modified. We now illustrate the optimal modification of the criteria evaluations opt for the influencing strategy applied by the stakeholder. Having a set of criteria weights $w_c = (0.5; 0.4; 0.1)$ and criteria evaluations of alternatives $\sigma_c = [s \mapsto (1, 1, N), f \mapsto (0, 2, N)]$, we have $\rho(s, w_c, \sigma_c) = 1$ and $\rho(f, w_c, \sigma_c) = 0$. In other words, the stakeholder wants to influence the decision maker towards selecting a more secure Wi-Fi s.

Let us consider now that the decision maker has different criteria weights $w_u = (0.3; 0.5; 0.2)$ and different criteria evaluations $\sigma_u = [s \mapsto (1, 1, N), f \mapsto (1, 2, N)]$, which leads to the decision maker choosing a faster network $\pi(w_u, \sigma_u) = f$. Since $\mathcal{M} = \{l\}$, only the color criterion value can be modified. Table 2 details all the possible cases, where, for the sake of compactness, we write σ_{xy} for the evaluation function $\sigma_{xy} = [s \mapsto (1, 1, x), f \mapsto (1, 2, y)]$. We also consider that when s and f have the same value, the decision maker selects f by default.

In other words, $\text{opt}(w_u, \sigma_u, w_c, \sigma_c) = [s \mapsto (1, 1, G), f \mapsto (1, 2, R)]$, i.e., changing the color of s to green, and that of f to red, results in the influencing effect making the decision maker to swap his/her choice and to select an alternative preferred by the stakeholder.

However, note the impact of this modification depends on the set of non-modifiable criteria $\{t, r\}$. For instance, if the utility of f is null and of s is maximal, there is no effect that will make a decision maker with a weight on $l \neq 1$ and additive aggregation function to change its decision from s to f. Similarly, if the decision maker has a weight equal to 0 on the criterion t, then all effects have no impact. See details of the last case in Table 3 for the same set of criteria evaluations $\sigma_u = [s \mapsto (1, 1, N), f \mapsto (1, 2, N)]$ of the decision maker as in the previous example, but different set of weights $w_u = (0; 0.8; 0.2)$. This case demonstrates that if decision makers do not care about the trust of the network, there is no chance to make them to select a more secure alternative whatever modifications are applied to the modifiable criteria.

Table 3. Impact of all modifications to the color criterion for initial alternatives evaluations $\sigma_u = [s \mapsto (1, 1, N), f \mapsto (1, 2, N)]$ and criteria weights $w_u = (0; 0.8; 0.2)$ of the decision maker with $\rho(f, w_c, \sigma_c) = 1$ and $\rho(s, w_c, \sigma_c) = 0$

σ_{xy}	$v(s, w_u, \sigma_{xy})$	$v(f, w_u, \sigma_{xy})$	$a = \pi(w_u, \sigma_{xy})$	$\rho(a, w_c, \sigma_c)$
σ_{NN}	0.5	0.9	f	0
σ_{NR}	0.5	0.8	f	0
σ_{NG}	0.5	1	f	0
σ_{GN}	0.6	0.9	f	0
σ_{RN}	0.4	0.9	f	0
σ_{GR}	0.6	0.8	f	0
σ_{RG}	0.4	1	f	0
σ_{RR}	0.4	0.8	f	0
σ_{GG}	0.6	1	f	0

5   Influencing Population

In all previous sections, we have considered a deterministic decision maker by default. To allow modeling groups of users rather than single users, we may consider a *probabilistic* decision maker. We model this aspect by considering a probability distribution over weights, such that given a weight function w, $\psi(w)$

represents the probability of w. From a statistical point of view, $\psi(w)$ represents the percentage of the population with the weight distribution w.

The policy of the entire population can therefore be defined as given a MAUT model $M = (\mathcal{A}, \mathcal{G}, \Sigma_{\mathcal{G}}, n, \psi(w))$:

$$\pi(\psi, \sigma, a) = \sum_{w \in W} \{\psi(w) \mid \pi(w, \sigma,) = a\}. \tag{2}$$

For influencing a population of users, a stakeholder needs to look for an alternative (or subset of alternatives) with highest impact and a subset of modifiable criteria that makes this alternative (preferred by the stakeholder) to be selected by the majority of population.

$$\mathsf{opt}(w_c, \sigma_c, \psi_u, \sigma_u) = \arg \max_{\sigma' \in P_M(\sigma_u)} \sum_{a \in \mathcal{A}} \rho(a, w_c, \sigma_c) \pi(\psi_u, \sigma'_u, a).$$

5.1 Population in Case Study: Selection of a Wi-Fi Network

As an example of population modeling, we can consider examples of three types of decision makers with the same criteria evaluation functions $\sigma_u = [s \mapsto (1, 1, N), f \mapsto (1, 2, N)]$, but different criteria weights $w_1 = (0.3; 0.5; 0.2)$, $w_2 = (0; 0.8; 0.2)$, and $w_3 = (0.8; 0; 0.2)$. Let us also consider a probability distribution ψ such that $\psi(w_1) = \psi(w_2) = \psi(w_3) = 1/3$. We can calculate that $\pi(w_1, \sigma_u) = f$, $\pi(w_2, \sigma_u) = f$, $\pi(w_3, \sigma_u) = s$, and therefore, following Equation 2, we have $\pi(f, \sigma, \psi) = 2/3$ and $\pi(s, \sigma, \psi) = 1/3$.

If a stakeholder wants to shift choices of a population of users, he/she may consider similar strategy as one proposed for influencing choice of individual decision makers, but taking into account weights of different groups of users.

6   Conclusion

In this work, we have proposed a model for influencing human decision making in security context. We have illustrated the approach with a case study of a public Wi-Fi selection, and have shown how optimal influence may be selected. Even though the resented multi-criteria model is simplified, when compared to possible real-life scenario, however, it establishes a basis for developing a more complex framework, which we consider as our future work.

The first step will be to consider more than two alternatives to select from. Moreover, it will be interesting to investigate more complex impact functions (e.g. non-monotonic), which may lead to a backfire of influencing, with decision maker selecting a worse alternative when compared to his/her initial intention. Another interesting aspect is related to studying different normalization functions and their interpretation by different decision makers. For instance, it was observed in [23] that a padlock sign, usually assigned to trusted Wi-Fi's, may be perceived as blocking of access by some users, who misinterpret, and, consequently, normalize differently evaluation of Wi-Fi's. Interesting aspects to study are dependence between different security decisions and applying a sequence or combinations of influencing effects. For instance, in [23], it was shown that the color effect has a higher impact when applied in combination with ordering effect of different networks presented to the decision maker by default.

Taking into account complexity of different criteria, such as 'trust' criterion, MAUT contribution may be further investigated by modeling more complex shapes of marginal utility functions, such as convex or concave utility functions corresponding to risk (risk-prone or risk-averse) attitude of decision maker, when compared to the linear marginal utility functions modeled here. Moreover, the quantities obtained through MAUT can be used to characterize the strength of the effect applied, following, for instance, the recent approach in the context of quantitative access control policies [18].

Finally, the Wi-Fi scenario provides an interesting basis for future work. The importance of name when choosing a Wi-Fi was studied in the context of trust in [9]. The 'trust' criterion is interesting as it may take into account various information, e.g. about decision maker's location, to avoid situations, where the most trusted network for a researcher located in a coffee shop far away from universities appears to be the 'eduroam' Wi-Fi, an international network for all university staff of universities provided within campuses of universities only.

Acknowledgements. The authors acknowledge funding for Choice Architecture for Information Security (ChAISe) project EP/K006568/1 from Engineering and Physical Sciences Research Council (EPSRC), UK, and Government Communications Headquarters (GCHQ), UK, as a part of Cyber Research Institute. We gratefully acknowledge the support and contribution of our colleagues on the ChAISe project from Newcastle University: James Turland and Northumbria University: Debora Jeske, Lynne Coventry, Pam Briggs and Christopher Laing who worked with us to identify issues and solutions in this project. We are also thankful to unknown reviewers who helped to improve this paper.

References

1. Aime, M., Calandriello, G., Lioy, A.: Dependability in wireless networks: Can we rely on WiFi? IEEE Security Privacy 5(1), 23–29 (2007)
2. Belton, V., Stewart, T.: Multiple Criteria Decision Analysis: An Integrated Approach. Kluwer Academic Publishers, Dordrecht (2002)
3. Bishop, M.A.: The Art and Science of Computer Security. Addison-Wesley Longman Publishing Co., Inc., Boston (2003)
4. Bovens, L.: The ethics of nudge. In: Grüne-Yanoff, T., Hansson, S. (eds.) Preference Change: Approaches from Philosophy, Economics and Psychology. Philosophy and Methodology of Social Sciences, vol. 42, pp. 207–219. Springer, Theory and Decision Library (2009)
5. Chismon, D., Carter, T., Ruks, M., Hoggard, H.: Mobile devices: Guide for implementers. White paper, MWRInfoSecurity and Center for the Protection of National Infrastructure (CPNI), Basingstoke, UK (February 2013)

6. Choe, E.K., Jung, J., Lee, B., Fisher, K.: Nudging people away from privacy-invasive mobile apps through visual framing. In: Kotzé, P., Marsden, G., Lindgaard, G., Wesson, J., Winckler, M. (eds.) INTERACT 2013, Part III. LNCS, vol. 8119, pp. 74–91. Springer, Heidelberg (2013)
7. Clarke, J., Hidalgo, M.G., Lioy, A., Petkovic, M., Vishik, C., Ward, J.: Consumerization of IT: Top risks and opportunities. ENISA deliverables, European Network and Information Security Agency (ENISA), European Network and Information Security Agency (ENISA) report (2012)
8. Farnham, G., Leune, K.: Tools and standards for cyber threat intelligence projects. Technical report, SANS Institute (2013)
9. Ferreira, A., Huynen, J.-L., Koenig, V., Lenzini, G., Rivas, S.: Socio-technical study on the effect of trust and context when choosing WiFi names. In: Accorsi, R., Ranise, S. (eds.) STM 2013. LNCS, vol. 8203, pp. 131–143. Springer, Heidelberg (2013)
10. Goodwin, P., Wright, G.: Decision Analysis for Management Judgment, 4th edn. J. Wiley (2009)
11. Heilmann, C.: Success conditions for nudges: A methodological critique of libertarian paternalism. European Journal for Philosophy of Science 4(1), 75–94 (2014)
12. AIDC worldwide mobile worker population 2010-2015 forecast. Technical report, IDC Australia (2012)
13. Kahneman, D.: Thinking, fast and slow. Farrar, Straus & Giroux, New York (2011)
14. Kahneman, D., Tversky, A.: Prospect theory: An analysis of decision under risk. Econometrica 47(2), 263–291 (1979)
15. Keeney, R., Raiffa, H.: Decisions with Multiple Objectives: Preferences and Value Tradeoffs. J. Wiley, New York (1976)
16. Kennedy, D., O'Gorman, J., Kearns, D., Aharoni, M.: Metasploit: The Penetration Tester's Guide, 1st edn. No Starch Press, San Francisco (2011)
17. Morisset, C., Groß, T., van Moorsel, A., Yevseyeva, I.: Formalization of influencing in information security. Technical Report CS-TR-1423, Newcastle University (May 2014)
18. Morisset, C., Groß, T., van Moorsel, A., Yevseyeva, I.: Nudging for quantitative access control systems. In: Tryfonas, T., Askoxylakis, I. (eds.) HAS 2014. LNCS, vol. 8533, pp. 340–351. Springer, Heidelberg (2014)
19. Seigneur, J.-M., Kölndorfer, P., Busch, M., Hochleitner, C.: A survey of trust and risk metrics for a BYOD mobile worker world. In: Proceedings of SOTICS 2013, pp. 82–91. IARIA (2013)
20. Servin, A., Kudenko, D.: Multi-agent reinforcement learning for intrusion detection: A case study and evaluation. In: Bergmann, R., Lindemann, G., Kirn, S., Pěchouček, M. (eds.) MATES 2008. LNCS (LNAI), vol. 5244, pp. 159–170. Springer, Heidelberg (2008)
21. Applying behavioural insights to reduce fraud, error and debt. Policy paper: Transforming government services to make them more efficient and effective for users, Cabinet Office, Behavioural Insights Team, UK (February 2012)
22. Thaler, R.H., Sunstein, C.R.: Nudge: Improving Decisions About Health, Wealth, and Happiness. Yale University Press, New Haven (2008)
23. Turland, J., Coventry, L., Jeske, D., Briggs, P., Laing, C., Yevseyeva, I., van Moorsel, A.: Nudging towards security: Developing an application for wireless network selection for android phones (in preparation, 2014)
24. Yevseyeva, I., Morisset, C., Turland, J., Coventry, L., Groß, T., Laing, C., van Moorsel, A.: Consumerization of IT: Mitigating risky user actions and improving productivity with nudging. In: Proceeding of CENTERIS 2014 - Conference on ENTERprise Information Systems. Springer (accepted, 2014)

Automated Capacity Planning for PEPA Models

Christopher D. Williams and Jane Hillston

LFCS, School of Informatics, University of Edinburgh, Edinburgh, Scotland, UK
jane.hillston@ed.ac.uk
http://www.quanticol.eu

Abstract. Capacity planning is concerned with the provisioning of systems in order to ensure that they meet the demand or performance requirements of users. Currently for PEPA models, a modeller who wishes to solve a capacity planning problem has to either carry out a manual search for an optimal configuration or work outside the provided tool suite. We present a new extension to the Eclipse Plug-in for PEPA which integrates automated capacity planning into the functionality of the tool, thus allowing optimal configurations of large scale PEPA models to be found.

1 Introduction

Performance analysis occurs in many guises during system development. One important role of a performance analyst is as a capacity planner, helping the system developers to appropriately dimension their system in order to meet user demand in a satisfactory manner. This involves choosing the appropriate number of copies for each type of component within the system, for both software components (e.g. threads) and hardware components (e.g. database servers). This can involve running models of the system with many different configurations to thoroughly explore the parameter space to find an optimum. There is often a tension between the number of components and the user perceived performance, as system managers wish to limit the number of components for a variety of reasons, including economic, efficiency and maintainability considerations.

The Eclipse PEPA Plug-in is a mature analysis tool supporting the PEPA modelling language, and offering a number of different solution techniques. Whilst it does offer an experimentation facility, this is intended for varying one or two parameters within a model and all results are returned to the modeller. In contrast, in capacity planning the modeller will typically want to thoroughly explore a multi-dimensional parameter space but only be presented with results for those points in parameter space which are optimal with respect to some modeller-defined performance and population target. Currently a modeller who wishes to conduct such an exploration must manually search parameter space, or export their model to another format such as Matlab and then code up a search algorithm themselves.

In this paper we present an extension of the Eclipse PEPA Plug-in which incorporates an automated capacity planning tool which addresses this problem.

A. Horváth and K. Wolter (Eds.): EPEW 2014, LNCS 8721, pp. 209–223, 2014.

The modeller can specify their optimisation problem and let the tool search for a model configuration which best matches the performance target whilst also keeping the number of components minimal. In the current implementation this is aimed at scalable PEPA models amenable to solution based on ordinary differential equations, making interactive capacity planning possible. But the developed framework is flexible and now that the feasibility of the approach has been established, could be readily extended to a broader class of PEPA models and other solution techniques.

1.1 Related Work

Hillston *et al.* suggested the feasibility of a capacity planning tool for PEPA in [1]. Within that paper, the parameter space of a moderately-sized example is explored by hand to find a configuration which matches a target average response time, when one population size is fixed and all others are allowed to vary. This paper provided inspiration for our current work. Genetic algorithms and genetic programming metaheuristics have previously been used in conjunction with PEPA and Bio-PEPA models by Marco *et al.* [2,3]. In that work they sought to find model parameters which give optimal fit of model output to a given time series of biological data. Both activity rates and model structure make up the search space for the metaheuristic. Similarly Karaman *et al.* use genetic algorithms to construct a process algebra model satisfying a path optimisation property, again focussing on the time series view of the process algebra model output [4]. In contrast our work identifies emergent global properties of the process algebra model as the goal. Like Karaman *et al.* our primary focus is on investigating model with different structures, i.e. different numbers of components in this case. The work of Geisweiller also sought to match to given performance characteristics but by finding optimal rate parameters for a PEPA model with fixed structure, using expectation-maximisation techniques [5]. More generally, in [6], Cerotti *et al.* present a general capacity planning tool for dimensioning in Cloud systems, based on simple queueing abstractions.

1.2 Structure of the Paper

The rest of this paper is organised as follows. In Section 2, we present the necessary background. In Section 3 we describe the basic functionality of the tool and how a user can specify a search based on a performance target and cost requirements. The search procedure is sensitive to the settings of the algorithm, so we additionally offer a *driven* search in which another, simpler, metaheuristic, is used to find the best settings for the intended search. This is explained in detail in Section 4. In the following section, Section 5 we present some results from the tool, run on a number of different models, and in particular show how the Particle Swarm Optimisation (PSO) search compares with simpler metaheuristics such as hill-climbing and a genetic algorithm. In Section 6, the paper concludes with a summary of the results and a discussion of how the work can be developed further.

2   Background

In this section we give a brief overview of the background to the project, reviewing the PEPA modelling language, and specifically the fluid approximation which allows large scale models to be rapidly solved via a set of ordinary differential equations. In this work we choose this solution technique to underlie our capacity planning because we envisage a tool which the user will engage and experiment with. However, the same framework could be used with alternative solution methods albeit with slower response time and the capacity planning would no longer be interactive. Finally in this section we describe the metaheuristic that is the principal focus of our work.

2.1   PEPA

PEPA is a CSP-like process calculus extended with the notion of exponentially distributed activities [7]. A PEPA model consists of a collection of components or *processes*, which undertake actions. A component may perform an action autonomously, *independent actions*, or in synchronisation with other components, *shared actions*. PEPA models are generated by the following two-level grammar:

$$S ::= (\alpha, r).S \quad | \quad S + S \quad | \quad A_S, A_S \stackrel{def}{=} S$$

$$C ::= S \quad | \quad C \underset{L}{\bowtie} C \quad | \quad C/L \quad | \quad A_C, A_C \stackrel{def}{=} C$$

The first production defines *sequential components*, i.e., processes which only exhibit sequential or branching behaviour (by means of prefix, ".", or choice, "+", respectively). The second production defines *model components*, in which the interactions between the sequential components are expressed through the cooperation (" $\underset{L}{\bowtie}$ ") and hiding ("/") operators. Within a cooperation, the set L specifies which action types must be shared; components can proceed independently and concurrently on other action types. A *system equation* specifies all the components within a system and how they must interact.

Typically, each sequential component corresponds to a component of the system and the performance of the system is constrained by the interactions between components as imposed by the cooperations. For example for a client-server system, some number of clients may compete for access to a limited number of servers. This may be written as the system equation

$$Client[N_c] \underset{\{request\}}{\bowtie} Server[N_s]$$

where $Client[N_c]$ is shorthand for $Client \underset{\emptyset}{\bowtie} \cdots \underset{\emptyset}{\bowtie} Client$ for a population of N_c clients, and similarly for $Server[N_s]$.

The capacity planning problem is to find appropriate population sizes for the components in the system equation which allow the system to meet a performance target. For example this might be *response time should be on average less than 2s when there are 100 clients in the system*. Some populations, such as the clients, may be fixed as they are specified by the requirement, whereas others

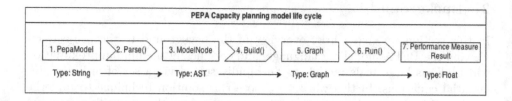

Fig. 1. The model life cycle in the capacity planning tool. The square boxes represent classes of models, the arrowed boxes represent a conversion method.

may vary, allowing the modeller to explore a parameter space. In this simple case, the parameter space is one-dimensional and capacity planning amounts to finding the number of servers which is sufficient to meet the response time target. However, in general the search space will be multi-dimensional as the system will be made up of many different interacting populations of components.

The original structured operational semantics for PEPA [7], gives rise to a continuous time Markov chain (CTMC) via a labelled multi-transition system. In [8], an alternative symbolic semantics in terms of generating functions is presented, which allows a fluid approximation of large scale PEPA models to be derived automatically. This derivation is incorporated in the tool which supports PEPA modelling, the PEPA Eclipse Plug-in tool [9]. Thus the PEPA Eclipse Plug-in supports numerical solution of the CTMC, stochastic simulation and ODE numerical simulation.

Our capacity planning currently focuses on this latter approach: the scalability and efficiency of the ODE-based fluid approximation means that many model instances can be solved relatively quickly, providing an interactive experience for the user. Moreover, the large scale models amenable to this approach are typically those for which it is difficult to predict the relative influence of one individual over all the interacting populations.

To evaluate a PEPA Model using ODEs, the PEPA Eclipse Plug-in first converts the String model class (`PepaModel` — the model as input by the modeller) into a Graph model class (`Graph`), via an abstract syntax tree termed the `ModelNode`. Once the model is a graph, the ODEs can be evaluated, returning performance measure results as an array of floats. Fig. 1 shows the PEPA Eclipse Plug-in methods used by the capacity planning tool, and the life cycle of a model.

The intervention of the capacity planning tool, compared with a regular ODE solution of a PEPA model, is that the capacity planning tool manipulates model configurations during step 3 of the lifecycle. Using a Visitor pattern and a Java class called `ASTHandling`, the capacity planning tool can operate on models, and change the population value of one or more components. This updated AST is then built into a `Graph` (step 5) and evaluated using ODEs. This will be explained in more detail in Section 3.

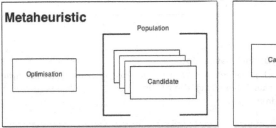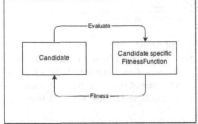

Fig. 2. Schematic view of a metaheuristic

2.2 Particle Swarm Optimization (PSO)

Metaheuristics are a class of general algorithms which may be used to find optimised solutions to a large class of problems. Examples include hill-climbing, genetic algorithms, simulated annealing, ant colony optimisation and particle swarm optimisation (PSO). These can be thought of as strategies for guiding a search, and they do not typically guarantee convergence or optimality. But in many practical situations they have been found to perform well, both with respect to optimality and efficiency compared with brute-force or manual search. During the course of developing the capacity planning tool we implemented and experimented with a number of different metaheuristics: hill-climbing, a genetic algorithm and PSO. The experimentation suggested that the PSO was the most successful in the sense of both efficiency and optimality. Thus in the final version of the tool this is the offered algorithm, and in this paper we focus on that due to space limitations, although we will present some of the experimental results in Section 5.

```
Initiation of variables
generation = user defined value          //number of iterations
candidatePopulation = user defined value   //how many candidates
localBest = user defined value           //weight of local best in new velocity
globalBest = user defined value          //weight of global best in new velocity
originalVelocity = user defined value    //weight of original velocity in new velocity

Scattering of candidates
bestCandidate = null
arrayOfCandidates = []

for candidatePopulation do:
    newCandidate = (new candidate) //candidate random position and velocity
    arrayOfCandidates = arrayOfCandidates ∪ newCandidate
```

Fig. 3. PSO initiation

```
Iterative search
for generation do:

   //find fittest
   for each candidate in arrayOfCandidates do:
      fitnessFunction(candidate) //use system equation fitness function to assess fitness
      if(candidate 'is better than' bestCandidate) do:
         bestCandidate = candidate

   //update each candidate
   for each candidate in arrayOfCandidates do:
      currentPosition ← candidate's position // the system equation
      previousVelocity ← candidate's velocity
      localBestPosition ← candidate's previous best position
      globalBestPosition ← bestCandidate's best position
      newVelocity = (originalVelocity * previousVelocity) +
         localBest * (localBestPosition - currentPosition) +
         globalBest * (globalBestPosition - currentPosition)
      candidate's velocity ← newVelocity  //the candidate gets a new velocity vector

   //move each candidate
   for each candidate in arrayOfCandidates do:
      candidate's current position ←
         floor(candidate's current position + candidate's velocity)
```

Fig. 4. PSO search

PSO is a stochastic optimisation algorithm modelled after flocking or swarming agents [10]. A PSO works by scattering a number of candidates (or particles) in some search space and providing each with a random velocity. Each generation, or iteration, of the optimisation method, each candidate moves according to its velocity. Then the best candidate of that iteration, called the *global best*, is found and its position is made known to all other candidates. Each agent then uses this global best, its own velocity, and its own best position historically, to create a new velocity vector which it uses in the next step. After a number of iterations the PSO should converge on an optimum position in the search space. (See Figs. 3 and 4 for pseudocode inspired by Luke [11].)

3   Simple Search

In the basic use-case for our tool, the modeller uses the capacity planning tool to set up a search directly. In this case the modeller establishes a fitness function with the aid of wizard in the Eclipse Plug-in. As will be explained later in this section, the fitness function is constructed of a number of components, allowing a target performance measure, and the population sizes to be taken into consideration. The wizard also allows the modeller to choose the setting for the PSO algorithm, such as how much weight should be given to the global best position in the definition of a new velocity for each candidate. In this simple search

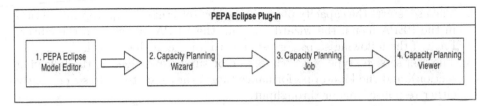

Fig. 5. Steps in the use of the capacity planning tool. Input is a PEPA model from the Editor, the output is displayed in a Viewer pane within Eclipse.

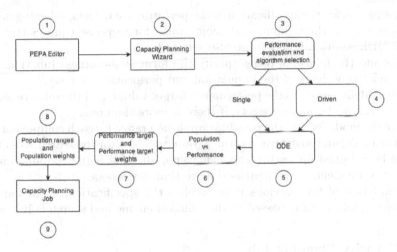

Fig. 6. A high level picture of the capacity planning wizard

each candidate in the PSO corresponds to a system equation, or configuration of the system. Once the search is fully specified, one run of the PSO algorithm is used to explore parameter space and return the candidate which gives the best value of the fitness function. The parameter space is defined by the range of populations considered for each of the component types in the PEPA model.

The steps in the use of the capacity planning tool are depicted in Fig. 5. We will focus on steps 2, 3 and 4.

3.1 Capacity Planning Wizard

An Eclipse wizard is provided to support the modeller in the set up and initiation of a capacity planning job. A `Wizard` is a Java class and is used to present a logical ordering of more Java classes called `WizardPages`. `WizardPages` are used to guide the user in entering the parameters, the input and settings, required for a capacity planning search. Fig. 6 shows the steps of the capacity planning wizard pages.

1. Input is a model created in the PEPA editor.

2. The user starts the capacity planning tool by selecting the appropriate action in the PEPA menu: the wizard picks up the PEPA model from the editor. Each of the following steps corresponds to a page in the wizard.
3. The user sets the type of search, driven or single (explained in the following section), and the kind of performance target they seek to address: currently either response time or throughput.
4. Driven or single search are specified separately. Here the user can change the number of experiments (explained in the following section) and algorithm settings.
5. This page is for the specification of the performance targets, selecting actions in the case of throughput, and agent states for response time. Settings for the ODE solution function are also selected.
6. This and the following pages specify the form of the fitness function. Here the relative weights of the population and performance are set.
7. The modeller specifies the performance target values, and the relative weighting of the performance targets, if there is more than one.
8. Here the modeller sets the possible population range for each component type in the model and so determines the parameter space for the search. Weighting can be assigned for each component type, allowing higher populations for some components to be penalised more than for others.
9. Completion of the previous page finishes the specification of the capacity planning job. Data is passed to the solution engine and search is initiated.

3.2   Capacity Planning Job

General Design. The capacity planning wizard passes all data to the capacity planning Job, and then starts the search. A capacity planning job is created as a separate thread and so runs separately from the Eclipse user interface. Fig. 7 presents a schematic view of a capacity planning job. As the metaheuristics which we have considered are stochastic, in order to increase the likelihood of finding a good result, a number of searches are run serially; each run is termed an *experiment*, and each set of experiments is termed a *Lab*. Each experiment is a run of the metaheuristic with a randomly seeded *candidate population*.

Each candidate corresponds to a point in the parameter space, i.e. an instantiation of the system equation for the model, and each will have a corresponding value of the fitness function. Since the fitness function has an element corresponding to the performance target, each model instance must be solved. Currently the existing ODE solver within the PEPA Eclipse Plug-in is used for this, but the architecture has been designed so that a metaheuristic can run on any type of candidate, and with any of the solvers present in the tool.

Fitness Function Evaluation. The fitness function, built by the modeller with the aid of the wizard, determines the search that is carried out. The PSO seeks to minimise the value of the fitness function, by seeking candidates which have the following characteristics:

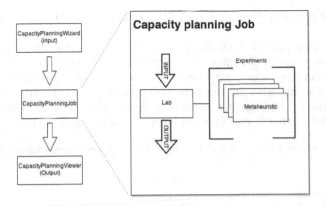

Fig. 7. Schematic view of a capacity planning job

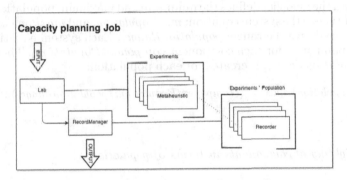

Fig. 8. Recording output

- Good performance with regards to the target. This will be based on the result of the performance measures from the ODE function.
- Minimal total component population (*componentPopulation*), i.e. the number of components in the system equation.
- There will typically be a trade-off between performance and population so the modeller gives a relative weight to these objectives (*performanceWeight* vs. *populationWeight*).
- Component weighting (*componentPopulation$_i$*): similarly within a system, it might be more important to minimise the population size of some component types than others, for example due to cost or other considerations.
- Performance target weighting (*performanceTargetWeight$_i$*): when the modeller has specified a performance target with multiple elements they can give relative weights to those elements.

These five elements collectively build up the fitness value, which determines how good one system equation candidate is relative to another[1]. The fitness is better the *lower* it is. Thus for each iteration of the PSO we construct:

---

[1] Here we use a weighted sum, but of course, a weighted product could also be used.

1. The result of the performance evaluations from the ODE function

The performance results, $ODEResult_i$, is returned by the ODE solver after it has evaluated a model instance. There will be one $ODEResult_i$ for each user defined performance target, $performanceTarget_i$, for each performance measure specified by the modeller in the capacity planning wizard. Any result that is better than the target is given a value of 0. A $scaledPerformanceValue$ is created for each target so that it can be used later in the fitness function:

$$scaledPerformanceValue_i = |(100 - (ODEResult_i/performanceTarget_i) \times 100)|$$

2. The number of components in a system equation

In the wizard the modeller defines the minimum and maximum population for each component type in the system equation: $minPopulation_i$ and $maxPopulation_i$, and from these we derive the range: $populationRange_i$. Each system equation candidate has a population for each component $componentPopulation_i$. Then a value, $scaledPopulationValue_i$, is created for each population:

$$scaledPopulationValue_i = |((componentPopulation_i/populationRange_i) \times 100)|$$

3. The weighting of components in terms of population

The modeller also specifies the weight for each component, $componentWeight_i$, which gives the fitness function some notion of cost per component. From these values we derive the total population weight ($totalCWeight = \sum_0^n(componentWeight_i)$) and the contribution of the population of this candidate to the fitness function:

$$weightedPopResult = \sum_0^n (scaledPopulationValue_i \times (\frac{componentWeight_i}{totalCWeight}))$$

4. The weighting of performance targets across all selected performance targets

Similarly we construct the contribution of the performance target to the fitness function by defining $totalPWeight = \sum_0^n(performanceTargetWeight_i)$ and $weightedPerfResult$:

$$weightedPerfResult = \sum_0^n (scaledPerformanceValue_i \times \frac{performanceTargetWeight_i}{totalPWeight})$$

This allows the user to put more importance on finding one performance target over any others.

5. The balance between performance targets and population size

Finally we use the weights for population and performance entered by the modeller, *populationWeight* and *performanceWeight* to combine these values in the final fitness value. Here *totalWeight = populationWeight + performanceWeight*

$$fitnessValue = weightedPopResult \times (populationWeight/totalWeight)$$
$$+ weightPerfResult \times (performanceWeight/totalWeight)$$

In summary, this creates a fitness value such that the smaller the value, the lower the number of components used, and the closer the performance evaluation will be to the user defined target.

3.3 Capacity Planning Viewer

The capacity planning viewer is a viewer pane in the Eclipse environment. At the end of each run of the metaheuristic it is passed the ten top candidates by fitness value and displays them in order in the view pane.

To keep track of all the experiments and candidates we use a Java class `RecorderManager`. A `RecorderManager` is created with every Lab, and every experiment has a `Recorder`. It is the `Recorder`'s role to track the progress of an experiment, and it is the task of the `RecorderManager` to collect all `Recorders` and pass the results to the capacity planning `Viewer`.

4   Driven Search

As will be discussed in the next section the simple search proved effective for finding a model configuration satisfying the performance targets, whilst minimising the number of components. But in testing users had difficulties in choosing the best settings for the PSO algorithm to achieve the best results. These settings include the initial population and velocity of components, weightings for global and local best as well as the number of generations and experiments. Therefore we experimented with using another metaheuristic to find the settings for a metaheuristic search over the parameter space. We term this *driven search*: a *driving* metaheuristic is used to find the best settings for the second *driven* metaheuristic. After some investigation we found that this works well when a hill-climbing algorithm is used to find the best settings for a PSO algorithm.

In a driven search the candidate is itself a Lab, as defined for single search; this is termed a `Lab candidate`. Each experiment consists of a single search `Lab`, and so one driven experiment, consists of many Lab experiments. This is represented schematically in Fig. 9.

Lab Fitness Function. In order to evaluate the fitness of a Lab candidate, we construct a Lab fitness function. This Lab fitness function calls the `RecordManager` from the underlying single search (Fig. 10) to return four values;

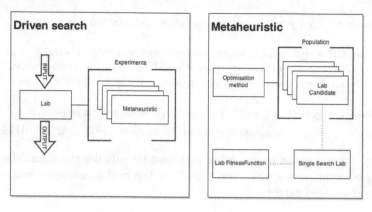

Fig. 9. Schematic view of a driven search

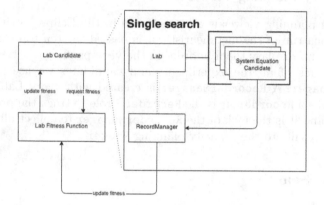

Fig. 10. Lab candidate calling a fitness update

- Top fitness from the underlying experiments (*topFitness*);
- Mean fitness of all experiments in this Lab candidate (*meanFitness*);
- Standard deviation of the best values found for this Lab candidate (*standardDeviation*);
- Average response time of the underlying experiments (*averageResponseTime*)[2].

The lower a lab fitness value is, the better the underlying single search will be at finding the optimal candidate. Getting the best fitness has the highest priority and is reflected in the fitness function by having a weight of 0.6, next priority is finding a single search that on average returns a high value and therefore it uses a weight of 0.2. In order to break any ties between Lab candidates, accuracy (standard deviation) and response time each are given weights of 0.1. These weights were arrived at through experimentation.

---

[2] How long an experiment took to run, not to be confused with the average response time performance measure evaluated on the model.

$$fitnessValue = (0.6 \times topFitness) + (0.2 \times meanFitness)$$
$$+ (0.1 \times standardDeviation) + (0.1 \times averageResponseTime)$$

5 Evaluation

As mentioned previously, our initial implementation offered three different meta-heuristics. In addition to PSO we implemented a stochastic hill climbing (HC) and a genetic algorithm (GA). Table 1 shows the results of experimentation with these different search algorithms over a variety of models[3]. The model sizes ranged from 2 populations (Simple) to 9 populations (E University), with components from 2 local states (Traffic) up to 18 local states (E University).

Table 1. Evaluation results for Top fitness (to 6 s.f.) and Response time in milliseconds (RT). Best values highlighted in each row.

Model	GA Top fitness	RT	HC Top fitness	RT	PSO Top fitness	RT
Brewery	**21.1459**	31.49	23.7710	46.15	22.6990	**29.075**
Brewery2	18.8109	220.39	16.3775	217.16	**14.1821**	**208.05**
E University	11.8022	1723.17	10.5644	**1327.02**	**5.22160**	1653.36
Example System	8.26005	46.14	7.33108	46.75	**3.91710**	**43.98**
Example System2	6.58843	75.75	7.65840	75.67	**1.75361**	**71.2**
Large-t	**2**	120.71	**2**	122.68	**2**	**119.95**
Simple	8.00494	**29.17**	7.50368	33.02	**7.25008**	30.10
Simple2	9.03451	33.82	11.0040	28.45	**4.86591**	**26.70**
Traffic	34.2665	33.05	34.4215	33.32	**32.9015**	**30.87**

It can be seen from Table 1 that the PSO algorithm is generally achieving the best result and often in the shortest time. This is the reason why we decided to only include PSO in the final implementation of the tool.

Figs. 11(a) and 11(b) show what happens to fitness over generations with PSO and GA. Each algorithm was run 1000 times on the same model and the average fitness of each generation was calculated. An effective algorithm should improve (decrease) the average fitness through the generations. Figs. 11(a) shows the average fitness and the variance of fitness over generations of the PSO on our example model, `Example.pepa`. On average the PSO has converged after 8 generations — there is no significant improvement in average fitness after that. In contrast, GA convergence appears to happen around the 5th generation, but there is much wider variance, indicating that there are better candidates found but not consistently. The width of the variance shows GA has different behaviour on each run, whereas PSO has much less variance in the final generations. The results for HC exhibited even greater variability and much less convergence. These results show that the PSO has the same behaviour on average independently of how the algorithm was started. Thus it is more predictable.

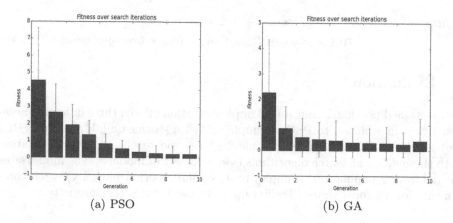

(a) PSO   (b) GA

Fig. 11. Graph showing the average fitness and variance of fitness over generations of PSO (left) and GA (right)

Table 2. The top fitness results of the driven PSO against the single PSO

Model	Driven fitness	Single fitness
Simple ab	**5.00523**	6.30427
Simple a	**7.25008**	7.25008
Brewery ab	18.6424	**15.1921**
Brewery a	**22.3579**	23.1876
E University	**5.11754**	6.06217
Example system a	**3.82410**	5.21521
Example system ab	**2.78335**	4.90980

Finally in Table 2 we show the results of a comparison of a single PSO search, against a driven PSO search on each of our example models. Note that the driven search achieves a better fitness result in the majority of models.

6   Conclusions

In this paper we have presented the new capacity planning facility within the Eclipse Plug-in for PEPA. This tool offers both a fast *single* search and a slower *driven* search. The former requires the modeller to understands the heuristic and how to appropriately choose settings, the latter has only two settings. Experimentation suggests that the PSO metaheuristic offers the best compromise between speed of convergence and satisfaction of requirements. The heuristic's optimisation method manages the candidate search, how the candidates communicate, how the candidates are created and mutated, but the fitness function

---

[3] More details of this evaluation and the models used can be found in [12].

defines the search. The capacity planning wizard enables and supports the user to define an appropriate fitness function.

Future work can proceed in a number of directions. For example, we aim to generalise the tool to work with other solvers within the PEPA Eclipse Plug-in tool suite. Whilst this is likely to be much more computationally expensive, it will be suitable for a broader range of models. Currently activity rates can only be searched through cloning subpopulations of components with different rates, but in the future we will extend the support for activity rates in the search parameter space. Furthermore, we will also allow more general specification of performance targets to be logic-based.

Acknowledgement. This work is partially supported by the EU project QUANTICOL, 600708.

References

1. Hillston, J., Tribastone, M., Gilmore, S.: Stochastic Process Algebras: From Individuals to Populations. Computer Journal 55(7), 866–881 (2012)
2. Marco, D., Cairns, D., Shankland, C.: Optimisation of process algebra models using evolutionary computation. In: IEEE Congress on Evolutionary Computation, pp. 1296–1301 (2011)
3. Marco, D., Scott, E., Cairns, D., Graham, A., Allen, J., Mahajan, S., Shankland, C.: Investigating co-infection dynamics through evolution of Bio-PEPA model parameters: A combined process algebra and evolutionary computing approach. In: Gilbert, D., Heiner, M. (eds.) CMSB 2012. LNCS, vol. 7605, pp. 227–246. Springer, Heidelberg (2012)
4. Karaman, S., Shima, T., Frazzoli, E.: A process algebra genetic algorithm. IEEE Transactions on Evolutionary Computation 16(4), 489–503 (2012)
5. Geisweiller, N.: Finding the Most Likely Values inside a PEPA Model According to Partially Observable Executions. PhD thesis, LAAS (2006)
6. Cerotti, D., Gribaudo, M., Piazzolla, P., Serazzi, G.: Asymptotic behaviour and performance constraints of replication policies. In: Proceedings of PASM 2014 (2014)
7. Hillston, J.: A Compositional Approach to Performance Modelling. Cambridge University Press (2005)
8. Tribastone, M., Gilmore, S., Hillston, J.: Scalable differential analysis of process algebra models. IEEE Transactions on Software Engineering 38(1), 205–219 (2012)
9. Tribastone, M., Duguid, A., Gilmore, S.: The PEPA Eclipse Plugin. Performance Evaluation Review 36(4), 28 (2009)
10. Poli, R., Kennedy, J., Blackwell, T.: Particle swarm optimization; an overview. Swarm Intelligence (1), 33 (2007)
11. Luke, S.: Essentials of metaheuristics, Lulu (2011)
12. Williams, C.: A capacity planning tool for PEPA. Master's thesis, School of Informatics, University of Edinburgh (2014)

Stochastic Approximation of Global Reachability Probabilities of Markov Population Models*

Luca Bortolussi[1] and Roberta Lanciani[2]

[1] DMG, University of Trieste, Italy
CNR/ISTI, Pisa, Italy
luca@dmi.units.it
[2] IMT Lucca, Italy
roberta.lanciani@imtlucca.it

Abstract. Complex computer systems, from peer-to-peer networks to the spreading of computer virus epidemics, can often be described as Markovian models of large populations of interacting agents. Many properties of such systems can be rephrased as the computation of time bounded reachability probabilities. However, large population models suffer severely from state space explosion, hence a direct computation of these probabilities is often unfeasible. In this paper we present some results in estimating these probabilities using ideas borrowed from Fluid and Central Limit approximations. We consider also an empirical improvement of the basic method leveraging higher order stochastic approximations. Results are illustrated on a peer-to-peer example.

Keywords: Stochastic model checking, reachability, hitting times, fluid approximation, central limit approximation, linear noise approximation.

1 Introduction

Recent years have seen an impressively growing complexity of computer systems being engineered and deployed, from cloud computing to smart cities, resulting in a stringent need for model-based design. However, describing such systems and analysing the so-obtained models in a mathematically sound and computationally efficient way is extremely challenging. These systems are often described as *population models*, which naturally capture the complexity in terms of interactions between relatively simple heterogeneous sets of components. Examples include peer-to-peer networks [20], epidemic spreading in computer networks [21], but also models of smart city scenarios like bike sharing [12]. Population models are usually given a semantics in terms of Continuous Time Markov Chains, and many tools have been developed in the past years for their analysis. Here we recall *model checkers* [17, 18], taking as input complex queries specified as temporal logic formulae and returning their satisfaction probability, using numerical routines to compute transient and steady state probabilities [3, 4].

* This research has been partially funded by the EU-FET project QUANTICOL (nr. 600708) and by FRA-UniTS.

A. Horváth and K. Wolter (Eds.): EPEW 2014, LNCS 8721, pp. 224–239, 2014.

These analysis techniques, however, suffer from scalability issues caused by the *state space explosion* of population models, which is severe even for populations of the order of few hundreds. A growing trend to tackle this challenge is the use of stochastic approximations to simplify model complexity, predominantly in the form of *Fluid* (or *Mean-Field*) *Approximation* [8]. These methods have been employed successfully to capture mean transient behaviour [8], but also to estimate passage times [15,16] and for stochastic model checking [6,9]. In particular, Fluid Approximation has been used to estimate passage times [15,16] and for model checking [6, 7] of properties of individual agents in large population models. These local specifications have been lifted to the collective level by considering the fraction of agents satisfying a local specification, expressed as a deterministic automaton [15] or a deterministic timed automaton [9], and bounding [15,16] or approximating their probability by exploiting a second order Gaussian approximation [9].

In this paper we continue along this direction, applying stochastic approximation ideas to a different class of global properties, namely *reachability properties* at the collective population level. Specifically, we are interested in the fast computation of the (approximate) probability with which the system can reach a certain region of the state space, defined by a non-linear inequality of population variables, within a given time horizon $T < \infty$. Examples of such properties are the probability that a large fraction of users of a peer-to-peer network has an updated piece of information, or the probability with which a given fraction of computers in a LAN becomes infected by a virus. Reachability queries are important in many respects: safety properties are of this kind and they constitute the core subroutine to check time-bounded CSL properties [4].

The main idea of our approach to approximate such probabilities is to exploit second order stochastic approximations of hitting time probabilities [11]. Consider the Fluid Approximation, which is a deterministic process described by a set of ODE, and assume their solution enters the region $\mathcal{R}$ of interest at a given time $t_{\mathcal{R}}$. Then, $t_{\mathcal{R}}$ is an estimate of the hitting time of $\mathcal{R}$ also for the stochastic model with a very large population. However, for populations of the order of hundreds of individuals, a typical size in heterogeneous models, this approximation is too crude, as stochastic effects cannot be easily neglected. Hence, our idea is to exploit the *Central Limit Approximation* [9]: by replacing the CTMC with a Gaussian process, we can obtain a Gaussian approximation of the time to reach $\mathcal{R}$, and therefore of the reachability probability. The effectiveness of this approach will be illustrated in the paper discussing an example of software update in a peer to peer network, inspired by [14].

We consider also improvements of this estimate. The idea is to define a *higher order approximation* of the moments of the hitting time distribution, leveraging a recently developed approach for chemical reaction networks [13], and use this information to reconstruct a plausible distribution, within a moment reconstruction scheme based on the maximum entropy principle [1]. In this paper, we illustrate this idea focusing on the first two moments, but higher order approximations can be easily obtained as well.

The paper is organised as follows. In Section 2, we introduce the modelling language for Markov population models, the running example, and the Fluid and Central Limit Approximations. In Section 3, we present the global reachability problem, and its Gaussian approximation. Section 4 discusses the higher order approximation, while Section 5 shows the method in practice on the peer-to-peer example. The final discussion can be found in Section 6.

2 Background

2.1 Markov Population Models

A *Markov population model* is a system comprised of a large number of interacting entities, or *agents*. Each agent is an instance of an *agent class*, which defines its (finite) state space and the set of its possible actions (the *local transitions*).

Definition 1 (Agent class). *An* agent class $\mathscr{A}$ *is a pair* (S, E) *in which* $S = \{1, \ldots, n\}$ *is the state space and* $E = \{\epsilon_1, \ldots, \epsilon_m\} \subseteq S \times A \times S$ *is the set of local transitions of the form* $\epsilon_i = s_i \xrightarrow{\alpha_i} s_i'$, *where* $s_i, s_i' \in S$ *are the initial and arrival states, and* $\alpha_i \in A$ *is an action label belonging to the action set* A.

The evolution in time of an agent belonging to a class $\mathscr{A} = (S, E)$ is described by a random variable $Y(t) \in S$, that denotes the state of the agent at time t.

To ease the notation, in this work, we consider populations whose agents Y_j, $j = 1, 2, \ldots$ all belong to the same class $\mathscr{A} = (S, E)$ with $S = \{1, \ldots, n\}$. The extension of the method to multiple classes is straightforward. We further assume that agents in the same state are indistinguishable and we define the *collective variables* $X_i(t) = \sum_j \mathbb{1}\{Y_j(t) = k\}$, $X_i \in \{0, 1, \ldots\}$, $i = 1, \ldots, n$, which count how many agents are in each state $1, \ldots, n$ at time t. In this way, the vector of collective variables $\mathbf{X} = (X_1, \ldots, X_n)$ can be used to describe the state of the population. Formally, we define a *population model* in the following way.

Definition 2 (Population model). *A population model* $\mathcal{X}$ *is a tuple* $\mathcal{X} = (\mathscr{A}, \mathcal{T}, \boldsymbol{x}_0)$, *where:*

- $\mathscr{A}$ *is an agent class, as in Definition 1;*
- $\mathcal{T} = \{\tau_1, \ldots, \tau_\ell\}$ *is the set of global transitions of the form* $\tau_i = (\mathbb{S}_i, f_i)$, *where:*
  - $\mathbb{S}_i = \{s_1 \xrightarrow{\alpha_1} s_1', \ldots, s_p \xrightarrow{\alpha_p} s_p'\}$ *is the (finite) set of local transitions synchronized by* τ_i;
  - $f_i : \mathbb{R}^n \longrightarrow \mathbb{R}_{\geq 0}$ *is the (Lipschitz continuous) global rate function.*
- $\boldsymbol{x}_0 = \mathbf{X}(0)$ *is the initial state.*

The set $\mathcal{T}$ of global transitions identifies all the events that can change the state $\mathbf{X}$ of the population model. Intuitively, when a global transition $\tau_i = (\mathbb{S}_i, f_i)$ with $\mathbb{S}_i = \{s_1 \xrightarrow{\alpha_1} s_1', \ldots, s_p \xrightarrow{\alpha_p} s_p'\}$ occurs, the local transitions $s_1 \xrightarrow{\alpha_1} s_1', \ldots, s_p \xrightarrow{\alpha_p} s_p'$ fire and p agents change state accordingly. The expected frequency of τ_i is given by global rate f_i as a function of the state vector $\mathbf{X}$.

To describe the evolution in time of a population model $\mathcal{X} = (\mathcal{A}, \mathcal{T}, \mathbf{x}_0)$, we define the associated CTMC as follows. For each global transition $\tau \in \mathcal{T}$, $\tau = (\mathbb{S}_\tau, f_\tau)$ with $\mathbb{S}_\tau = \{s_1 \xrightarrow{\alpha_1} s'_1, \ldots, s_p \xrightarrow{\alpha_p} s'_p\}$, we encode the net change in the state $\mathbf{X}$ due to τ in the *update vector* $\mathbf{v}_\tau = \sum_{i=1}^p (\mathbf{e}_{s'_i} - \mathbf{e}_{s_i})$, where $\mathbf{e}_{s_i}$ is the unit vector equal to 1 in position s_i and zero elsewhere. Then, the *CTMC* $\mathbf{X}(t)$ *associated with* $\mathcal{X}$ has state space $\mathcal{S} \subset \mathbb{Z}^n$, initial probability distribution concentrated on $\mathbf{x}_0$, and is uniquely characterised by the *infinitesimal generator matrix* $\mathbf{Q}$ whose component $q_{\mathbf{x},\mathbf{x}'}$ for $\mathbf{x}, \mathbf{x}' \in \mathcal{S}$, $\mathbf{x} \neq \mathbf{x}'$, is defined by $q_{\mathbf{x},\mathbf{x}'} = \sum_{\tau \in \mathcal{T} \mid \mathbf{v}_\tau = \mathbf{x}' - \mathbf{x}} f_\tau(\mathbf{x})$.

2.2 Running Example

To illustrate the method of the paper, we consider a simple variant of the peer-to-peer software update process introduced in [14]. In the modelled network, a node can be *old*, meaning that it has an old version of the software, or *updated*, when it has been able to receive the update. In both cases, the node can be switched *ON* and *OFF*, and an old node can update only when it is on. The search for the update in the network lasts until a certain timeout is reached, after which the old node gives up and reaches a *oldOUT* state from which it can be eventually switched off. Finally, we mimic also the possibility that an *oldOUT* node obtains the update from an external source (ext_O) and that the license of the updated version of the software eventually expires or a new version is released ($\exp_U$).

The agent class $\mathcal{A}_{node} = (S_{node}, E_{node})$ of the network nodes can be easily derived from the automaton representation depicted in Figure 1. The population model $\mathcal{X}_{network} = (\mathcal{A}_{node}, \mathcal{T}, \mathbf{x}_0)$ is described by the vector of counting variables $\mathbf{X} = (X_{oldOFF}, X_{oldON}, X_{oldOUT}, X_{updatedOFF}, X_{updatedON})$ and the set of global transition is given by $\mathcal{T} = \{\tau_{\text{on}_O}, \tau_{\text{off}_O}, \tau_{\text{out}_O}, \tau_{\text{off}_T}, \tau_{\text{ext}_O}, \tau_{\text{off}_U}, \tau_{\text{on}_U}, \tau_{\text{update}}, \tau_{\exp_U}\}$. For example, the switching on of an old node is described by $\tau_{\text{on}_O} = \{\{oldOFF \xrightarrow{\text{on}_O} oldON\}, f_{\text{on}_O}\}$, where the synchronisation set specifies that only one (old) node is involved and changes state from *oldOFF* to *oldON* at an expected rate given by the function $f_{\text{on}_O}(\mathbf{X}) = \lambda_{\text{on}_O} X_{oldOFF}$, in which λ_{on_O} is the constant indicating the rate of switching on of old nodes per single unit. The global transitions $\tau_{\text{off}_O}, \tau_{\text{out}_O}, \tau_{\text{off}_T}, \tau_{\text{ext}_O}, \tau_{\text{off}_U}, \tau_{\text{on}_U}, \tau_{\exp_U}$ have a similar form (with λ_{ext_O} and $\lambda_{\exp_U}$ having low values to implement the fact that ext_O and $\exp_U$ happen on a much lower time-scale than the others). The global transition τ_{update}, instead, synchronises two local transitions. In particular, $\tau_{\text{update}} = \{\{oldON \xrightarrow{\text{update}_O} updatedON, updatedON \xrightarrow{\text{update}_U}$ $updatedON\}, f_{\text{update}}\}$ and involves an *oldON*-node and an *updatedON*-node. In this case, we assume that an *updatedON*-node sends the update to an *oldON*-one at an instantaneous rate given by λ_{update} and the rate function has the classical mass action form $f_{\text{update}}(\mathbf{X}) = \lambda_{\text{update}} X_{oldON} X_{updatedON}$, depending on the number of pairs of nodes that are ready to communicate [2].

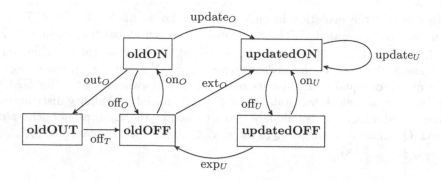

Fig. 1. The automaton representation of the peer-to-peer software update process of Section 2.2

2.3   Fluid and Central Limit Approximations

To introduce the Fluid and Central Limit approximations, we first need to define a notion of *size* $\gamma_\mathcal{X}$ associated with a population model $\mathcal{X} = (\mathcal{A}, \mathcal{T}, \mathbf{x}_0)$. In the context of this paper, $\gamma_\mathcal{X}$ will be the *initial number of agents* in the population, i.e. $\gamma_\mathcal{X} = \sum_i X_i(0)$, but in general, the size is just a suitable constant positive integer number $N > 0$, which is by no means limited to the population size. For example, in queueing models, $\gamma_\mathcal{X}$ can be the arrival rate or, in other context, the size can refer to an environmental factor (like the volume of the container in a biochemical mixture model). For a more detailed discussion on the notion of size, see [8].

Given a population model $\mathcal{X} = (\mathcal{A}, \mathcal{T}, \mathbf{x}_0)$ with population size $\gamma_\mathcal{X} = N$, the *Fluid Approximation* of $\mathcal{X}$ [6–8] provides a deterministic estimation of its dynamics, exact in the limit of an *infinite* population. To define such approximation, we consider a sequence $(\mathcal{X}^{(N)})_{N \in \mathbb{N}}$ of population models that have the same structure of $\mathcal{X}$ for increasing values of population size N. To compare the dynamics of the models in the sequence, we define the *normalised counting variables* $\hat{\mathbf{X}} = \frac{1}{N}\mathbf{X}$ (also known as *population densities* or *occupancy measures*, see [8] for further details) and we consider the *normalized population models* $\hat{\mathcal{X}}^{(N)} = (\mathcal{A}, \hat{\mathcal{T}}^{(N)}, \hat{\mathbf{x}}_0^{(N)})$, obtained by making the rate functions $f_\tau^{(N)}(\hat{\mathbf{X}})$, $\tau \in \hat{\mathcal{T}}^{(N)}$, depend on the normalised variables and rescaling the initial state, i.e. $\hat{\mathbf{x}}_0^{(N)} = \frac{1}{N}\mathbf{x}_0$. We assume that there exist a compact set $\mathcal{D} \subseteq [0,1]^n$ such that $\bigcup_N \hat{S}^{(N)} \subseteq \mathcal{D}$ and a point $\hat{\mathbf{x}}_0 \in \mathcal{D}$ such that $\hat{\mathbf{x}}_0 = \lim_{N \to \infty} \hat{\mathbf{x}}_0^{(N)}$. For each transition $\tau \in \hat{\mathcal{T}}^{(N)}$, we require that the *density dependent condition* $\frac{1}{N}f_\tau^{(N)}(\hat{\mathbf{X}}) = f_\tau(\hat{\mathbf{X}})$ holds true for some Lipschitz function $f_\tau : \mathcal{D} \longrightarrow \mathbb{R}_{\geq 0}$. Furthermore, we define the *drift* $\mathbf{F}$ of $\mathcal{X}^{(N)}$ to be the mean instantaneous net change in the normalised counting variables, i.e. $\mathbf{F}(\hat{\mathbf{X}}) = \sum_{\tau \in \hat{\mathcal{T}}^{(N)}} \mathbf{v}_\tau f_\tau(\hat{\mathbf{X}})$. Then, the unique solution[1] $\boldsymbol{\Phi} : \mathcal{D} \longrightarrow \mathbb{R}^n$ (*independent of N*) of the differential equation

---

[1] The drift F is Lipschitz continuous, since we assume that every f_τ is. Hence, the solution $\boldsymbol{\Phi} : \mathbb{R}_{\geq 0} \to \mathbb{R}^n$ exists and is unique.

$$\frac{d\boldsymbol{\Phi}(t)}{dt} = \mathbf{F}(\boldsymbol{\Phi}(t)), \qquad \text{with} \ \ \boldsymbol{\Phi}(0) = \hat{\mathbf{x}}_0, \tag{1}$$

is the *Fluid Approximation* of the CTMC $\hat{\mathbf{X}}^{(N)}(t)$ associated with $\hat{\mathcal{X}}^{(N)}$. A limit theorem by Kurtz [8,11] states that, for any $T < \infty$, $\sup_{t \in [0,T]} \|\hat{\mathbf{X}}^{(N)}(t) - \boldsymbol{\Phi}(t)\|$ converges to zero (almost surely) as N goes to infinity, hence the approximation is exact in the limit of an infinite number of agents.

The Fluid Approximation can be successfully implemented to estimate the dynamics of large population models [8]. However, whenever $\mathcal{X}^{(N)}$ is a *mesoscopic* population model, meaning that its population is in the order of hundreds of agents, the behaviour of $\mathcal{X}^{(N)}$ remains intrinsically *probabilistic*, hampering the accuracy of the *deterministic* estimation $\boldsymbol{\Phi}(t)$ of $\hat{\mathbf{X}}^{(N)}(t)$. For this reason, in this work, we consider an alternative probabilistic approximation, the *Central Limit Approximation* (CLA) [9], which takes into account the probabilistic fluctuations $\hat{\mathbf{X}}^{(N)}(t)$ around the deterministic average behaviour described by the fluid limit $\boldsymbol{\Phi}(t)$, and can be successfully applied also in the case of mesoscopic populations.

Consider the (normalised) stochastic process $\mathbf{Z}^{(N)}(t) := N^{\frac{1}{2}}(\hat{\mathbf{X}}^{(N)}(t) - \boldsymbol{\Phi}(t))$, which captures the noise of $\hat{\mathbf{X}}^{(N)}(t)$ around the deterministic fluid estimation $\boldsymbol{\Phi}(t)$. Let $\{\mathbf{Z}(t) \in \mathbb{R}^n \mid t \in \mathbb{R}\}$ be the Gaussian process (*independent of N*) with mean $\mathbf{E}(t)$ and covariance $\mathbf{C}(t)$ given by

$$\frac{\partial \mathbf{E}(t)}{\partial t} = \mathbf{J_F}(\boldsymbol{\Phi}(t))\mathbf{E}(t), \qquad \text{with} \ \ \mathbf{E}(0) = 0, \tag{2}$$

and

$$\frac{\partial \mathbf{C}(t)}{\partial t} = \mathbf{J_F}(\boldsymbol{\Phi}(t))\mathbf{C}(t) + \mathbf{C}(t)\mathbf{J_F}^T(\boldsymbol{\Phi}(t)) + \mathbf{G}(\boldsymbol{\Phi}(t)), \qquad \text{with} \ \ \mathbf{C}(0) = 0, \tag{3}$$

where $\mathbf{J_F}(\boldsymbol{\Phi}(t))$ is the Jacobian of $\mathbf{F}$ calculated along the deterministic fluid limit $\boldsymbol{\Phi} : \mathcal{D} \longrightarrow \mathbb{R}^n$, and $\mathbf{G}(\hat{\mathbf{X}}) = \sum_{\tau \in \hat{\mathcal{T}}^{(N)}} \mathbf{v}_\tau \mathbf{v}_\tau^T f_\tau(\hat{\mathbf{X}})$ is called the *diffusion* term. Then, the following result holds true [11].

Theorem 1. *Assume that* $\lim_{N \to \infty} \mathbf{Z}^{(N)}(0) = \mathbf{Z}(0)$. *Then,* $\mathbf{Z}^{(N)}(t)$ *converges in distribution to* $\mathbf{Z}(t)$ *($\mathbf{Z}^{(N)}(t) \Rightarrow \mathbf{Z}(t)$).*

The stochastic process given by

$$\boldsymbol{\Phi}(t) + N^{-\frac{1}{2}}\mathbf{Z}(t) \tag{4}$$

is the *Central Limit Approximation* of the CTMC $\hat{\mathbf{X}}^{(N)}(t)$ and Theorem 1 guarantees that the approximation is exact in the limit of an infinite population size.

3 Stochastic Approximation of Reachability Probabilities

3.1 The Reachability Problem

In this paper, we introduce a model checking procedure for the validation of *reachability properties*. In particular, we consider instances of *global* reachability properties, describing the dynamics of the system at the population level, i.e. characterising the collective behaviour of all agents. In order to verify such requirements, we compute the probability of reaching, within a given time horizon T, a specific *target region* $\mathcal{R} \subset \mathcal{S}$ of the state space $\mathcal{S}$ of the population model, starting from the initial state $\mathbf{x}_0$:

$$\mathbb{P}_{\mathcal{R}}(T) = \mathbb{P}\{\mathbf{X}(t) \in \mathcal{R} \mid t \in [0, T]\}. \tag{5}$$

Reachability is a fundamental notion in the analysis and verification of complex systems and it has been widely studied in many disciplines, including physics, biology and computer science. In the latter community, the investigation of reachability has been usually motivated by the *safety verification problem*, that checks the performance of a model by computing the probability associated with its failure, i.e. with those trajectories that end up in a dead-lock or error state. This type of analysis is indeed fundamental for a sound and reliable verification of software and hardware systems, and in recent years great variety of stochastic model checking techniques have been developed in order to efficiently tackle the problem [10].

The standard *stochastic model checking* procedures address the reachability problem (5) by making *absorbing* the states in the target region $\mathcal{R}$ and computing the transient probability of being in such states at time T. However, all these methods severely suffer by the *state space explosion* of population models, which hampers the computability of transient and steady-state probabilities. In this paper, we introduce a model checking procedure, which tackles the problem of the state space explosion by considering scalable approximations of the population dynamics.

In the following, we will reformulate the reachability problem (5) as a *hitting time problem*. In particular, instead of computing the probability of reaching the target region $\mathcal{R}$ before time T, we consider the *hitting time* $t_{\mathcal{R}}$, the instant in which the trajectory of the population model enters $\mathcal{R}$, and we compute the probability that $t_{\mathcal{R}} < T$:

$$\mathbb{P}_{\mathcal{R}}^{hit}(T) = \mathbb{P}\{t_{\mathcal{R}} \leq T\} \qquad \text{with} \qquad t_{\mathcal{R}} = \inf\{t > 0 \mid \mathbf{X}(t) \in \mathcal{R}\}. \tag{6}$$

It is straightforward to prove that $\mathbb{P}_{\mathcal{R}}(T) = \mathbb{P}_{\mathcal{R}}^{hit}(T)$.

To compute the reachability probability $\mathbb{P}_{\mathcal{R}}(T)$, we will define a Gaussian estimation of the hitting time $t_{\mathcal{R}}$, starting from the Central Limit Approximation of the dynamics of the population model $\mathcal{X}$. Hence, following the procedure illustrated in Section 2.3, we define the sequence of population models $(\mathcal{X}^{(N)})_{N \in \mathbb{N}}$ and, normalising with respect to the population size N, we consider the following reachability probability:

$$\mathbb{P}_{\mathcal{R}}^{(N)}(T) = \mathbb{P}\{t_{\mathcal{R}}^{(N)} \leq T\} \qquad \text{with} \qquad t_{\mathcal{R}}^{(N)} = \inf\{t > 0 \mid \hat{\mathbf{X}}^{(N)}(t) \in \mathcal{R}\}.$$

Moreover, we assume that the (normalised) target region $\mathcal{R}$ is defined by an inequality on population variables. Formally, we introduce a suitable *target function* $\rho : \mathcal{D} \to \mathbb{R}$, such that $\mathcal{R}$ is the subset of $\mathcal{D}$ where ρ is negative. The function ρ is defined on the compact set $\mathcal{D} \subseteq [0,1]^n$ such that $\bigcup_N \hat{\mathcal{S}}^{(N)} \subseteq \mathcal{D}$ (see Section 2.3) and comes in the form of a nonlinear differentiable function of the normalised counting variables $\hat{X}_1(t), \ldots, \hat{X}_n(t)$. Hence, the final form of the *reachability problem* we want to solve is given by

$$\mathbb{P}_{\mathcal{R}}^{(N)}(T) = \mathbb{P}\{t_{\mathcal{R}}^{(N)} \leq T\} \quad \text{with} \quad t_{\mathcal{R}}^{(N)} = \inf\{t > 0 \mid \rho(\hat{\mathbf{X}}^{(N)}(t)) < 0\}. \quad (7)$$

Example 1. As an example, consider the peer-to-peer software update process described in Section 2.2 in a network with 100 nodes, i.e. the population size is N=100. We can validate the performance of the model by considering the simple reachability property which controls the time in which 95% of the nodes have been updated. In this case, the target region $\mathcal{R}$ is that in which the number of agents that have received the update, i.e. $X_{updatedOFF} + X_{updatedON}$, is greater or equal to 95% of the population, i.e. $0.95 \cdot 100$. Hence,

$$\mathcal{R} := \{\mathbf{X}(t) \in \mathcal{S} \mid X_{updatedOFF}(t) + X_{updatedON}(t) \geq 0.95 \cdot 100\},$$

and the target function $\rho : \mathcal{D} \to \mathbb{R}$ is given by

$$\rho(\hat{\mathbf{X}}^{(N)}(t)) := 0.95 - \hat{X}_{updatedOFF}(t) - \hat{X}_{updatedON}(t).$$

3.2 Central Limit Approximation of the Hitting Time Distribution

To compute the cumulative probability distribution associated with the reachability problem (7), the model checking procedure that we are presenting exploits a Corollary of Theorem 1, which provides a Gaussian estimation of $t_{\mathcal{R}}^{(N)}$ based on the Fluid and Central Limit Approximations of $\hat{\mathbf{X}}^{(N)}(t)$. In this section, we review how to define this estimation.

The Fluid Approximation of $\hat{\mathbf{X}}^{(N)}(t)$ provides a deterministic approximation $t_{\mathcal{R}}$ (*independent of* N) of the hitting time $t_{\mathcal{R}}^{(N)}$ of the reachability problem (7), namely

$$t_{\mathcal{R}} = \inf\{t > 0 \mid \Phi(t) \in \mathcal{R}\}. \quad (8)$$

As a direct consequence of Kurtz's Theorem on the fluid limit, such an approximation is exact in the limit of an infinite population. Consider now the (normalised) stochastic variable $\varepsilon^{(N)} := \sqrt{N}(t_{\mathcal{R}}^{(N)} - t_{\mathcal{R}})$, which captures the noise of $t_{\mathcal{R}}^{(N)}$ around the deterministic fluid estimation $t_{\mathcal{R}}$. Let $\{\mathbf{Z}(t) \in \mathbb{R}^n \mid t \in \mathbb{R}\}$ be the Gaussian noise of the Central Limit Approximation with mean and covariance given by (2) and (3) respectively, and let ε be the random variable given by

$$\varepsilon := -\frac{\nabla\rho(\Phi(t_{\mathcal{R}})) \cdot \mathbf{Z}(t_{\mathcal{R}})}{\nabla\rho(\Phi(t_{\mathcal{R}})) \cdot \mathbf{F}(\Phi(t_{\mathcal{R}}))},$$

where $\rho : \mathcal{D} \to \mathbb{R}$ is the target function identifying the (normalised) target region $\hat{\mathcal{R}}$ in (7), ∇ is the gradient, and $\cdot$ is the Euclidean scalar product. Then, the following result holds true ([11], Ch 11, Theorem 4.1).

Theorem 2. *Assume that* $\lim_{N \to \infty} \boldsymbol{Z}^{(N)}(0) = \boldsymbol{Z}(0)$ *as in Theorem 1. If* $t_{\mathcal{R}} < \infty$ *and* $\nabla\rho(\boldsymbol{\Phi}(t_{\mathcal{R}})) \cdot \boldsymbol{F}(\boldsymbol{\Phi}(t_{\mathcal{R}})) < 0$, *then* $\varepsilon^{(N)}(t)$ *converges in distribution to* $\varepsilon(t)$.

In conclusion, the Gaussian approximation of the hitting time $t_{\mathcal{R}}^{(N)}$ that we will consider in our model checking procedure is given by

$$t_{\mathcal{R}} - \frac{1}{\sqrt{N}} \frac{\nabla\rho(\boldsymbol{\Phi}(t_{\mathcal{R}})) \cdot \boldsymbol{Z}(t_{\mathcal{R}})}{\nabla\rho(\boldsymbol{\Phi}(t_{\mathcal{R}})) \cdot \boldsymbol{F}(\boldsymbol{\Phi}(t_{\mathcal{R}}))} \qquad (9)$$

and Theorem 2 guarantees that the estimation is exact in the limit of an infinite population size.

3.3   The Algorithm

The algorithm of our model checking procedure for the verification of reachability properties over Markov population models has the following form.

Input:

- an agent class $\mathscr{A} = (S, E)$ as described in Definition 1;
- a Markov population model $\mathcal{X} = (\mathscr{A}, \mathcal{T}, \mathbf{x}_0)$ as described in Definition 2;
- a global reachability property with target region $\mathcal{R}$ identified by a target function $\rho : \mathcal{D} \to \mathbb{R}$.

Steps:

1. *Integration of the Fluid and Central Limit differential equations.* Numerically solve the ODE systems (1) for the Fluid Approximation $\boldsymbol{\Phi}(t)$, and (2) and (3) for the mean $\mathbf{E}(t)$ and covariance $\mathbf{C}(t)$ of the Gaussian noise of the Central limit Approximation;
2. *Computation of the fluid estimation* $t_{\mathcal{R}}$. Compute the fluid estimation $t_{\mathcal{R}}$ of the hitting time by solving $t_{\mathcal{R}} = \inf\{t > 0 \mid \rho(\boldsymbol{\Phi}(t)) < 0\}$;
3. *Computation of the mean and covariance of the Gaussian approximation.* Identify the mean μ_{hit} and variance σ_{hit}^2 of the Gaussian approximation of the hitting time defined in (9) by solving

$$\mu_{hit} = t_{\mathcal{R}} - \frac{1}{\sqrt{N}} \frac{\nabla\rho(\boldsymbol{\Phi}(t_{\mathcal{R}})) \cdot \mathbf{E}(t_{\mathcal{R}})}{\nabla\rho(\boldsymbol{\Phi}(t_{\mathcal{R}})) \cdot \mathbf{F}(\boldsymbol{\Phi}(t_{\mathcal{R}}))}$$

and

$$\sigma_{hit}^2 = -\frac{1}{\sqrt{N}} \frac{\nabla\rho(\boldsymbol{\Phi}(t_{\mathcal{R}})) \cdot \operatorname{diag}(\mathbf{C}(t_{\mathcal{R}}))}{\nabla\rho(\boldsymbol{\Phi}(t_{\mathcal{R}})) \cdot \mathbf{F}(\boldsymbol{\Phi}(t_{\mathcal{R}}))},$$

where $\operatorname{diag}(\mathbf{C}(t_{\mathcal{R}}))$ is the vector of diagonal elements of $\mathbf{C}(t_{\mathcal{R}})$.

4. *Computation of the reachability probability.* Let $f(t \mid \mu, \sigma^2)$ be the probability density function of a Gaussian distribution in t with mean μ and variance σ^2. Approximate the global reachability probability $\mathbb{P}_{\mathcal{R}}^{(N)}(T)$ by

$$\mathbb{P}_{\mathcal{R}}^{(N)}(T) \sim \int_{-\infty}^{T} f(t \mid \mu_{hit}, \sigma_{hit}^2) dt. \tag{10}$$

The asymptotic correctness of the approximation of the reachability probability $\mathbb{P}_{\mathcal{R}}(T)$ is guaranteed by the following result, which is a straightforward corollary of Theorem 2.

Theorem 3. *Let $\mathbb{P}_{\mathcal{R}}^{(N)}(T)$ be the exact value of the global reachability probability defined in (7), and let $\widetilde{\mathbb{P}}_{\mathcal{R}}^{(N)}(T) = \int_{0}^{T} f(t \mid \mu_{hit}, \sigma_{hit}^2) dt$ be the Gaussian approximation computed in (10). Then, under the assumptions of Theorem 2, it holds that $\lim_{N \to \infty} \| \mathbb{P}_{\mathcal{R}}^{(N)}(T) - \widetilde{\mathbb{P}}_{\mathcal{R}}^{(N)}(T) \| = 0.$*

4 Higher-Order Approximation

As discussed in Section 2.3, the Central Limit Approximation (CLA) is an estimation of the behaviour of a population model which is exact in the limit of an infinite population size, but can be efficiently applied even when considering mesoscopic systems. Indeed, the CLA provides a Gaussian estimation of the stochastic fluctuations of the dynamics of the population model around the (deterministic) average behaviour described by the fluid limit. Again, such Gaussian approximation is asymptotically correct, but in the case of mesoscopic populations it can happen that even the CLA fails to properly describe the dynamics of the population model: the fluid limit itself may indeed fail to accurately describe the average behaviour of the system and/or the stochastic fluctuations around the fluid estimation could be not normally distributed. In these cases, to tackle the error in the estimation of the CLA, *higher-order approximations* of the system behaviour have been proposed and applied in the literature [13]. In this section, we exploit these higher-order approximations to improve the accuracy of Kurtz's Gaussian estimation of the hitting time (9).

Let us first introduce the higher-order approximation of the CTMC $\mathbf{X}^{(N)}(t)$ of a population model $\mathcal{X}^{(N)}$. To ease the presentation, we will describe just the fundamental steps of the definition, leaving aside most of the mathematical prone details (the interested reader can refer to [13] and [22]). As in the case of the CLA, we are interested in estimating the (normalised) process $\mathbf{Z}^{(N)}(t) := N^{\frac{1}{2}}(\hat{\mathbf{X}}^{(N)}(t) - \boldsymbol{\Phi}(t))$, capturing the noise of $\hat{\mathbf{X}}^{(N)}(t)$ around the deterministic Fluid Approximation $\boldsymbol{\Phi}(t)$ given by (1). To achieve this, we write $\hat{\mathbf{X}}^{(N)}(t) = \boldsymbol{\Phi}(t) + N^{-\frac{1}{2}}\mathbf{Z}^{(N)}(t)$ and we substitute this formula in an expansion in powers of N of the Master Equation associated with $\hat{\mathbf{X}}^{(N)}(t)$ [19] , describing the evolution in time of the probability $\Pi(\mathbf{Z}^{(N)}(t))$ of being in state $\hat{\mathbf{X}}^{(N)}(t) = \boldsymbol{\Phi}(t) + N^{-\frac{1}{2}}\mathbf{Z}^{(N)}(t)$ at time t. By dropping high order terms in N in the so obtained form of the Master Equation, we can control the level of accuracy and define different higher-order approximations of $\hat{\mathbf{X}}^{(N)}(t)$. The simplest correction to the CLA, following [13], defines a stochastic process $\mathbf{Z} * (t)$ whose first and second moments are given by:

$$\frac{\partial \mathbf{E}^*(t)}{\partial t} = \mathbf{J}(\boldsymbol{\Phi}(t))\mathbf{E}^*(t) + N^{-1/2}\boldsymbol{\Delta}(\mathbf{C}^*(t)) + O(N^{-1}), \quad \mathbf{E}^*(0) = 0, \quad (11)$$

and

$$\frac{\partial \mathbf{C}^*(t)}{\partial t} = \mathbf{J}(\boldsymbol{\Phi}(t))\mathbf{C}^*(t) + \mathbf{C}^*(t)\mathbf{J}^T(\boldsymbol{\Phi}(t)) + \mathbf{G}(\boldsymbol{\Phi}(t)) + O(N^{-\frac{1}{2}}), \quad \mathbf{C}^*(0) = 0, \quad (12)$$

where $\mathbf{F}$ and $\mathbf{G}$ are the drift and diffusion respectively, $\mathbf{J}$ is the Jacobian of $\mathbf{F}$, and $\boldsymbol{\Delta}(\mathbf{C}^*(t))$ is the vector whose i-th component is given by:

$$\Delta_i(\mathbf{C}^*(t)) = -\frac{1}{2}\left(\sum_{j,k}\frac{\partial^2}{\partial\Phi_j\partial\Phi_k}F_i(\boldsymbol{\Phi})C_{ij}^* - \sum_j \Phi_j\frac{\partial^2}{\partial\Phi_j{}^2}F_i(\boldsymbol{\Phi})\right). \quad (13)$$

Notice that $\mathbf{Z}^*(t)$ depends on N, and moreover, if in Equation (11) we drop the term that is $O(N^{-1/2})$, Equations (11) and (12) describe the mean and covariance of the Central Limit Approximation (i.e. they correspond to the ODE systems (2) and (3)).

The higher-order approximation $\mathbf{Z}^*(t)$ of the noise around the fluid limit $\boldsymbol{\Phi}(t)$ of $\hat{\mathbf{X}}^{(N)}(t)$ can be used to define the *higher-order approximation* $\widetilde{t}_{\mathcal{R}}^{(N)}$ of the hitting-time $t_{\mathcal{R}}^{(N)}$ given by

$$\widetilde{t}_{\mathcal{R}}^{(N)} = t_{\mathcal{R}} - \frac{1}{\sqrt{N}}\frac{\nabla\rho(\boldsymbol{\Phi}(t_{\mathcal{R}}))\cdot\mathbf{Z}^*(t_{\mathcal{R}})}{\nabla\rho(\boldsymbol{\Phi}(t_{\mathcal{R}}))\cdot\mathbf{F}(\boldsymbol{\Phi}(t_{\mathcal{R}}))}. \quad (14)$$

While in (9) the CLA guarantees that $\mathbf{Z}(t)$ is a Gaussian process [11], in (14) there is no limit result that characterises the nature of the distribution of the higher-order approximation $\mathbf{Z}^*(t)$ (and, thus, of the stochastic variable $\widetilde{t}_{\mathcal{R}}^{(N)}$ defined in (14)). To tackle this problem and construct a plausible probability density function for $\widetilde{t}_{\mathcal{R}}^{(N)}$, we leverage an advanced information theoretic *moment-reconstruction* technique based on the *maximum entropy principle* [1]. This computational method is used to design smooth approximations of multidimensional probability distributions by maximising the Shannon entropy subject to taking the moments as constraints. Hence, to improve the estimation of the hitting time $t_{\mathcal{R}}$ given by (9), we consider the first and second moments of $\widetilde{t}_{\mathcal{R}}^{(N)}$, which are given by

$$\mathbf{E}\left[\widetilde{t}_{\mathcal{R}}^{(N)}\right] = t_{\mathcal{R}} - \frac{1}{\sqrt{N}}\frac{\nabla\rho(\boldsymbol{\Phi}(t_{\mathcal{R}}))\cdot\mathbf{E}^*(t_{\mathcal{R}})}{\nabla\rho(\boldsymbol{\Phi}(t_{\mathcal{R}}))\cdot\mathbf{F}(\boldsymbol{\Phi}(t_{\mathcal{R}}))} \quad (15)$$

and

$$\mathbf{C}\left[\widetilde{t}_{\mathcal{R}}^{(N)}\right] = t_{\mathcal{R}} - \frac{1}{\sqrt{N}}\frac{\nabla\rho(\boldsymbol{\Phi}(t_{\mathcal{R}}))\cdot\mathrm{diag}(\mathbf{C}^*(t_{\mathcal{R}}))}{\nabla\rho(\boldsymbol{\Phi}(t_{\mathcal{R}}))\cdot\mathbf{F}(\boldsymbol{\Phi}(t_{\mathcal{R}}))}. \quad (16)$$

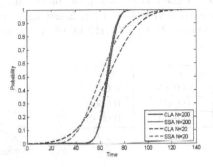 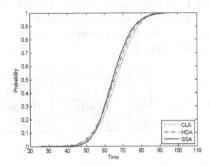

Fig. 2. Results of the experimental analysis of the running example with $\rho(\hat{\mathbf{X}}) = 0.95 - \hat{X}_{oldON} - \hat{X}_{upON}$ and $\hat{\mathbf{x}}_0 = (0.9, 0, 0, 0.1, 0)$. Left: Comparison of reachability probabilities obtained by Central Limit Approximation (CLA) and Gillespie's statistical algorithm (SSA) $N = 20$ and $N = 200$. Right: Comparison of reachability probabilities obtained by CLA, SSA and the higher-order approximation of Section 4 (HOA) for $N = 100$

Then, the moment-reconstruction maximum entropy principle states that the best approximation (in terms of the Shannon entropy) for a one-dimensional probability distribution given its first and second moments, μ and σ^2 respectively, is the Gaussian distribution $\mathcal{N}(\mu, \sigma^2)$. Hence, we conclude that $\tilde{t}_{\mathcal{R}}^{(N)} \sim \mathcal{N}(\mu_{hit}^*, \sigma_{hit}^{2*})$, where $\mu_{hit} = \mathbf{E}[\tilde{t}_{\mathcal{R}}^{(N)}]$ and $\sigma_{hit}^{2*} = \mathbf{C}[\tilde{t}_{\mathcal{R}}^{(N)}]$ are the mean (15) and the variance (16), respectively.

In conclusion, if we consider the higher-order approximation $\tilde{t}_{\mathcal{R}}^{(N)}$ of the hitting time $t_{\mathcal{R}}^{(N)}$ defined in (14), the algorithm of our model checking procedure keeps the form described in Section 3.3, substituting the occurrences of the mean $\mathbf{E}(t)$ and covariance $\mathbf{C}(t)$ of the Central Limit Approximation with the mean $\mathbf{E}^*(t)$ and covariance $\mathbf{C}^*(t)$ of the higher-order approximation given by (12) and (11).

Remark 1. Since by definition $\mathbf{C}(t) = \mathbf{C}^*(t)$, in practice the higher-order approximation (14) improves the estimation given by the Central Limit Approximation (9) by subtracting an N-dependent correction term to its mean, which was actually equal to the Fluid estimation $t_{\mathcal{R}}$ (indeed, by integration of (2), we have that $\mathbf{E}(t) \equiv 0$).

5   Experimental Analysis

In the following, we describe the experimental results obtained on the peer-to-peer software update process introduced in Section 2.2 where we set $\lambda_{on_O} = \lambda_{off_O} = \lambda_{off_T} = \lambda_{on_U} = \lambda_{off_U} = 0.4$, $\lambda_{out_O} = 0.9$, $\lambda_{ext_O} = 0.008$, $\lambda_{exp_U} = 0.001$ and $\lambda_{update} = 1$.

Table 1. Maximum and mean absolute error on the reachability probability estimations obtained by the Central Limit Approximation (max(errCLA), $\mathbb{E}$[errCLA]) and the higher-order approximation of Section 4 (max(errHOA), $\mathbb{E}$[errHOA]) in the experiments of Figure 2

N	max(errCLA)	$\mathbb{E}$[errCLA]	max(errHOA)	$\mathbb{E}$[errHOA]
20	0.1453	0.0443	0.0908	0.0313
100	0.0931	0.0128	0.0415	0.0066
200	0.0623	0.0063	0.0171	0.0023

First, we considered reachability properties characterised by *linear* target functions. In particular, we chose $\rho(\hat{\mathbf{X}}) = 0.95 - \hat{X}_{oldON} - \hat{X}_{upON}$, which is used to check whenever the number of updated nodes reaches 95% of the network size (see Example 1). Figure 2 shows the reachability probabilities $\mathbb{P}_{\mathcal{R}}^{(N)}(T) = \mathbb{P}\{t_{\mathcal{R}}^{(N)} \leq T\}$ for three population sizes: $N = 20, 100, 200$. On the left of Figure 2, the Gaussian estimation (9) obtained by the Central Limit Approximation (CLA) is compared with a statistical estimation (the Gillespie's Stochastic Simulation Algorithm (SSA)) computed over 10000 simulation runs. As expected the accuracy of the estimation increases with N and is already very good for $N = 200$. On the right of Figure 2, for the case $N = 100$, we show also the higher-order approximation defined in Section 4, which corrects the mean obtained by the CLA, improving the quality of the approximation. In Table 1, we report the maximum and mean absolute errors obtained by the CLA and the higher-order approximation. Again as expected, the results improve with N and the quality is already quite good for $N = 100$. Moreover, when $N = 100$ and $N = 200$, the higher-order approximation reduces the errors of more than 50%.

In the second set of experiments, instead, we considered reachability properties identified by *non-linear* target functions. We chose to verify the efficiency of the communication across the network by checking whenever the throughput $X_{oldON} * X_{upON}$ gets below a certain N-dependent threshold. In particular, we set $\rho(\hat{\mathbf{X}}) = \hat{X}_{oldON} * \hat{X}_{upON} - 0.006$. In this case, the experimental results showed that, to reach a good level of accuracy in the approximation, much larger population sizes have to be considered (probably due to the fact that the error in the estimation of gets amplified by the product $\hat{X}_{oldON} * \hat{X}_{upON}$, hence larger N are needed for the method to converge). Indeed, Figure 3 compares the results obtained by the CLA, the SSA and the higher-order approximation for two population sizes: $N = 1000$ and $N = 10000$. When $N = 1000$, the CLA performs poorly in estimating the reachability probability and the quality of the approximation slightly improves when we compute the higher-order approximation of Section 4, which however still fails to capture the slope of the cumulative distribution function obtained by the SSA, due to an inaccuracy in the prediction of the variance. Only when we consider population sizes of the order of 10000, the method starts to converge and the higher-order approximation is finally able to efficiently predict the (estimated) true reachability probability. Including higher-order terms in the method of Section 4 may improve this scenario, too.

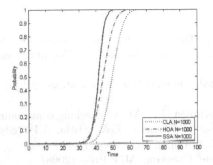 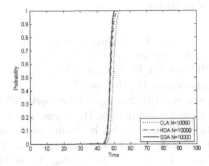

Fig. 3. Results of the experimental analysis of the running example with $\rho(\hat{\mathbf{X}}) = \hat{X}_{oldON} * \hat{X}_{upON} - 0.006$ and $\hat{x}_0 = (0, 0.9, 0, 0, 0.1)$. Comparison of reachability probabilities obtained by Central Limit Approximation (CLA), Gillespie's simulation algorithm (SSA) and the higher-order approximation of Section 4 (HOA) for $N = 1000$ (left) and $N = 10000$ (right).

6 Conclusion

In this paper, we presented a model checking procedure for global reachability properties of Markov population models based on stochastic approximations of the system behaviour. In particular, we exploited the CLA as in [9], but for a considerably larger class of global reachability properties. Indeed, in [9] we considered queries based on counting how many agents satisfy a local specification, resulting in a reachability problem in which the target region $\mathcal{R}$ is guaranteed to be absorbing. Here, instead, we consider arbitrary regions $\mathcal{R}$, defined by differentiable functions on collective variables, which cannot be made absorbing in a consistent way with the CLA. Hence, we relied on a different mathematical machinery, based on a Gaussian approximation of the time instant in which the trajectory of the population model enters $\mathcal{R}$. Moreover, we improved the accuracy of the estimation considering higher-order approximations of the (first two) moments of the reachability probability distribution. The method was experimentally validated on a peer-to-peer software update process.

The main limitation of our methodology is that it requires the fluid limit trajectory to enter the target region $\mathcal{R}$ associated with the reachability constraint. And even when this happens, the quality of our approximation is correlated with the unimodality of the hitting time distribution: if the true distribution is multimodal, then the accuracy of our method will be hampered [5]. This can happen if the fluid trajectory passes close to the boundary of $\mathcal{R}$ without crossing it. We are currently investigating possible ways of overcoming these limitations.

Other directions for future work are the release of an implementation, the investigation and characterisation of the effect of higher-order approximations on the estimate of the reachability probability, and the application of the framework on larger case studies.

References

1. Abramov, R.: A practical computational framework for the multidimensional moment-constrained maximum entropy principle. Journal of Computational Physics 211(1), 198–209 (2006)
2. Andersson, H., Britton, T.: Stochastic epidemic models and their statistical analysis, vol. 151. Springer, New York (2000)
3. Baier, C., Boudewijn, H., Hermanns, H., Katoen, J.P.: Model checking continuous-time Markov chains by transient analysis. In: Emerson, E.A., Sistla, A.P. (eds.) CAV 2000. LNCS, vol. 1855, pp. 358–372. Springer, Heidelberg (2000)
4. Baier, C., Katoen, J.P.: Principles of Model Checking. MIT Press (2008)
5. Bortolussi, L.: Hybrid behaviour of Markov population models. CoRR, abs/1211.1643 (2012)
6. Bortolussi, L., Hillston, J.: Fluid model checking. In: Koutny, M., Ulidowski, I. (eds.) CONCUR 2012. LNCS, vol. 7454, pp. 333–347. Springer, Heidelberg (2012)
7. Bortolussi, L., Hillston, J.: Checking Individual Agent Behaviours in Markov Population Models by Fluid Approximation. In: Bernardo, M., de Vink, E., Di Pierro, A., Wiklicky, H. (eds.) SFM 2013. LNCS, vol. 7938, pp. 113–149. Springer, Heidelberg (2013)
8. Bortolussi, L., Hillston, J., Latella, D., Massink, M.: Continuous approximation of collective system behaviour: A tutorial. Performance Evaluation 70(5), 317–349 (2013)
9. Bortolussi, L., Lanciani, R.: Model checking markov population models by central limit approximation. In: Joshi, K., Siegle, M., Stoelinga, M., D'Argenio, P.R. (eds.) QEST 2013. LNCS, vol. 8054, pp. 123–138. Springer, Heidelberg (2013)
10. Bujorianu, L.M.: Stochastic Reachability Analysis of Hybrid Systems. In: Communications and Control Engineering. Springer (2012)
11. Ethier, S.N., Kurtz, T.G.: Markov Processes: Characterization and Convergence. Wiley (2005)
12. Fricker, C., Gast, N.: Incentives and redistribution in homogeneous bike-sharing systems with stations of finite capacity. EURO Journal on Transportation and Logistics, 1–31 (2012)
13. Grima, R.: An effective rate equation approach to reaction kinetics in small volumes: Theory and application to biochemical reactions in nonequilibrium steady-state conditions. The Journal of Chemical Physics 133(3), 035101 (2010)
14. Hayden, R.A.: Mean field for performance models with deterministically-timed transitions. In: Proceedings of QEST 2012, pp. 63–73 (2012)
15. Hayden, R.A., Bradley, J.T., Clark, A.: Performance specification and evaluation with unified stochastic probes and fluid analysis. IEEE Trans. Software Eng. 39(1), 97–118 (2013)
16. Hayden, R.A., Stefanek, A., Bradley, J.T.: Fluid computation of passage-time distributions in large Markov models. Theor. Comput. Sci. 413(1), 106–141 (2012)
17. Katoen, J.P., Khattri, M., Zapreevt, I.S.: A Markov reward model checker. In: Proceedings of QEST 2005, pp. 243–244 (2005)
18. Kwiatkowska, M., Norman, G., Parker, D.: Prism 4.0: Verification of probabilistic real-time systems. In: Gopalakrishnan, G., Qadeer, S. (eds.) CAV 2011. LNCS, vol. 6806, pp. 585–591. Springer, Heidelberg (2011)
19. Norris, J.R.: Markov Chains. Cambridge University Press (1997)

20. Qiu, D., Srikant, R.: Modeling and performance analysis of BitTorrent-like peer-to-peer networks. In: Proceedings of ACM SIGCOMM 2004, pp. 367–378 (2004)
21. Stamatiou, Y.C., Spirakis, P.G., Komninos, T., Vavitsas, G.: Computer Network Epidemics: Models and Techniques for Invasion and Defense. CRC Press (2012)
22. Van Kampen, N.G.: Stochastic Processes in Physics and Chemistry. Elsevier (1992)

Explicit State Space and Markov Chain Generation Using Decision Diagrams

Junaid Babar and Andrew S. Miner*

Department of Computer Science, Iowa State University, Ames IA 50011, USA
{junaid,asminer}@iastate.edu

Abstract. Various forms of decision diagrams have been successfully used for quite some time to generate the state space and Markov chain from models expressed in some high–level formalism. A variety of efficient, "symbolic" algorithms, which manipulate sets of states instead of individual states, are known for this purpose. However, there are cases where explicit generation algorithms are still used. This paper seeks to efficiently use decision diagrams as replacement data structures within an existing explicit generation implementation. The necessary decision diagram algorithms are presented, and small changes to the explicit generation algorithm are suggested to improve the overall generation process. The efficiency of the new algorithms is illustrated using several models.

Keywords: Quantitative model checking, Markov chain generation, Markov chain storage, decision diagrams

1 Introduction

Generating the state space and underlying Markov chain from a model described in a high–level formalism, such as a Petri net, is a necessary first step for many types of analysis. Important applications that utilize this information include performance evaluation (e.g., [22,23]) and stochastic model checking (e.g., [3]). The well–known *state explosion problem*, in which a "small" high–level model may describe a huge state space and Markov chain, effectively limits the size and complexity of systems that may be analyzed in this manner.

A successful approach to help manage the state explosion problem has been the use of so–called "symbolic" or implicit algorithms, which utilize *decision diagrams* (DDs) (e.g., [4,16]), a structure that represents finite sets of integer vectors compactly for many (but not all) practical sets. As the symbolic algorithms work directly with DDs, by building a transition relation from the high level model and constructing the state space by a sequence of operations on DDs, they are limited by the sizes of the DDs generated, which can remain small even for huge state spaces. Once this was realized [5], many other researchers adopted the idea and expanded upon it; for example, by applying it to Petri nets [24], by developing generation algorithms [9], or by proposing variants of DDs

---

* This work was supported by the National Science Foundation, grant CNS-0546041.

A. Horváth and K. Wolter (Eds.): EPEW 2014, LNCS 8721, pp. 240–254, 2014.

[12,17,20,30]. In particular, several variants [15,19,26] have been proposed for representing Markov chains by including the rate information in the transitions relations.

While symbolic generation algorithms have enjoyed widespread success, they have not *completely* replaced explicit algorithms. For Markov chain analysis in particular, there are cases where explicit algorithms are often still used in practice. Notably, algorithms that exploit symmetry, such as the generation algorithms for Stochastic Well–Formed Nets [8] that immediately build a lumped process, remain explicit, despite preliminary work to make them symbolic [13]. For models containing immediate events, it is non-trivial to symbolically construct a Markov chain over the "tangible" states (e.g., [20]). Finally, even when DDs compactly represent large Markov chains, since the numerical solution of the Markov chain usually requires one or more *explicit* solution vectors [18], in practice, explicit generation algorithms are often considered "good enough".

In this paper, we propose the use of DDs as data structures to represent the state space and underlying Markov chain while using *explicit* generation algorithms. This approach is not intended to compete with symbolic generation algorithms (although, for certain models, symbolic generation can be slow and explicit generation is a viable alternative); rather, the intention is to allow cases that cannot (or for various reasons, do not) use symbolic generation, such as those discussed above, to utilize DDs. As DDs are usually much more compact than traditional explicit data structures, the goal is to reduce the memory requirements during generation, and to do so without greatly increasing the generation times. Once the state space and underlying Markov chain have been generated as DDs, the numerical solution algorithms in [18], using explicit solution vectors and the DD representation of the Markov chain, may be invoked.

In addition to numerical solution, there are other instances when explicit algorithms are used with DDs. Algorithms for classifying the states of a Markov chain, stored using DDs, into recurrent and transient classes, are partly explicit: the initial "seed" states are chosen, one at a time, at random; each seed state and potentially a large set of other states may then be classified using symbolic operations [29]. The "symblicit" approach for analyzing Markov Decision Processes presented in [27] is a combination of explicit and symbolic approaches. Similarly, the lumping algorithm in [14] contains explicit loops over states.

However, for state space and Markov chain or reachability graph generation, most algorithms are either completely explicit or completely symbolic. We are only aware of a few exceptions to this rule, and these previous works are most relevant to this paper. In [19], a type of DD called *matrix diagram* is used to store the Markov chain, and an explicit algorithm is used to generate the Markov chain. However, this work assumes that the state space is already known and is represented using a DD. Recently, in [1], an explicit state space generation algorithm was integrated into GreatSPN [6]; however, this uses a fairly simple algorithm. This paper builds upon these works, with extensions and generalizations over [19], and algorithm improvements over [1].

The remainder of this paper is organized as follows. Section 2 briefly recalls the traditional, explicit generation algorithms and motivates the data structure requirements. Section 3 formally defines DDs and some variants, and reviews some necessary, known algorithms. Section 4 describes how DDs can be efficiently integrated into explicit generation algorithms, and discusses some improvements to make this more efficient. Section 5 gives experimental results for our new algorithms, and section 6 concludes the work.

2   Background

Instead of a specific high–level formalism, we use an abstract model definition, and treat the model as a "black box". We assume that a model consists of:

- A set of L state variables, $x_L, \ldots, x_1$. Each state variable may assume a finite number of possible values; for simplicity, we assume these are the first naturals, i.e., $x_l \in \mathcal{S}_l = \{0, 1, \ldots, n_l - 1\}$. The possible states of the model are therefore $\hat{\mathcal{S}} = \mathcal{S}_L \times \cdots \times \mathcal{S}_1$; a finite set. While we assume the existence of bounds $n_L, \ldots, n_1$, we do not require that these are known a priori.
- An initial distribution, $\pi_0 : \hat{\mathcal{S}} \to [0, 1]$, with the restriction $\sum_{\mathbf{s} \in \hat{\mathcal{S}}} \pi_0(\mathbf{s}) = 1$.
- A next–state function, $\mathcal{N} : \hat{\mathcal{S}} \to 2^{\hat{\mathcal{S}} \times \mathbb{R}^+}$, where $\mathbb{R}^+$ denotes the set of positive reals. The meaning of $(\mathbf{s}', r) \in \mathcal{N}(\mathbf{s})$ is that the model can change from state $\mathbf{s}$ to state $\mathbf{s}'$ in one step, with rate r.

The state space of the model, $\mathcal{S}$, is the set of all states that can be reached from the initial states, when zero or more actions occur. Formally, the set of initial states may be defined as $\mathcal{I} = \{\mathbf{s} : \pi_0(\mathbf{s}) > 0\}$, and $\mathcal{S}$ is the smallest superset of $\mathcal{I}$ that satisfies $\mathbf{s} \in \mathcal{S} \Rightarrow \forall (\mathbf{s}', r) \in \mathcal{N}(\mathbf{s}), \mathbf{s}' \in \mathcal{S}$. The (reachable) Markov chain of the model, $\mathbf{R} : \mathcal{S} \times \mathcal{S} \to \mathbb{R}^{\geq 0}$, where $\mathbb{R}^{\geq 0}$ denotes the set of non-negative reals, is defined as $\mathbf{R}(\mathbf{s}, \mathbf{s}') = \sum_{(\mathbf{s}', r) \in \mathcal{N}(\mathbf{s})} r$.

A traditional algorithm to generate the state space and Markov chain from a model is shown in Figure 1(a). Sometimes, it may be useful to generate the Markov chain after the state space is known (for example, by collecting information while generating the state space, a more compact data structure may be used for the Markov chain); this may be done using the two–pass algorithm shown in Figure 1(b). Both algorithms require data structures for $\mathcal{S}$, the currently–known state space; $\mathcal{U}$, a set of states that still need to be explored; and $\mathbf{R}$, the currently–known Markov chain. Note that each algorithm terminates if and only if the state space is finite. After termination, $\mathcal{S}$ and $\mathbf{R}$ will respectively hold the complete state space and Markov chain, and $\mathcal{U}$ will be empty.

From a data structure perspective, the critical operations for $\mathcal{S}$ are adding states (cf. line 8) and determining if a given state is contained in the set (cf. line 7). The critical operations for $\mathcal{U}$ are checking if the set is empty (cf. line 4), adding states (cf. line 9), and removing *some* state (cf. line 5). Note that if $\mathcal{U}$ removes states in FIFO order (i.e., $\mathcal{U}$ is a queue), then the generation algorithm uses breadth–first search; however, this is not required. A critical operation applicable only to the two–pass algorithm is enumeration (cf. line 13). The only critical

1. $\mathcal{S} \leftarrow \{s : \pi_0[s] > 0\}$;
2. $\mathcal{U} \leftarrow \mathcal{S}$;
3. $\mathbf{R} \leftarrow 0$;
4. **while** $\mathcal{U} \neq \emptyset$ **do**
5.   Remove some s from $\mathcal{U}$;
6.   **for all** $(s', r) \in \mathcal{N}(s)$ **do**
7.   **if** $s' \notin \mathcal{S}$ **then**
8. $\mathcal{S} \leftarrow \mathcal{S} \cup \{s'\}$;
9. $\mathcal{U} \leftarrow \mathcal{U} \cup \{s'\}$;
10.   **end if**
11. $\mathbf{R}(s, s') \leftarrow \mathbf{R}(s, s') + r$;
12.   **end for**
13. **end while**

(a) One–pass algorithm

1. $\mathcal{S} \leftarrow \{s : \pi_0[s] > 0\}$;
2. $\mathcal{U} \leftarrow \mathcal{S}$;
3. $\mathbf{R} \leftarrow 0$;
4. **while** $\mathcal{U} \neq \emptyset$ **do**
5.   Remove some s from $\mathcal{U}$;
6.   **for all** $(s', r) \in \mathcal{N}(s)$ **do**
7.   **if** $s' \notin \mathcal{S}$ **then**
8. $\mathcal{S} \leftarrow \mathcal{S} \cup \{s'\}$;
9. $\mathcal{U} \leftarrow \mathcal{U} \cup \{s'\}$;
10.   **end if**
11.   **end for**
12. **end while**
13. **for all** $s \in \mathcal{S}$ **do**
14.   **for all** $(s', r) \in \mathcal{N}(s)$ **do**
15. $\mathbf{R}(s, s') \leftarrow \mathbf{R}(s, s') + r$;
16.   **end for**
17. **end for**

(b) Two–pass algorithm

Fig. 1. Algorithms to generate the state space and Markov chain from a model

operation for $\mathbf{R}$ is to add edges; note that this operation should allow for the destination state to appear more than once, i.e., we could have a model with $(s', r_1), (s', r_2) \in \mathcal{N}(s)$.

3 Decision Diagrams

There are many possible data-structures that can be used for storing $\mathcal{S}, \mathcal{U}$, and $\mathbf{R}$, and the design choice for one usually affects the others. A traditional strategy for the generation of $\mathcal{S}$ and $\mathbf{R}$, is:

- store an explicit (but compressed) representation of the states so that each state has a unique index that coincides with the discovery order of the states;
- use a classical dictionary data structure (such as a splay tree or hash table) to determine if states are already contained in $\mathcal{S}$;
- use an integer u for $\mathcal{U}$, to specify that all states with an index greater than or equal to u are yet to be explored;
- and use a dynamic data structure for directed, weighted, sparse graphs, such as adjacency lists, for $\mathbf{R}$.

See, for example, [7] for a survey of efficient techniques for explicit generation.

This paper proposes the use of Decision Diagrams (DDs) instead of the explicit structures mentioned above to store $\mathcal{S}, \mathcal{U}$ and $\mathbf{R}$. A DD is a directed acyclic graph used to represent a function on a finite number of variables, where each variable can assume a finite number of values. To simplify the discussion, we assume that the DD variables exactly match the state variables of the model. However, this is not required; it is possible (and often desired) to split a single model variable into several DD variables (e.g., [15]), or conversely to group several model variables into a single DD variable (e.g., [9]).

Each DD node is either *terminal*, with no outgoing edges, or *non-terminal*, labeled with a variable x_l and containing n_l outgoing edges to other nodes. For a given non-terminal node p, we use $p[i]$ to denote the ith outgoing edge. Ordered DDs require a *total ordering* $\succ$ on the variables such that any outgoing edge from a node labeled x_l must go to either a terminal node or to a node labeled x_k with $x_l \succ x_k$. In this paper, we assume that all DDs are Ordered.

DDs employ *reduction rules* to eliminate duplicate nodes, and to determine cases in which nodes are eliminated; done carefully, this ensures that each DD variant is a *canonical* representation. Two non-terminal nodes p and q are *duplicates* if they have the same variable label x_l, and the same outgoing edges: $p[i] = q[i], \forall\, 0 \le i < n_l$. All DDs in this paper eliminate duplicate nodes. Additional reduction rules, combined with different variable types and ranges, produce a variety of DDs:

- Multiway Decision Diagrams (MDDs) are used to represent functions of the form $f : \hat{S} \to \{0, 1\}$, and are typically *fully–reduced*, eliminating the pattern of a variable that does not matter. If each variable is boolean ($n_l = 2, \forall\, l$) then these become classical BDDs [4]. An MDD can represent a set of model states, by encoding the characteristic function of the set.
- Matrix Diagrams (MxDs) are used to represent functions of the form $f : \hat{S} \times \hat{S} \to \{0, 1\}$, using an "interleaved" ordering $x_L \succ x'_L \succ \cdots \succ x_1 \succ x'_1$. These are typically *identity–reduced*, eliminating the pattern of a variable that does not change (e.g., $x_l = x'_l$) [20].
- Multi–terminal Matrix Diagrams (MTMxDs) are used to represent functions of the form $f : \hat{S} \times \hat{S} \to \mathbb{R}$, and are essentially the same as MxDs except the terminal nodes are labeled with real values.
- Multiplicative Edge-Valued Matrix Diagrams (EV*MxDs) also represent functions of the form $f : \hat{S} \times \hat{S} \to \mathbb{R}$, but store real values along each edge in the DD, and values are multiplied along the path from the root to the terminal node to obtain the encoded value. A *normalization rule* (e.g., "the first non-zero edge value must be 1.0") is necessary to preserve canonicity. EV*MxDs are essentially EV*MDDs [26] that utilize the notions of interleaved variable ordering and identity reductions from MxDs.
- Canonical Matrix Diagrams (CMDs) [19] are an earlier version of EV*MxDs, where each node is a *matrix* of outgoing edges; effectively, a node for variable x_l in a CMD corresponds (roughly) to nodes for variables x_l and x'_l in an EV*MxD. Note that the CMDs of [19] did *not* utilize the identity reduction.

As a simple example, consider an open network of 3 bounded queues, where arrivals enter queue 3, queue 3 feeds into queue 2, queue 2 feeds into queue 1, and departures from queue 1 exit the system. If each queue can contain at most two customers, then the model has $S_3 = S_2 = S_1 = \{0, 1, 2\}$. Figure 2 shows the MTMxD, EV*MxD, and CMD representations for the underlying Markov chain

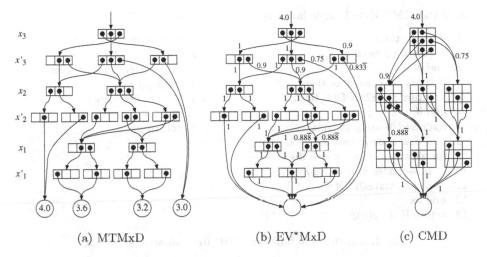

(a) MTMxD   (b) EV*MxD   (c) CMD

Fig. 2. Comparison of DDs encoding the same Markov chain

for this system, using an arrival rate of 3.0, and service rates of 4.0, 3.2, and 3.6 for queues 3, 2, and 1, respectively. In the illustrations, edges to terminal node 0 are omitted for clarity. For the CMD illustration, edges with the same edge value are clustered together, with only the common edge value drawn. EV*MxD and CMD use the normalization rule: "the largest edge value out of a node is 1.0".

There are a variety of operations that may be performed on DDs. Typically these operations traverse the DDs recursively, and utilize a compute table to avoid duplication of work since the recursion could visit the same node several times. For the purpose of this work, we note that for MDDs encoding sets of states, there are efficient operations to take the union or difference of two sets, determine the cardinality of a set, or enumerate the elements of a set. For an EV*MxD, MTMxD, or CMD representation of a Markov chain, it is possible to compute measures of interest (e.g., the steady–state distribution) using an *explicit* solution vector of size $|\mathcal{S}|$. For more details, please refer to [18].

4   Explicit Generation with Decision Diagrams

Our goal is to use either the one–pass or two–pass algorithm shown in Figure 1, with MDDs for $\mathcal{S}$ and $\mathcal{U}$, and EV*MxDs, MTMxDs, or CMDs for **R**. To do so, we must address the critical operations discussed in Section 2, namely: adding elements, removing elements, checking if $\mathbf{s} \in \mathcal{S}$, and checking if $\mathcal{U} \neq \emptyset$.

Because any given DD node can have several incoming edges, most DD algorithms do not modify DDs in place, but rather construct entirely new DDs. Thus, a statement such as "$\mathcal{S} \leftarrow \mathcal{S} \cup \{\mathbf{s}'\}$" would require constructing an MDD encoding the set $\{\mathbf{s}'\}$ (which can be done using procedure state2MDD shown in Figure 3), and then invoking a union operation on two MDDs. Since the union operation can be rather heavy, we instead use a fixed–size *state buffer* to collect node states2MDD(var L, state buffer b)

1. **if** $L = 0$ **then**
2.   **return** terminal node 1;
3. **end if**
4. **if** b has only one state **then**
5.   **return** state2MDD(L, b[0]);
6. **end if**
7. RadixSort b using variable x_L;
8. $p \leftarrow$ new node at level L;
9. **for all** values v for x_L in b **do**
10.   b$' \leftarrow$ states in b with $x_L = v$;
11. $p[v] \leftarrow$ states2MDD($L - 1$, b$'$);
12. **end for**
13. **return** Reduce(p);

node state2MDD(var L, state s)

1. $p \leftarrow$ terminal node 1;
2. **for all** $l \in \{1, \ldots, L\}$ **do**
3. $q \leftarrow$ new node at level l
4. $q[s[l]] \leftarrow p$;
5. $p \leftarrow$ Reduce(q);
6. **end for**
7. **return** p;

Fig. 3. Algorithm to build an MDD from an array of states

states, so that we can reduce the number of union operations. Thus, we need an algorithm to efficiently build an MDD encoding a set $\{s_1, s_2, \ldots, s_b\}$ where each state s_i is stored explicitly in an array. This can be done using procedure "states2MDD" shown in Figure 3. The procedure is recursive by variable number. The state array is sorted by the current variable (c.f. line 7), and the node at the current level is constructed by building the children recursively based on the sub-arrays grouped by variable value (c.f. lines 10 and 11). Note that these states are contiguous because of the sort, so in practice b$'$ can be built from b inexpensively. Recursion continues until either the bottom level is reached, or the array of states shrinks down to a single state. It can be shown that, because the states are sorted, MDD nodes are created bottom–up without creating any intermediate nodes. In other words, every node that is created, will appear in the "final" MDD, unless it is eliminated by the reduction step (c.f., line 13). The algorithm can be modified as appropriate to add matrix elements to an EV*MxD or MTMxD; we call that procedure "rates2MxD".

One way to implement the statement "Remove some s from $\mathcal{U}$", is to find a state $s \in \mathcal{U}$ by modifying the algorithm to enumerate elements of an MDD (stopping after the first element), build an MDD for the set $\{s\}$ using "state2MDD", and invoke the set difference operation. Taking this approach, and combining it with the "batch addition" idea discussed above, we can modify the one–pass algorithm to obtain the "Single Removal" algorithm shown in Figure 4(a). "$\mathcal{U} \neq \emptyset$" can be determined easily due to the fact that MDDs are a canonical representation, and "$s \in \mathcal{S}$" can be determined by following the appropriate path through the MDD for $\mathcal{S}$, until a terminal node is reached, 0 for $s \notin \mathcal{S}$, and 1 for $s \in \mathcal{S}$. Note that the state buffer, b_S, must be flushed whenever the set $\mathcal{U}$ becomes empty (c.f. lines 22—24), otherwise the algorithm may terminate prematurely. Similarly, the rate buffer b_R must be flushed before terminating (c.f. line 27).

Performing a set difference operation for each element of $\mathcal{U}$ (i.e. every discovered state) could be expensive. We might save time by removing states in

(a) Single Removal

1. $\mathcal{S} \leftarrow \{\mathbf{s} : \pi_0[s] > 0\}$;
2. $\mathcal{U} \leftarrow \mathcal{S}$; $\mathbf{b}_S \leftarrow \emptyset$; $\mathbf{b}_R \leftarrow \mathbf{0}$;
3. **while** $\mathcal{U} \neq \emptyset$ **do**
4. $\mathbf{s} \leftarrow$ FindFirstElement($\mathcal{U}$);
5. $\mathcal{U} \leftarrow \mathcal{U} \setminus$ state2MDD($L, \mathbf{s}$);
6.   **for all** $(\mathbf{s}', r) \in \mathcal{N}(\mathbf{s})$ **do**
7.   **if** $\mathbf{s}' \notin \mathcal{S}$ **then**
8.   Add $\mathbf{s}'$ to $\mathbf{b}_S$;
9.   **if** $\mathbf{b}_S$ is full **then**
10. $\mathcal{S}' \leftarrow$ states2MDD($L, \mathbf{b}_S$);
11. $\mathcal{S} \leftarrow \mathcal{S} \cup \mathcal{S}'$; $\mathcal{U} \leftarrow \mathcal{U} \cup \mathcal{S}'$;
12. $\mathbf{b}_S \leftarrow \emptyset$;
13.   **end if**
14.   **end if**
15.   Add $(\mathbf{s}, \mathbf{s}', r)$ to $\mathbf{b}_R$;
16.   **if** $\mathbf{b}_R$ is full **then**
17. $\mathbf{R} \leftarrow \mathbf{R} +$ rates2MxD($L, \mathbf{b}_R$);
18. $\mathbf{b}_R \leftarrow \mathbf{0}$;
19.   **end if**
20.   **end for**
21.   **if** $\mathcal{U} = \emptyset$ **then**
22. $\mathcal{S}' \leftarrow$ states2MDD($L, \mathbf{b}_S$);
23. $\mathcal{S} \leftarrow \mathcal{S} \cup \mathcal{S}'$; $\mathcal{U} \leftarrow \mathcal{U} \cup \mathcal{S}'$;
24. $\mathbf{b}_S \leftarrow \emptyset$;
25.   **end if**
26. **end while**
27. $\mathbf{R} \leftarrow \mathbf{R} +$ rates2MxD($L, \mathbf{b}_R$);

(b) Batch Removal

1. $\mathcal{S} \leftarrow \{\mathbf{s} : \pi_0[s] > 0\}$;
2. $\mathcal{U} \leftarrow \mathcal{S}$; $\mathbf{b}_S \leftarrow \emptyset$; $\mathbf{b}_R \leftarrow \mathbf{0}$;
3. **while** $\mathcal{U} \neq \emptyset$ **do**
4. $\mathcal{U}' \leftarrow \mathcal{U}$; $\mathcal{U} \leftarrow \emptyset$;
5.   **for all** $\mathbf{s} \in \mathcal{U}'$ **do**
6.   **for all** $(\mathbf{s}', r) \in \mathcal{N}(\mathbf{s})$ **do**
7.   **if** $\mathbf{s}' \notin \mathcal{S}$ **then**
8.   Add $\mathbf{s}'$ to $\mathbf{b}_S$;
9.   **if** $\mathbf{b}_S$ is full **then**
10. $\mathcal{S}' \leftarrow$ states2MDD($L, \mathbf{b}_S$);
11. $\mathcal{S} \leftarrow \mathcal{S} \cup \mathcal{S}'$; $\mathcal{U} \leftarrow \mathcal{U} \cup \mathcal{S}'$;
12. $\mathbf{b}_S \leftarrow \emptyset$;
13.   **end if**
14.   **end if**
15.   Add $(\mathbf{s}, \mathbf{s}', r)$ to $\mathbf{b}_R$;
16.   **if** $\mathbf{b}_R$ is full **then**
17. $\mathbf{R} \leftarrow \mathbf{R} +$ rates2MxD($L, \mathbf{b}_R$),
18. $\mathbf{b}_R \leftarrow \mathbf{0}$;
19.   **end if**
20.   **end for**
21.   **end for**
22. $\mathcal{S}' \leftarrow$ states2MDD($L, \mathbf{b}_S$);
23. $\mathcal{S} \leftarrow \mathcal{S} \cup \mathcal{S}'$; $\mathcal{U} \leftarrow \mathcal{U} \cup \mathcal{S}'$;
24. $\mathbf{b}_S \leftarrow \emptyset$;
25. **end while**
26. $\mathbf{R} \leftarrow \mathbf{R} +$ rates2MxD($L, \mathbf{b}_R$);

Fig. 4. One–pass algorithm, modified to use batch addition

batches. As described earlier, the single–removal algorithm removes a state $\mathbf{s}$ from $\mathcal{U}$ and uses it to discover new states. We need a procedure to obtain $\mathbf{s}$ from $\mathcal{U}$ without performing a set difference operation on $\mathcal{U}$, while permitting $\mathcal{U}$ to be updated within the loop (Figure 4(a), line 11). Our solution is to copy $\mathcal{U}$ into *another* MDD $\mathcal{U}'$ and then enumerate the elements in $\mathcal{U}'$. This gives us all the elements in $\mathcal{U}$ (via $\mathcal{U}'$), eliminates the set difference operation, and allows $\mathcal{U}$ to be modified as needed. Note that copying an MDD as described above is a trivial operation. The "Batch Removal" algorithm is shown in Figure 4(b).

We now make several observations about these algorithms. First, note that the choice of removing the "first element" from $\mathcal{U}$ in the single–removal algorithm is arbitrary; any element will work. The order of the traversal of $\mathcal{U}'$ in the batch–removal algorithm is also arbitrary, but efficient algorithms exist [18] for

Table 1. Markov chain sizes for the models

Model	$|\mathcal{S}|$	$|\mathbf{R}|$	Model	$|\mathcal{S}|$	$|\mathbf{R}|$
Courier 3	999,450	5,544,240	Kanban 5	2,546,432	24,460,016
Courier 4	5,133,600	31,392,540	Kanban 6	11,261,376	115,708,992
FMS 8	4,459,455	42,302,007	Kanban 7	41,644,800	450,455,040
FMS 9	11,058,190	108,084,295	QN 60	11,510,083	74,544,210
FMS 10	25,397,658	234,523,289	QN 80	58,214,137	384,196,600

lexicographical or reverse lexicographical order. Changing these could affect the order in which states are explored. Second, the order in which states are discovered and added to the batches could affect both the "intermediate" MDDs representing $\mathcal{S}$ (it does not affect the final MDD for $\mathcal{S}$ due to MDD canonicity), and the overall computational cost, since the union operations will be performed on different sets. Thus, while the procedure "states2MDD" is optimal for building an MDD from a set of states, the same *cannot* be said for the single–removal and batch–removal algorithms. More precisely: there is at least one order of state exploration that will be optimal in terms of size of the largest intermediate MDD, and it is unknown whether either of the algorithms in Figure 4 uses such an order. Third, note that it is possible to add the same unexplored state to the batch $\mathbf{b}_S$ more than once; this happens when a state is discovered again before the batch is added to $\mathcal{S}$. Since each reachable state is guaranteed to be explored exactly once, the overhead of duplication is somewhat contained, and we suspect that this overhead is much smaller than the overhead of a mechanism to detect duplicate states in a batch. Finally, Figure 4 gives the modifications for single removal and batch removal for the one–pass algorithm only; the analogous two–pass algorithms easily follow from this, and are omitted for space.

5   Experimental Results

The DD algorithms discussed in Section 4 are implemented in Meddly [2], and our explicit generation algorithms based on DDs are implemented in SMART [10]. Experiments are based on version 3.2.1411 of SMART and version 0.12.526 of Meddly, and were run on a typical desktop machine: a 2.7 GHz Intel Core i5 processor with 8 Gb of 1333 MHz DDR3 memory, running Mac OS 10.8.5.

5.1   Models

We tested our algorithms on a variety of models, mostly taken from the literature, described below. Table 1 shows, for each model, the size of the Markov chain described by the model, where $|\mathcal{S}|$ denotes the number of states in the Markov chain, and $|\mathbf{R}|$ denotes the number of non-zero entries in the rate matrix.

The Courier model [28] describes a communications protocol, where model parameter N corresponds to the window size and transport space parameters in [28] (i.e., we use $M = N$). This Petri Net contains 45 places, and we use a single place for each MDD variable. Some places may contain up to N tokens, while other places may contain only up to 1 token.

The FMS model [11] describes a flexible manufacturing system where model parameter N specifies the number of pallets to move parts. The model is partitioned into 19 MDD variables, where the number of possible values for most of the state variables grows with N, while the rest are fixed. The Petri Net for this system utilizes marking–dependent arc cardinalities, marking–dependent firing rates, and immediate transitions.

The Kanban model [25] describes a manufacturing system. The model parameter, N, determines the number of parts circulating in the system. This Petri net contains 16 places, and we assign each place to its own MDD variable.

QN is a classless, closed queueing network model. Initially, N customers are in an infinite–server queue, "pool". From this queue, customers arrive into a single–server "dispatcher" queue, which quickly sends customers to a "back end" queue with shortest queue length. Each of the 5 single–server "back end" queues process customers, and send them back into the "pool". All queues have exponentially-distributed service, with different rates, and process customers on a first come, first served policy. We use a decomposition of 7 MDD variables, one for each queue.

We also present some preliminary results obtained by incorporating the one-pass algorithm into GreatSPN [6]. The following colored nets were used: (a) CSRepititions [21]: a client/server application where communications from clients to servers is unreliable and requests are buffered, (b) DrinkVendingMachine [21]: a symmetric net modeling a hot drink vending machine, and (c) GlobalResAllocation [21]: a resource management model that prevents deadlocks.

5.2   Results

Table 2 shows the generation times and memory required to generate the state space $\mathcal{S}$ and Markov chain $\mathbf{R}$ for each model, using the one–pass algorithm, which constructs $\mathcal{S}$ and $\mathbf{R}$ simultaneously. Table 3 shows similar information for the two–pass algorithm, which constructs $\mathcal{S}$ first, and then constructs $\mathbf{R}$. Generation times are based on CPU usage and are reported in seconds. Memory requirements are based on "peak" sizes of the data structure, and are reported in appropriate units ("k" for 1024 bytes, and "m" for 1024^2 bytes).

In both tables, "BST" uses a balanced binary search tree (splay tree) for $\mathcal{S}$, and an explicit sparse matrix representation for $\mathbf{R}$; both are typical choices in practice. "EV*" uses an MDD for $\mathcal{S}$, and an EV*MxD for $\mathbf{R}$, as implemented in Meddly. "MT" uses an MDD for $\mathcal{S}$, and an MTMxD for $\mathbf{R}$, as implemented in Meddly. "CMD" uses an MDD for $\mathcal{S}$ as implemented in Meddly, and a Canonical Matrix Diagram for $\mathbf{R}$, using the implementation from [19] with some changes (discussed below). Finally, "br" stands for "batch removal", and "sr" stands for "single removal". We use a batch size of 1024 for all experiments.

In Table 2, the CMD implementation was modified to construct CMD nodes from a batch of entries, namely, an array of source states, destination states, and rates. This modification allows the use of CMDs without knowledge of $\mathcal{S}$,

Table 2. CPU and peak memory requirements using the **one–pass algorithm** for state space and Markov chain generation

	Generation times (seconds)							Peak memory required (k=1024 bytes, m=1024k)													
	BST	EV*		MT		CMD		BST		br EV*		sr EV*		br MT		sr MT		br CMD		sr CMD	
Model		br	sr	br	sr	br	sr	S	R	S	R	S	R	S	R	S	R	S	R	S	R
Courier 3	10	34	54	47	56	39	56	36m	114m	408k	2.2m	626k	2.2m	422k	4.0m	2.1m	943k	588k	3.4m	463k	1.8m
Courier 4	57	205	284	277	300	244	313	194m	638m	662k	13m	738k	2.2m	664k	5.8m	2.0m	1017k	1.1m	6.2m	473k	2.1m
FMS 8	72	116	157	211	183	141	1653	157m	769m	326k	2.8m	542k	619k	378k	5.5m	1.9m	1.4m	441k	14m	116k	7.5m
FMS 9	193	303	397	603	497	379	425	390m	1977m	452k	4.4m	547k	851k	569k	11m	3.8m	2.5m	538k	28m	120k	16m
FMS 10	—	731	939	1540	1224	954	999	—	—	574k	5.9m	554k	1.1m	693k	14m	7.6m	4.3m	817k	54m	121k	31m
Kanban 5	22	35	60	50	66	45	64	80m	486m	120k	1.3m	332k	211k	131k	1.2m	1005k	376k	170k	1.4m	151k	845k
Kanban 6	110	165	275	244	315	217	290	358m	2294m	179k	5.1m	346k	253k	196k	1.5m	1019k	442k	302k	2.0m	163k	1014k
Kanban 7	—	650	1052	1006	1229	866	1104	—	—	260k	17m	362k	291k	309k	2.6m	1.0m	499k	374k	2.9m	178k	1.2m
QN 60	56	102	207	673	473	170	251	356m	1506m	209k	1.9m	595k	517k	301k	3.8m	932k	898k	268k	8.4m	151k	3.1m
QN 80	—	531	1175	4425	3327	1031	1463	—	—	401k	5.9m	643k	902k	628k	8.5m	1.9m	1.7m	533k	20m	195k	4.6m

Table 3. CPU and peak memory requirements using the **two–pass algorithm** for state space and Markov chain generation

	Generation of S						Generation of R											
	Time (seconds)			Peak memory			Time (seconds)			CMD			Peak memory			CMD		
		MDD			MDD									EV*	MT			
Model	BST	br	sr	BST	br	sr	BST	EV*	MT	bat.	f = 3	f = 8	BST			bat.	f = 3	f = 8
Courier 3	9	29	109	36m	947k	508k	10	13	15	17	38	18	114m	327k	352k	767k	186k	197k
Courier 4	54	165	595	194m	1.2m	518k	64	76	85	95	218	104	638m	378k	674k	1.6m	290k	303k
FMS 8	66	133	404	157m	428k	126k	88	62	82	68	64	71	769m	712k	1.6m	7.5m	1.5m	4.4m
FMS 9	173	344	1069	390m	611k	131k	239	160	225	176	168	194	1977m	1.1m	3.8m	16m	2.1m	6.6m
FMS 10	—	840	2585	—	745k	133k	—	378	557	424	404	485	—	1.3m	5.6m	30m	2.8m	9.8m
Kanban 5	20	10	44	80m	170k	151k	27	20	26	21	24	23	486m	100k	172k	311k	145k	2.9m
Kanban 6	101	48	202	358m	302k	163k	166	92	126	106	116	127	2294m	115k	191k	351k	198k	5.9m
Kanban 7	—	187	763	—	374k	178k	—	359	521	391	419	648	—	133k	215k	397k	262k	12m
QN 60	47	37	166	356m	268k	151k	73	60	266	102	102	183	1506m	519k	1.2m	2.5m	1.9m	51m
QN 80	—	188	957	—	533k	195k	—	310	2085	587	581	1001	—	948k	2.3m	4.2m	3.8m	66m

Table 4. Generation using GreatSPN (time in seconds, peak memory in MB)

| Model | One-Pass EV* | | | | | | WNRG |
| | No Batches | | sr | | br | | |
	Time	Memory	Time	Memory	Time	Memory	Time
CSRepetitions	13	10.92	4	2.09	1	1.40	3
DrinkVendingMachine	28	77.22	5	5.97	2	5.45	4
GlobalResAllocation	144	109.07	38	14.49	9	4.86	8

as required in [19]. However, as the CMD implementation is rather specialized, it requires a known bound for each MDD variable a priori. In contrast, the EV*MxD and MTMxD implementations in Meddly are designed to be more robust and general–purpose; these instead allow the MDD variable bounds to expand as needed. Looking at the generation times, we see that "BST" is always the fastest, followed by "EV*" and then "CMD" using batch removal, which are quite close except for large models. Batch removal tends to be faster than single removal, except for some models when MTMxDs are used. Recall from the discussion in Section 4, that the order in which states are explored can affect the computational cost of the batch additions in the DDs; it must be the case for these models that single removal happens to result in a significantly better order for MTMxDs, enough to offset the performance improvement of removing states in batch. Looking instead at memory usage, we see that "BST" always uses orders of magnitude more memory than "EV*", "MT", or "CMD". In particular, for several models, "BST" required more memory than was available on the machine, and could not complete; these cases are marked with the entries "—". Also, we note that, with some exceptions, "EV*" tends to have the smallest memory requirement, followed by "MT" and then "CMD". However, assuming the ultimate goal is to analyze the Markov chain numerically, it is perhaps unnecessary to compare *peak* memory usage between "EV*", "MT", and "CMD", since any of these should require a fraction of the memory needed during numerical solution. For example, the largest memory requirement is 54m for model FMS 10 using "br CMD"; in this case, one double–precision solution vector will require about 203m. Using MDDs for S and EV*MxDs, MTMxDs, or CMDs for R, the bottleneck becomes the memory required for solution vectors.

In Table 3, we split generation of S and R into separate steps. When building only S, we compare BSTs with MDDs. Perhaps surprisingly, there are a few cases where MDDs are faster than splay trees. For all cases, using batch removal is faster than using single removal. Again, MDDs require orders of magnitude less memory than BSTs. The memory requirements for batch removal are higher than for single removal; one cause for this is the fact that batch removal keeps two MDDs for unexplored states, while single removal keeps only one. Once S is built, we construct R by examining each state in S and determining its outgoing edges based on the model. For BSTs, we examine states of S in the order in which they were originally discovered; for MDDs instead, we efficiently enumerate the MDD which will visit states in lexicographical order. We believe that this leads to a good, but probably not optimal, order for adding rates to **R**. For "EV*" and "MT", we add entries in batch as usual. For CMDs, we can either use our new batch–based implementation, or the original implementation of [19]. The table shows results for both: "bat." is the batch–based implementation, and "$f =$" is the original implementation, which merges the CMDs every time the MDD enumeration crosses level f. For all cases, "EV*" is faster than "CMD", and batch–based CMD is faster than MT. Note that setting the f parameter can be tricky and model–dependent, while a fixed batch size of 1024 gives fairly consistent results. Also note that, except for the Courier model, "EV*" is faster than BST, and for models FMS and Kanban, BST is actually the *slowest* to generate **R**. For many of the models and for EV*s, MTMxDs, and CMDs, the total time for the two–pass algorithm is *lower* than the total time for the one–pass algorithm. As for memory, the trends noted in the "one–pass" table are also seen in the "two–pass" table: BST requires orders of magnitude more memory than the others, with "EV*" requiring the least, followed by "MT". "—" are cases where BST needed more memory than was available, and did not complete.

In Table 4, we report our preliminary results obtained by incorporating the one–pass algorithms into GreatSPN. There is a clear improvement in CPU and peak memory usage as we progress from "No batches" to "sr" and then to "br". Note that the CPU usage of "br" for EV* is close to GreatSPN's BST implementation (WNRG). The peak memory usage of WNRG was not available but we expect it to be similar to the BST implementation used for Table 2.

6   Conclusion

In this paper, we have examined the use of DDs for efficiently storing the state space and Markov chain of a high–level model. We have presented algorithms that improve the performance of DDs when used by traditional explicit generation algorithms. We have also presented small changes to the traditional explicit generation algorithms to take advantages of efficient DD operations.

Our experimental results show that DDs can greatly reduce the peak memory requirements of traditional explicit generation algorithms. There is often the typical time–space tradeoff between a classical BST–based implementation and the DD–based implementations. For the models we tested, BSTs can be nearly a factor of 4 faster than the fastest DD implementation, but at the expense of using on the order of 1000 times as much memory. Models for which traditional explicit implementations exceed the available memory are prime candidates for the DD–based explicit generation. Additionally, for applications where a two–pass generation algorithm is used anyway, we find that DDs are quite competitive with BSTs, sometimes even outperforming them in generation times.

An unexpected result is that EV*MxDs tend to be faster than MTMxDs. It is possible that the cost of the expensive EV*MxD reduction rules is insignificant compared to the cost of accessing more memory (EV*MxDs tend to be more compact than MTMxDs); further analysis of this phenomenon is needed.

Our preliminary tests with GreatSPN using the one–pass algorithm, indicate a clear improvement over previous work in [1]. Once fully integrated we plan on further experiments with generating lumped Markov chains, where the computational cost of determining the next (abstract) state from the high–level model is significantly higher.

Another direction for further work is to study different state exploration orders and develop heuristics for selecting an order that is good in terms of the union operations, and efficient in enumerating an MDD of unexplored states.

Acknowledgment. The authors thank Elvio Amparore and Marco Beccuti for implementing the one–pass algorithms in GreatSPN and running experiments.

References

1. Babar, J., Beccuti, M., Donatelli, S., Miner, A.: GreatSPN enhanced with decision diagram data structures. In: Lilius, J., Penczek, W. (eds.) PETRI NETS 2010. LNCS, vol. 6128, pp. 308–317. Springer, Heidelberg (2010)
2. Babar, J., Miner, A.: Meddly: Multi–terminal and Edge–valued Decision Diagram Library. In: 7th Int. Conf. on Quantitative Evaluation of Systems (QEST 2010), pp. 195–196 (September 2010)
3. Baier, C., Haverkort, B., Hermanns, H., Katoen, J.P.: Model checking algorithms for continuous-time Markov chains. IEEE Trans. Softw. Eng. 29(6), 524–541 (2003)
4. Bryant, R.E.: Graph–based algorithms for boolean function manipulation. IEEE Trans. Comp. C-35(8), 677–691 (1986)
5. Burch, J.R., Clarke, E.M., McMillan, K.L., Dill, D.L., Hwang, L.J.: Symbolic model checking: 10^{20} states and beyond. Information and Computation 98(2), 142–170 (1992)
6. Chiola, G., Franceschinis, G., Gaeta, R., Ribaudo, M.: GreatSPN 1.7: Graphical Editor and Analyzer for Timed and Stochastic Petri Nets. Perf. Eval. 24(1-2), 47–68 (1995)
7. Chiola, G.: Compiling techniques for the analysis of stochastic Petri nets. In: Proc. 4th Int. Conf. on Modelling Techniques and Tools for Performance Evaluation, pp. 13–27 (1989)
8. Chiola, G., Dutheillet, C., Franceschinis, G., Haddad, S.: Stochastic well-formed colored nets and symmetric modeling applications. IEEE Trans. Comp. 42(11), 1343–1360 (1993)
9. Ciardo, G., Lüttgen, G., Miner, A.S.: Exploiting interleaving semantics in symbolic state–space generation. Formal Methods in System Design 31(1), 63–100 (2007)
10. PCiardo, G., Miner, A.S., Wan, M.: Advanced features in SMART: the Stochastic Model checking Analyzer for Reliability and Timing. SIGMETRICS Perform. Eval. Rev. 36(4), 58–63 (2009)
11. Ciardo, G., Trivedi, K.S.: A decomposition approach for stochastic reward net models. Perf. Eval. 18(1), 37–59 (1993)
12. Couvreur, J.-M., Thierry-Mieg, Y.: Hierarchical decision diagrams to exploit model structure. In: Wang, F. (ed.) FORTE 2005. LNCS, vol. 3731, pp. 443–457. Springer, Heidelberg (2005)
13. Delamare, C., Gardan, Y., Moreaux, P.: Performance evaluation with asynchronously decomposable SWN: Implementation and case study. In: 10th Int. Workshop on Petri Nets and Performance Models (PNPM 2003), pp. 20–29. IEEE Comp. Soc. Press (September 2003)

14. Derisavi, S.: A symbolic algorithm for optimal Markov chain lumping. In: Grumberg, O., Huth, M. (eds.) TACAS 2007. LNCS, vol. 4424, pp. 139–154. Springer, Heidelberg (2007)

15. Hermanns, H., Kwiatkowska, M., Norman, G., Parker, D., Siegle, M.: On the use of MTBDDs for performability analysis and verification of stochastic systems. Journal of Logic and Algebraic Programming: Special Issue on Probabilistic Techniques for the Design and Analysis of Systems 56(1-2), 23–67 (2003)

16. Kam, T., Villa, T., Brayton, R., Sangiovanni-Vincentelli, A.: Multi–valued decision diagrams: theory and applications. Multiple-Valued Logic 4(1-2), 9–62 (1998)

17. Lai, Y.T., Pedram, M., Vrudhula, S.B.K.: Formal verification using edge–valued binary decision diagrams. IEEE Trans. Comp. 45, 247–255 (1996)

18. Miner, A., Parker, D.: Symbolic representations and analysis of large state spaces. In: Baier, C., Haverkort, B.R., Hermanns, H., Katoen, J.-P., Siegle, M. (eds.) AUTONOMY 2003. LNCS (LNAI), vol. 2925, pp. 296–338. Springer, Heidelberg (2004)

19. Miner, A.S.: Efficient solution of GSPNs using Canonical Matrix Diagrams. In: 9th Int. Workshop on Petri Nets and Performance Models (PNPM 2001), pp. 101–110. IEEE Comp. Soc. Press (September 2001)

20. Miner, A.S.: Implicit GSPN reachability set generation using decision diagrams. Perf. Eval. 56(1-4), 145–165 (2004)

21. Model Checking Contest at Petri Nets (2014), http://mcc.lip6.fr/models.php

22. Molloy, M.K.: Performance analysis using stochastic Petri nets. IEEE Trans. Comp. 31(9), 913–917 (1982)

23. Muppala, J.K., Ciardo, G., Trivedi, K.S.: Modeling using Stochastic Reward Nets. In: Proc. 1st Int. Workshop on Modeling, Analysis and Simulation of Computer and Telecommunication Systems (MASCOTS 1993), pp. 367–372. IEEE Comp. Soc. Press, San Diego (1993)

24. Pastor, E., Roig, O., Cortadella, J., Badia, R.M.: Petri net analysis using boolean manipulation. In: Valette, R. (ed.) ICATPN 1994. LNCS, vol. 815, pp. 416–435. Springer, Heidelberg (1994)

25. Tilgner, M., Takahashi, Y., Ciardo, G.: SNS 1.0: Synchronized Network Solver. In: 1st International Workshop on Manufacturing and Petri Nets, Osaka, Japan, pp. 215–234 (June 1996)

26. Wan, M., Ciardo, G., Miner, A.S.: Approximate steady–state analysis of large Markov models based on the structure of their decision diagram encoding. Perf. Eval. 68(5), 463–486 (2011)

27. Wimmer, R., Braitling, B., Becker, B., Hahn, E.M., Crouzen, P., Hermanns, H., Dhama, A., Theel, O.: Symblicit calculation of long–run averages for concurrent probabilistic systems. In: 7th Int. Conf. on Quantitative Evaluation of Systems (QEST 2010), pp. 27–36 (2010)

28. Woodside, C.M., Li, Y.: Performance Petri net analysis of communications protocol software by delay-equivalent aggregation. In: 4th Int. Workshop on Petri Nets and Performance Models (PNPM 1991), pp. 64–73. IEEE Comp. Soc. Press, Melbourne (1991)

29. Xie, A., Beerel, P.A.: Efficient state classification of finite state Markov chains. IEEE Trans. Computer-Aided Design 17(12), 1334–1339 (1998)

30. Yoneda, T., Hatori, H., Takahara, A., Minato, S.: BDDs vs. zero–suppressed BDDs: for CTL symbolic model checking of Petri nets. In: Srivas, M., Camilleri, A. (eds.) FMCAD 1996. LNCS, vol. 1166, pp. 435–449. Springer, Heidelberg (1996)

Non-Markovian Modeling of a BladeCenter Chassis Midplane

Salvatore Distefano[1], Francesco Longo[2], Marco Scarpa[2], and Kishor S. Trivedi[3]

[1] Dipartimento di Elettronica, Informazione e Bioingegneria, Politecnico di Milano,
20133 Milano, Italy
salvatore.distefano@polimi.it
[2] Dipartimento di Ingegneria DICIEAMA, Università di Messina,
98166 Messina, Italy
{flongo,mscarpa}@unime.it
[3] Department of Electrical and Computer Engineering, Duke University,
27708-0291 Durham, USA
ktrivedi@duke.edu

Abstract. In distributed contexts such as Cloud computing, the reliability and availability of the provided resources and services have to be assured in order to meet user requirements. At the infrastructure level, this specification is translated into tighter ones on the datacenter hosting physical resources. In this paper, starting from a real case study of the IBM BladeCenter, we provide a technique for the quantitative evaluation of datacenter infrastructure availability. The proposed technique allows one to take into account both aging phenomena and multiple operating conditions. In particular, one subsystem of the BladeCenter, the chassis midplane, is studied. Indeed, based on the stochastic characterization of the midplane reliability through statistic measurements, a model dealing with the non-exponential failure time distribution thus obtained is evaluated to demonstrate the suitability and the effectiveness of the proposed technique.

1 Introduction

New technologies and applications strongly impact on everyday life, aiming at improving quality standards. IT infrastructure lies at the heart of such technologies, acting as the engine of the digital "revolution". Blade server systems are becoming a de-facto standard architecture in distributed computing infrastructure. Indeed, they are used in many academic and business contexts, such as e-commerce, banking, financial, stock trading, and telephone communications as well as in research applications, in addition to several types of life-critical and safety-critical systems and services. Furthermore, blade server systems are the pillars on which Cloud, Web, and social network technologies are based.

A primary requirement a blade server has to fulfill is related to its availability (and associated downtime). A common practice is to consider server modules (blades) as stand-alone servers with shared services rather than the blade server

A. Horváth and K. Wolter (Eds.): EPEW 2014, LNCS 8721, pp. 255–269, 2014.

system as a whole. Several techniques to achieve high availability are known [1,2]. Analytical models can be used for quantifying computer system characteristics such as reliability and availability, specifically in the Cloud context [3]. Reliability block diagrams or fault trees are often used to formulate and solve blade server availability models because of their simplicity and efficiency [4]. But non-state-space methods cannot easily incorporate realistic system behavior such as (imperfect) fault coverage, multiple failure modes, hot-swap components, and so on [5]. By contrast, such dependencies and multiple failure modes can be easily captured by state-space models such as Markov chains, semi-Markov processes, and Markov regenerative processes [6,7]. However, the computational requirements for building, storing, and solving state space models for real systems can lead to a state space explosion, and could be mitigated by using symbolic techniques and Kronecker algebra [8,9,10]. But a more practical alternative is to use a hierarchical approach where a judicious combination of state space and non-state-space methods is utilized. In particular, an analytical approach has been applied in [11] to the evaluation of an IBM BladeCenter system availability. A two-level hierarchical technique was adopted, modeling each subsystem as a Markov chain and the entire system as a fault tree. This way, both the limitation of Markov chain (state space explosion) and that of fault trees (dependencies' representation) are overcome.

Even if the technique proposed in [11] is effective in dealing with complexity, it is based on the strong assumption that component times to failure are all exponentially distributed. However, actual behaviour, representing complex, even dependent, phenomena and composite workflow have by nature age-dependent hazard rates. In line with that, here, starting from an in depth analysis and regression of the statistical data related to the BladeCenter components' reliability, which identify Weibull distributions associated with their times to failure, we develop a non-Markovian model dealing with the issues related to age dependent behaviors. This way, though an in depth evaluation of the BladeCenter components and subsystems, we can numerically evaluate the quality of the exponential approximation in dealing with the BladeCenter system availability evaluation. More specifically, we investigate a specific component of the BladeCenter system, the chassis midplane, which provides interconnection paths to the blade servers, managed by a specific logic deployed on-board. The model allows us to take into account repair facilities and common mode faults. Furthermore, in the non-Markovian version of the model we also consider load sharing effects.

In order to evaluate the non-Markovian model some elaborations on the original state space model to deal with common mode faults and load sharing phenomena are required. Starting from a technique we specified in [12,13] for evaluating the longevity of a wireless sensor node subject to sleep-wakeup cycles, in this paper we extend and adapt it to the BladeCenter availability modelling, adding specific features able to deal with common mode faults and load-sharing effects.

2   Description of the System

The blade server technology have been widely adopted in recent years both in academia and in the industry for server deployment, thanks to its modular design, which is based on industry-standard racks and provides denser packaging thanks to the possibility to share services such as power and cooling among the servers housed in the same chassis. Integrated network switches provide additional space saving and significant reduction in cabling.

The IBM BladeCenter E chassis[1] is one of the most widely spread server infrastructure implementing the blade architecture. It supports up to 14 computing elements, known as *blade servers*. From the front side, access is provided to the control panel, removable media devices, and the blade servers. Network switch modules, power supplies, cooling devices, and management modules are located at the rear of the chassis. All these devices plug into a central midplane that provides power distribution, sideband management buses, and network interconnections. The midplane follows a redundant, fail-in-place design. Power domain 1 consists of power supplies 1 and 2. These power supplies are designed to provide redundancy, i.e., all devices attached to power domain 1 remain operational if one of the power supplies fails. Similarly, power domain 2 consists of the redundant power supplies 3 and 4 and supplies blades 7–14. Everything else in the BladeCenter chassis is supplied by power domain 1.

In [11], a typical configuration consisting of a chassis with redundant blowers, two power domains each containing two redundant power supply modules, redundant management modules, redundant Ethernet switches, and redundant Fiber Channel network switches along with 14 blade servers has been modeled from the availability point of view. However, the main assumption in such a work is that failure and repair times are exponentially distributed. Thus, the sub-models described in [11] are continuous time Markov chains. The advantage of such an approach is the possibility of analyzing the model through simple, well-known methods. However, the results provided by the models may not be so accurate especially if the distributions associated with the time of event occurrence are proved not to be exponential. In fact, experimental results demonstrate that failure time distribution stochastically characterising the components of a BladeCenter system are Weibull distributed.

Aim of this work is to take into account one of the BladeCenter subsystem, the chassis midplane, as represented in [11], modifying the model there specified to consider Weibull distributed failure time and to demonstrate that the simplifying assumption of exponentially distributed events leads to inaccurate results. Moreover, we show how the use of a non-Markovian approach allows us to deal with more complex phenomena such us load sharing and aging.

2.1   BladeCenter Chassis Midplane

A BladeCenter E chassis contains a midplane that provides the interconnection paths among the blade servers, the network switches, the management modules,

---

[1] http://www-03.ibm.com/systems/bladecenter/hardware/chassis/bladee/

and the other components. The midplane is designed to minimize the probability to cause a blade server outage since: (i) the midplane contains only a few active components; (ii) it provides two independent sets of interconnects to each blade; (iii) it has two independent connectors to each set of interconnects. Thus, the chassis midplane follows a fail-in-place design, allowing the chassis to tolerate failure of half of the midplane and the blade servers to continue functioning without an operational outage by utilizing the redundant communication paths. In the latter case, the still operating part has to deal with the whole load, which is initially shared with the other part of the system while operating. This could result in a kind of load sharing phenomenon, which could impact on the system reliability. In any case, as soon as a failure occurs, a midplane replacement is requested and a repairman is summoned to perform it. The time that is needed for the repairman to start the maintenance is considered as an uptime but, of course, while maintenance is in progress the whole BladeCenter experiences a downtime because the chassis must be taken out of service to replace the midplane. In practice, the midplane replacement is usually scheduled and the load is moved to other blade servers in a different chassis while the maintenance is performed to minimize such an outage time.

During the time required to start the maintenance, the redundant communication paths may still fail bringing the whole chassis to a downtime. Moreover, the failover of the communication path after the first failure may also be unsuccessful. In fact, any not-covered case, such as any common mode failure, may cause the entire chassis to fail and the blade servers to experience a downtime.

3 The State Space Model

In this section, we first report the midplane homogeneous continuous time Markov chain (CTMC) model as described in [11]. Then, we introduce our non-Markovian extensions pointing out the main differences against the homogeneous CTMC model. Both of them are characterized according to the state space model defined in [11] and generalized here as reported in Figure 1. The states there reported are tagged with a combination of two characters. The first character indicates the condition of the chassis midplane as *active* (A) if properly operating, *covered fault* (C) if still operating although a fault has been experienced, and *failed* (F) after an not-covered fault. The second character, is instead associated with the status of the repairman as not yet summoned (*unsummoned* - U), *summoned* (S), or while *repairing* (R).

This way, state AU represents the fault-free state of the midplane. State CS represents the midplane that is still operational after a fault has been detected. At this point, failover of a communication path, if necessary, was successful and midplane repair/replacement was requested and the repairmen summoned. This may trigger load sharing effects on the part of the system that is operating after this kind of fault, e.g., the working connector, since it has to deal with the whole load, before shared with the other connector. Indeed, the midplane has a transition from state AU to state CS for most of the faults (event e_{cov}), which

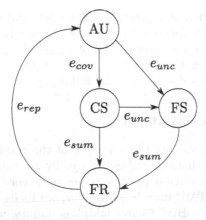

Fig. 1. State space model for the BladeCenter midplane

are mainly covered. Any not-covered case, such as any common mode failure, are represented by the transition to state FS (event e_{unc}). In both state CS and state FS, the system requires to be repaired but CS is a up state while FS is a down state considering the midplane functionalities. The midplane model shows a transition to state FR on the arrival of the repairman (event e_{sum}). State FR is a down state because the chassis must be taken out of service to replace the midplane. As soon as the replacement is performed the model transits back to state AU (event e_{rep}). The availability of the midplane is the probability that the system is in operational states AU or CS.

3.1 The Original Homogeneous CTMC Model

In [11], the following assumptions are taken into account with respect to the state space model reported in Figure 1:

i all the events in the model are associated with exponential distributions;
ii the midplane fails with a mean time to failure (MTTF) of $MTTF_{mp}$;
iii the transition rate to state FS is determined by a common mode factor fm so that the actual rate associated with event e_{unc} is given by $fm \cdot 1/MTTF_{mp}$;
iv alternatively, all covered cases are considered in event e_{cov} whose actual rate is $(1 - fm) \cdot 1/MTTF_{mp}$;
v no load sharing behaviors are taken into account in state AU, thus the same failure rate $(fm \cdot 1/MTTF_{mp})$ characterizes states AU and CS sojourn times;
vi thus in state CS, the midplane can experience additional faults that, if not covered, can bring the midplane in state FS with a rate $fm \cdot 1/MTTF_{mp}$;
vii the repairman arrives with a mean response time (MRT) of MRT_{sp};
viii the midplane is replaced with a mean time to repair (MTTR) of $MTTR_{mp}$.

Thus, in the original model proposed in [11] only exponentially distributed events are considered and the overall model is a CTMC. This brings to advantages in terms of simplicity of the model solution. However, experimental results through collection and regression of statistical data show that the failure events of the IBM BladeCenter midplane are characterised by Weibull distributions. This way, the state space model of Figure 1 has to be considered as non-Markovian, implying to adequately deal with the corresponding memory phenomena and effects as discussed in the following sections.

3.2   The Non-markovian Model

Aim of our work is to evaluate the accuracy of the results shown in [11], obtained by the Markov chain described above, by providing a more accurate, non-Markovian, model overcoming some issues and relaxing some assumptions of the homogeneous CTMC one. In particular, we focus on event e_f, i.e., the event associated with the BladeCenter midplane failure, which is not explicitly shown in the model of Figure 1 even if the statistical data collected are referred to it. Indeed, experiments conducted by IBM prove that the BladeCenter midplane failure time is associated with a Weibull distribution in the form:

$$F_f(t) = 1 - e^{-(t/\eta_f)^{\beta_f}}$$

In particular, IBM provides the parameters related to the BladeCenter midplane failure time both for the exponential and the Weibull distributions in the form of ranges (upper and lower limits) as reported in Table 1.

Table 1. BladeCenter midplane failure time parameters

Exponential distribution		Weibull distribution			
Low $MTTF_{mp}$	High $MTTF_{mp}$	Low β_f	High β_f	Low η_f	High η_f
310,000	420,000	0.424	0.574	4,237,335	5,732,865

Referring to the state space model of Figure 1, we split the midplane failure event into two derived events, the covered e_{cov} and not-covered e_{unc} fault events. This way, we have to characterize such events with the corresponding distribution functions starting from the common mode failure behavior. In the homogeneous CTMC model it is easy to represent common mode failure by just multiplying and splitting the rates according to the common mode factor fm. More complex is instead the case of non-Markovian model. In particular, being events e_{cov} and e_{unc} strictly related to event e_f, we associate with the corresponding state transitions three general distributions $F_{cov}(t)$, $F_{unc}(t)$, and $F'_{unc}(t)$, respectively. The latter distributions $F_{unc}(t)$, and $F'_{unc}(t)$, both characterize event e_{unc}, since the event can be associated with different distributions representing different operating conditions due to load sharing phenomena. More specifically, $F_{unc}(t)$ stochastically characterizes the unreliability of the BladeCenter midplane due to not-covered fault when all the midplane components, also the redundant ones such as the connection paths, are operating thus sharing the load. Then, $F'_{unc}(t)$

stochastically represents the unreliability of the BladeCenter midplane due to not-covered fault after some covered fault occurred, also including those related to the redundant parts. In the latter case, after a fault on a redundant component, the still operating component that performs the same function of the faulty one has to manage the whole load, before shared with the former. In terms of reliability, this may imply that, after a covered fault, the midplane could have a different probability to fault, represented by $F'_{unc}(t)$. Furthermore, keeping the assumption that events e_{sum} and e_{rep} are associated with exponential distributions, the other state transitions in the non-Markovian model have rates corresponding to the homogeneous CTMC ones.

In the following section, we provide insights about how to compute distributions $F_{cov}(t)$, $F_{unc}(t)$, and $F'_{unc}(t)$ starting from the knowledge of distribution $F_f(t)$ and of the common mode factor fm. Moreover, we show how distribution $F'_{unc}(t)$ could take into consideration load sharing phenomena between the two set of midplane interconnects. Finally, we provide few details about how the non-Markovian model can be solved through the use of continuous phase type (CPH) distributions and Kronecker algebra.

4 The Non-markovian Evaluation

In this section we discuss on how to deal with the specific problems raised by the BladeCenter midplane modelling when considering non-Markovian behaviors. In particular two main issues have to be addressed in the modelling, common mode faults and load sharing, also describing how to represent the aging/memory process and to solve the model thus obtained.

4.1 Computing Covered and Not-Covered Failure Distributions

In order to show how to compute distributions $F_{cov}(t)$, $F_{unc}(t)$, and $F'_{unc}(t)$ as functions of $F_f(t)$ and fm as discussed in Sections 3.2, we start by the case in which an event, whose duration is generally distributed, produces a transition into a *vanishing* state upon its occurrence. A vanishing state is a state of a state space model in which the sojourn time is null and the evolution of the process towards other states is instantaneous and probabilistically determined.

As an example we propose the stochastic process χ depicted in Figure 2a. When the process is in the state 0 only one event can stochastically occur according to the CDF $F(t)$. At the event occurrence the process transits into state 1. In the latter (vanishing) state no timed event is enabled and the state 2 or 3 can be reached with probability p and $1-p$, respectively. The time evolution of the process is completely known when the state probabilities over the time are known. Let $\pi_i(t)$, $i = 0, \ldots, 3$ be the state probabilities of χ and $\pi(t) = (\pi_0(t), \ldots, \pi_3(t))$ its probability vector. This way we have that $\pi_0(t) = 1 - F(t)$. Then, since state 1 is a vanishing state we have that $\pi_1(t) = 0$, and the following relations hold

$$\pi_2(t) = p \cdot (1 - \pi_0(t)) = p \cdot F(t) \tag{1}$$
$$\pi_3(t) = (1-p) \cdot (1 - \pi_0(t)) = (1-p) \cdot F(t) \tag{2}$$

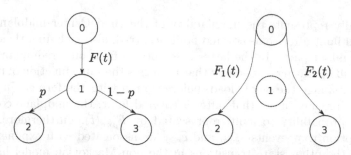

Fig. 2. Original (a) and equivalent (b) stochastic process

In order to reduce process χ into an equivalent one where the vanishing state 1 is removed, let us consider the process χ' depicted in Figure 2b, obtained by deleting the vanishing state 1 of χ and directly connecting its outgoing arcs to state 0. Let $\pi'(t)$ be the probability vector of χ'. We say χ' is time equivalent to χ if

$$\begin{cases} \pi_0(t) = \pi'_0(t) \\ \pi_2(t) = \pi'_2(t) \\ \pi_3(t) = \pi'_3(t) \end{cases} \tag{3}$$

According to this definition χ' could be used to derive the state probabilities of the non-vanishing states of χ. Indeed, χ' is characterized by two concurrent events, e_1 and e_2, in state 0, whose sojourn time random variables ξ_1 and ξ_2 are distributed according the CDFs $F_1(t)$ and $F_2(t)$, respectively. The process transits in either state 2 or state 3 depending on the first occurring event between e_1 and e_2. This way $\pi'_0(t) = (1 - F_1(t))(1 - F_2(t)) = 1 - F_1(t) - F_2(t) + F_1(t)F_2(t)$.

To compute the probability $\pi'_2(t)$ that χ' is in the state 2 at time t, two different cases have to be considered:

1. $\xi_2 > t$: in this case χ' transits in state 2 before time t when event e_1 occurs and the related probability is

$$\pi_2'^1(t) = F_1(t) \cdot (1 - F_2(t)) \tag{4}$$

2. $\xi_2 \leq t$: in this case event e_1 must happen before e_2 to be χ' in state 2; let us suppose $\xi_2 = x_2$ therefore

$$\pi_2'^2(t|\xi_2 = x_2) = F_1(x_2) \tag{5}$$

and by deconditioning on ξ_2

$$\pi_2'^2(t) = \int_0^t \pi_2'^2(t|\xi_2 = x_2)dF_2(x_2) = \int_0^t F_1(x_2)dF_2(x_2) = \int_0^t F_1(x_2)f_2(x_2)dx_2 \tag{6}$$

where $f_2(x) = \frac{dF_2(x)}{dx}$ is the probability density function of ξ_2.

Since the two cases are mutually exclusive the probability $\pi_2'(t)$ is computed by summing $\pi_2'^1$ and $\pi_2'^2$ by obtaining

$$\pi_2'(t) = F_1(t) \cdot (1 - F_2(t)) + \int_0^t F_1(x)dF_2(x) \qquad (7)$$

Following a similar reasoning we have that

$$\pi_3'(t) = F_2(t) \cdot (1 - F_1(t)) + \int_0^t F_2(x)dF_1(x) \qquad (8)$$

Since we are looking for $F_1(t)$ and $F_2(t)$ such that χ' is time equivalent to χ (in the sense before introduced), equation (3) must hold and the following ordinary differential (integral) equations' system can be written:

$$\begin{cases} 1 - F_1(t) - F_2(t) + F_1(t)F_2(t) = 1 - F(t) \\ F_1(t) \cdot (1 - F_2(t)) + \int_0^t F_1(x)f_2(x)dx = p \cdot F(t) \\ F_2(t) \cdot (1 - F_1(t)) + \int_0^t F_2(x)f_1(x)dx = (1 - p) \cdot F(t) \end{cases} \qquad (9)$$

with boundary conditions $F_1(0) = 0$, $F_2(0) = 0$. Since $F_1(t)$ and $F_2(t)$ are non-decreasing functions-CDFs the following property holds:

$$\lim_{t \to +\infty} F_1(t) = \lim_{t \to +\infty} F_2(t) = 1$$

The failure behavior of the BladeCenter midplane can be stochastically represented by the process depicted in Figure 2a, with $e_1 = e_{cov}$ and $e_2 = e_{unc}$, thus having $F_1(t) = F_{cov}(t)$, $F_2(t) = F_{unc}(t)$ and $p = fm$. Moreover, given that $F_f(t)$ is a Weibull distribution, it is easy to demonstrate by substitution in the system of equation (9) that F_{cov} and F_{unc} are both still Weibull distributions, with $\beta_{cov} = \beta_{unc} = \beta_f$ and $\eta_{cov} = \eta_f/fm^{(1/\beta_f)}$ and $\eta_{unc} = \eta_f/(1 - fm)^{(1/\beta_f)}$, respectively.

4.2   Modeling Aging and Load Sharing

The non-exponential distributions in the model of Figure 1 lead to address complex issues while dealing with the solution. In particular, during the transition from state AU to state C3, the wear out of the BladeCenter chassis midplane has to be taken into account at state changing. Moreover, since covered faults could imply higher load on the operating components, reflecting on its reliability, load sharing effects, if present, have to be adequately modeled.

Let us consider the lifetime of the BladeCenter midplane as stochastically modeled by a continuous random variable X. Moreover, let us assume the values of X also depend on a conditioning event characterizing and influencing the event associated with X according to different *operating conditions* assumed by the conditioning event random variable Y. In the BladeCenter chassis midplane two operating conditions can be identified, the fully operating (state AU) and, after a covered fault, the partially operating (state CS) ones. Then, X is characterized by a CDF

$$F_X^a(x) = Pr\{X \le x \mid The\ midplane\ is\ fully\ operating\} = 1 - R_X^a(x),$$

where $R_X^a(x)$ is the reliability of the fully operating midplane. In fact, $F_X^a(x) = F(x)$ as introduced in the previous section. In the partially operating condition, X is instead characterized by

$$F_X^f(x) = Pr\{X \le x \mid The\ midplane\ is\ partially\ operating\} = 1 - R_X^f(x).$$

In fact, $F_X^f(x) = F'(x)$ as introduced in the previous section. This way, we have a couple of CDFs for X, F_X^a and F_X^f, depending on the operating condition of the midplane. Since in our model we are specifically interested in evaluating such phenomenon for not-covered faults, we can just focus on the e_{unc} event thus obtaining $F_X^a = F_{unc}$ and $F_X^f = F'_{unc}$. Let us assume both F_{unc} and F'_{unc} are *continuous* and *strictly increasing*.

This way, the switching between F_{unc} and F'_{unc} is governed by the first covered fault event e_{cov}, associated with the random variable Y stochastically characterized by the CDF $F_{cov}(t)$. Y influences the behavior of X, i.e., X depends on Y. We are therefore interested in obtaining $F_X(x) = Pr\{X \le x\}$ that stochastically characterizes the event related to X when the operating condition varies. Since the condition changes are related to Y, F_X could be considered as a joint CDF of X and Y, i.e. $F_{X,Y}(x, y)$.

The main requirement we impose on $F_{X,Y}$ is that it must be a continuous function at changing point y. This choice implements the *conservation of reliability principle* [14]. As a consequence, at the condition changing points there must not be a probability mass for $F_{X,Y}$. The problem is to express $F_{X,Y}$ in terms of its marginal distributions F_{cov}, F_{unc} and F'_{unc}. Thus, assuming the midplane is initially fully operating, we have that

$$F_{X,Y}(x, y) = \begin{cases} F_{unc}(x) & x \le y \\ F'_{unc}(x + \tau) & x > y \end{cases} \tag{10}$$

where τ is a constant depending on the changing point, i.e. such that $F_{X,Y}(x, y)$ is continuous at y. It is therefore necessary to understand what happens at y in order to quantify τ. Given that F_{unc} and F'_{unc} must be continuous and strictly increasing (and thus invertible), by equation (10) at y we have that:

$$F_{unc}(y) = F'_{unc}(y + \tau) \Rightarrow \tau = F'^{(-1)}_{unc}(F_{unc}(y)) - y \tag{11}$$

where $F'^{(-1)}_{unc}$ is the inverse function of F'_{unc}.

By applying the law of total probability we can express $F_{X,Y}(x,y)$ in terms of a unique variable

$$F_{X,Y}(t) = \int_{-\infty}^{+\infty} Pr(X \leq t|Y = y)f_{cov}(y)dy$$

$$= \int_{0}^{t} Pr(X \leq t|Y = y)f_{cov}(y)dy + \int_{t}^{+\infty} Pr(X \leq t|Y = y)f_{cov}(y)dy =$$

$$= \int_{0}^{t} (1 - Pr(X > t|Y = y))f_{cov}(y)dy + Func(t)(F_{cov}(y)|_{t}^{\infty}) \qquad (12)$$

This way, considering $y \leq t$, $Pr(X > t|Y = y) = 1 - F'_{unc}(t + \tau) = 1 - F'_{unc}\left(t + F'^{(-1)}_{unc}(F_{unc}(y)) - y\right)$ and thus by equation (12) we can de-condition on Y obtaining

$$F_X(t) = F_{unc}(t)(1 - F_{cov}(t)) + \int_{0}^{t} F'_{unc}\left(t + F'^{(-1)}_{unc}(F_{unc}(y)) - y\right) f_{cov}(y)dy \qquad (13)$$

which represents the distribution of the midplane not-covered faults when the operating conditions change due to covered faults.

4.3 Solution Technique

As detailed in the previous sections, the state space model we propose to represent the BladeCenter midplane availability is not trivial to manage since it includes complex phenomena that should be dealt with through specific techniques. In particular, two main aspects should be addressed: i) representation of the Weibull distributed r.v., ii) memory management issues related to the aging phenomenon and the load sharing behavior. In order the overall distribution is continuous the evolution of the underlying stochastic process restarts in the new condition from the reliability/age level reached at changing point in the previous condition.

In [13], we proposed a solution technique for evaluating distributed systems affected by non-Markovian behaviors and changing conditions. This solution is based on a Markovization step, where we exploit the phase type (PH) approach taking into account the nature of the changing quantities as well as the way they are represented. In particular, starting from a state space model representing the system under exam, we propose a technique able to approximate non-exponentially distributed events through CHPs and to adequately manage memory conservation during changing conditions through an ad-hoc fitting algorithm.

Since the distribution or the function describing the observed phenomenon changes according to some events into different conditions it is necessary to know the lifetime CDFs of the observed system both in isolation or in the initial working condition (without dependencies or in the baseline environment), and in the new environments or after the application of a dependency. This way, by representing the lifetime CDFs as CPHs, it is possible to specify how the system jumps from one of such CPHs to the others by preserving, in the transition between two conditions, the lifetime or the reliability. The expansion technique can be thus used to numerically study the model, obtaining a CTMC where the states of the original stochastic process are expanded into set of states. The main drawback of this approach is the memory consumption since the space complexity grow up very quickly with the number of system states and the events there enabled. The adopted solution technique tackles this issue by using Kronecker algebra to generate the transition rates among the states of the expanded process. The system state space is described by using a two layer symbolic approach allowing a very compact representation of it. The Kronecker expressions involving CPHs matrices are similar to that presented in [9] and are not reported here for lack of space.

5   Analysis and Results

In order to demonstrate the effectiveness of our technique and to evaluate the quality of its accuracy, in this section we show the results obtained from the analysis of our BladeCenter midplane non-Markovian model comparing them against the exponential one obtained in [11]. The parameters adopted during the analysis that are related to the distribution associated with the midplane failure event have been reported in Table 1. Table 2 reports the coverage factor (fm) and the parameters related to the repairman arrival, the midplane maintenance, and the load-sharing behavior. In particular, we suppose the events associated with the repairman arrival and the midplane maintenance are exponentially distributed thus we only report the corresponding rates, i.e., $\lambda_{sum} = 1/MRT_{sp}$ and $\lambda_{rep} = 1/MTTR_{mp}$, respectively. On the other hand, the load-sharing behavior (if present) is modeled by assuming that the midplane failure time distributions $F(t)$ and $F'(t)$ can be framed into the accelerated life model (ALM) class such that $\eta'_f = l \cdot \eta_f$, where the η_f and η'_f are the shape parameters of the Weibull distributions as specified in Section 3.2 and $0 < l < 1$ is the load sharing parameter, quantifying the effect of load sharing in terms of the shape parameter. In case of Weibull distributions, this is equivalent to a proportional hazard model where $h'(t) = (1/l)^{\beta_f} h(t)$.

As a first step, we performed a transient analysis of our BladeCenter midplane non-Markovian model with the aim of computing the midplane point availability

Table 2. BladeCenter midplane non-Markovian model parameters

Parameter	Values
fm	[0.5,0.99]
λ_{sum}	0.4
λ_{rep}	0.4
l	0.7

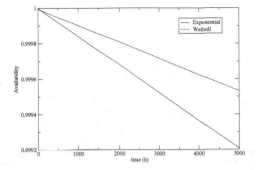

Fig. 3. Comparison of homogeneous CTMC model's and non-Markovian model's $A_{mid}(t)$

$A_{mid}(t)$. Such a metric can be calculated as the sum of the transient probabilities for the model to be in states AU and CS. The non-Markovian model has been solved comparing the cases in which the low parameters for the failure time distributions have been adopted with the purpose to analyzing the differences between exponentially and Weibull distributed failure times

Figure 3 shows $A_{mid}(t)$ in all the above described cases in the time interval $[0, 5000]$. From these results we can observe and quantify the quality of the exponential approximation against the Weibull one in transient evaluation. Both exponential and Weibull models have similar trends but different slopes, thus amplifying the gap between corresponding results by increasing the time. This trend is also highlighted by the steady state evaluation as shown in Figure 4 where the midplane steady state availability $A_{mid} = \lim_{t \to \infty} A_{mid}(t)$ is reported by varying the coverage factor fm. Such a measure could allow to understand and quantify the improvement in terms of system availability of empowering the software/hardware fault coverage. Figure 4a provides the results obtained by analyzing the model with exponential failure distributions against the Weibull one. Since we obtained very small differences when the exponential distributions characterized by the parameters of Table 1 are used, we depict only one of the two (Low and High) sets of results (Exponential curve). For similar reason, only a set of results related to the Weibull distributed failure model is reported (Weibull curve).

We can therefore argue that the exponential approximation under estimates the BladeCenter midplane availability, both in transient and steady state. As a consequence, it is necessary to carefully evaluate the quality of the approximation, the tolerance of the model when choosing to adopt exponential models. Indeed, the use of Weibull distributions allows a better approximation, showing some properties and behaviours of the system that are not caught by the homogeneous CTMC model. For example, from the non-Markovian model results we can observe that the system is highly available independently on the coverage factor value, since at least a 6-nine availability A_{mid} is obtained (Low Weibull and High Weibull curves). Moreover, when a load-sharing phenomena is experienced that modifies the Weibull distribution's shape parameters an improving

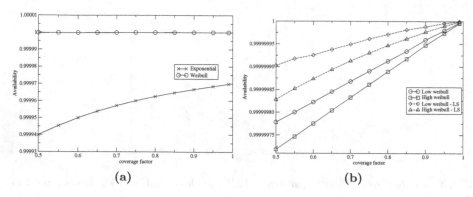

Fig. 4. A_{mid} in the homogeneous CTMC model and non-Markovian model (a) and (b), respectively

of the availability is obtained (Low Weibull - LS and High Weibull LS curves). This is mainly due to the decreasing failure rate (DFR) distributions considered in the model (Weibull with $\beta = 0.5 < 1$), since the load impacts on the system accelerating in time the aging process, as in ALM. In terms of failure rate this means that, being the Weibull DFR, this phenomenon accelerates the decreasing failure rate process. This is a common behavior for DFR distributions, which however could be considered as a bit strange. Basically, this is due to the choice of DFR Weibull distributions, which however have been obtained through statistical experiments.

6   Conclusions

In this paper, we proposed an analytical non-Markovian technique based on CPH distributions and Kronecker algebra for the modeling of an IBM Blade-Center system with particular attention to the chassis midplane evaluation. We compare the results of our model with the ones obtained from a homogeneous CTMC one, already proposed in a previous work. Results demonstrate how the capability to deal with non-exponentially distributed events and the corresponding aging phenomena allows our technique to be more accurate with respect to reality. Moreover, being able to model common mode faults and load sharing behaviors, our technique candidates itself as a powerful tool for assessing, ensuring, and enforcing reliability and availability of resources and services in distributed computing systems, both in transitory and steady-state.

References

1. Resnick, R.I.: A Modern Taxonomy of High Availability. Technical report, Interlog (1996),   http://www.generalconcepts.com/resources/reliability/resnick/HA.htm

2. Cisco Systems Inc.: Data Center High Availability Clusters Design Guide (2006), http://www.ciscocertified.info/en/US/docs/solutions/Enterprise/Data_Center/HA/_Clusters/HA/_Clusters.html
3. Ghosh, R., Longo, F., Frattini, F., Russo, S., Trivedi, K.: Scalable analytics for iaas cloud availability. IEEE Transactions on Cloud Computing 2(1), 57–70 (2014)
4. Zheng, J., Okamura, H., Dohi, T.: Component importance analysis of virtualized system. In: 2012 9th International Conference on Ubiquitous Intelligence Computing and 9th International Conference on Autonomic Trusted Computing (UIC/ATC), pp. 462–469 (September 2012)
5. Muppala, J., Malhotra, M., Trivedi, K.: Markov dependability models of complex systems: Analysis techniques. In: Özekici, S. (ed.) Reliability and Maintenance of Complex Systems. NATO ASI Series, vol. 154, pp. 442–486. Springer, Heidelberg (1996)
6. Fricks, R., Yin, L., Trivedi, K.: Application of semi-markov process and CTMC to evaluation of UPS system availability. In: RAMS 2002, January 28–31, pp. 584–591 (2002)
7. Mura, I., Bondavalli, A.: Markov regenerative stochastic petri nets to model and evaluate phased mission systems dependability. IEEE Transactions on Computers 50(12), 1337–1351 (2001)
8. Longo, F., Scarpa, M.: Applying symbolic techniques to the representation of non-markovian models with continuous PH distributions. In: Bradley, J.T. (ed.) EPEW 2009. LNCS, vol. 5652, pp. 44–58. Springer, Heidelberg (2009); Cited by (since 1996)6
9. Longo, F., Scarpa, M.: Two-layer symbolic representation for stochastic models with phase-type distributed events. International Journal of Systems Science, 1–32
10. Pérez-Ocón, R., Castro, J.E.R.: Two models for a repairable two-system with phase-type sojourn time distributions. Reliability Engineering & System Safety 84(3), 253–260 (2004)
11. Smith, W.E., Trivedi, K., Tomek, L., Ackaret, J.: Availability analysis of blade server systems. IBM Systems Journal 47(4), 621–640 (2008)
12. Bruneo, D., Distefano, S., Longo, F., Puliafito, A., Scarpa, M.: Reliability assessment of wireless sensor nodes with non-linear battery discharge (2010); Cited by (since 1996)5
13. Bruneo, D., Distefano, S., Longo, F., Puliafito, A., Scarpa, M.: Evaluating wireless sensor node longevity through markovian techniques. Comput. Netw. 56(2), 521–532 (2012)
14. Sedyakin, N.: On one physical principle in reliability theory. Tekhn. Kibernetika (in Russian - Technical Cybernetics) 3, 80–87 (1966)

Author Index